Operational Risk Modelling and Management

CHAPMAN & HALL/CRC FINANCE SERIES

Series Editor

Michael K. Ong

Stuart School of Business
Illinois Institute of Technology
Chicago, Illinois, U. S. A.

Aims and Scopes

As the vast field of finance continues to rapidly expand, it becomes increasingly important to present the latest research and applications to academics, practitioners, and students in the field.

An active and timely forum for both traditional and modern developments in the financial sector, this finance series aims to promote the whole spectrum of traditional and classic disciplines in banking and money, general finance and investments (economics, econometrics, corporate finance and valuation, treasury management, and asset and liability management), mergers and acquisitions, insurance, tax and accounting, and compliance and regulatory issues. The series also captures new and modern developments in risk management (market risk, credit risk, operational risk, capital attribution, and liquidity risk), behavioral finance, trading and financial markets innovations, financial engineering, alternative investments and the hedge funds industry, and financial crisis management.

The series will consider a broad range of textbooks, reference works, and handbooks that appeal to academics, practitioners, and students. The inclusion of numerical code and concrete real-world case studies is highly encouraged.

Published Titles

Decision Options®: The Art and Science of Making Decisions, **Gill Eapen**

Emerging Markets: Performance, Analysis, and Innovation, **Greg N. Gregoriou**

Introduction to Financial Models for Management and Planning, **James R. Morris and John P. Daley**

Pension Fund Risk Management: Financial and Actuarial Modeling, **Marco Micocci, Greg N. Gregoriou, and Giovanni Batista Masala**

Stock Market Volatility, **Greg N. Gregoriou**

Portfolio Optimization, **Michael J. Best**

Operational Risk Modelling and Management, **Claudio Franzetti**

Proposals for the series should be submitted to the series editor above or directly to:
CRC Press, Taylor & Francis Group
4th, Floor, Albert House
1-4 Singer Street
London EC2A 4BQ
UK

CHAPMAN & HALL/CRC FINANCE SERIES

Operational Risk Modelling and Management

Claudio Franzetti

SERV
Zurich, Switzerland

CRC Press
Taylor & Francis Group
Boca Raton   London   New York

CRC Press is an imprint of the
Taylor & Francis Group, an **informa** business

A CHAPMAN & HALL BOOK

Chapman & Hall/CRC
Taylor & Francis Group
6000 Broken Sound Parkway NW, Suite 300
Boca Raton, FL 33487-2742

© 2011 by Taylor and Francis Group, LLC
Chapman & Hall/CRC is an imprint of Taylor & Francis Group, an Informa business

No claim to original U.S. Government works

Printed in the United States of America on acid-free paper
10 9 8 7 6 5 4 3 2 1

International Standard Book Number: 978-1-4398-4476-2 (Hardback)

Library of Congress Cataloging-in-Publication Data

Franzetti, Claudio.
    Operational risk modelling and management / Claudio Franzetti.
        p. cm. -- (Chapman & Hall/CRC finance series)
    Summary: "In banking regulation, tools are needed to quantify risk and calculate the amount of capital reserve required to mitigate such risk. This book offers a complete model for the quantification of so-called operational risks. It offers a detailed discussion on the link between modeling approaches and management, which has been neglected in the literature, as well as the mathematical modeling of the loss distribution approach. With an emphasis on risk management and management fundamentals, the text presents a complete simulation model along with tested examples that can be replicated using R software. The author provides a broad view on managing risk using this mathematical model"-- Provided by publisher.
    Includes bibliographical references and index.
    ISBN 978-1-4398-4476-2 (hardback)
    1. Risk management. 2. Risk management--Mathematical models. I. Title.

HD61.F728 2010
658.15'5--dc22
                                                      2010032499

To Rita, Lorenzo, Bibiana, and Francesco

Frustra fit per plura quod potest fieri per pauciora.
(It is vain to do with more what can be done with less.)

William Ockham (1285-1349)

Far better an approximate answer to the right question,
than the exact answer to the wrong question, which can
always be made precise.

John W. Tukey

We move forward, ignoring the risks we create. From
time to time an accident shakes us up from our
numbness, and we glance into the precipice.

Ivar Ekeland

Contents

List of Figures

List of Tables

Preface

As writing a book takes quite a long time there is some chance that something important happens possibly affecting its content. As of this writing the financial crisis of 2008, which developed out of the sub-prime mortgage lending crisis of the previous years, is still ravaging and a stabilisation of the system is not in sight. Communication and leadership failure of governments has acerbated the turmoil and commentators were quick at pointing to bankers' greed, faulty models and inadequate regulation as the main culprit. Countries applying the new Basel 2 rules and others not having implemented them are scathed. It seems obvious that governments will call for "better" oversight when they think the crisis is tamed. It seems that governments would like to counteract the complexities of banking with simple, if not simplistic, rules. Implementing new regulation will take a long time, especially as there are cultural differences of views on the roles of free enterprise and state interference. The major emphasis is to be expected on credit risk and foremost liquidity risk. Therefore, we are quite confident that the rules governing operational risk being potentially the most sophisticated will survive and thus the ideas contained in this book are still most relevant. In the vein of its topic such an assertion is to be understood probabilistically.

This book does not advance any fundamental novelty of its own, but constructs a new edifice with known bricks. Its contribution is a complete, detailed description of one specific model for the Loss Distribution Approach, with auxiliary tools and methods, in the context of the new Basel Accord. The two main building blocks on which this text rests are the loss distribution approach (in the narrow sense) as a model for calculating the risk capital figure, and risk mitigation through management and management's actuations. Such mitigation must make risk figures more sensitive to internal and external environments, more representative of actual risk, more predictive and, one would hope, more controllable.

Basel 2 has taken several years to develop, engaging vast resources and engendering new services. The final documents of the Banking Committee on Banking Supervision (BCBS) cannot, and do not, prescribe in detail how to implement the rules. Nor can we expect legislators or supervisory bodies to do so. This is especially true for the most sophisticated approach toward operational risk, where much leeway is deliberately left for implementation.

Operational risk consideration has been added within this new framework. The Advanced Measurement Approach allows for the most sophisticated quantitative assessment of risk. To that end, specific models must be devised. Models are by their nature subjective, so regulators' requirements for potentially differing risk assessments may necessitate multiple models. However, we have tried to develop an "optimal" model that plays by the Basel rules, taking into account the methodological issues and the data situation, in order to provide the financial industry with a comprehensive and useful tool. It is our belief that the model outlined in this book complies with all the Basel 2 standards prescribed thus far.

We all know that operational risk management is perceived as a particularly confusing field. This may be because the many people involved bring specific and varied backgrounds to their practice of the discipline. For example, auditors tend to think of risk as a financial control issue, whereas lawyers focus on tort and liability, insurance experts on insurability, and strategists on business redesign. Not surprisingly, process experts see re-engineering as the solution, while software engineers perceive databases and applications as the key and financial experts devise ways to transfer risk to the financial markets. Last but not least, mathematicians, actuaries and statisticians persevere in their attempts to quantify risk.

Having said all that, we still believe that any book on operational risk quantification must broaden its scope. Therefore, a significant portion of this book examines a multitude of management issues that must be considered when adjusting the quantitative results of a comprehensive model. Naturally, some depth must be sacrificed for the sake of such breadth.

Let there be no mistake: this book was conceived and written in an "engineering" style. In the quantitative portions we offer key concepts and definitions without stating theorems or delving into mathematical proofs. Anyone interested in such background reading is welcome to peruse the literature and the references upon which we make it clear we draw. Our emphasis is on providing techniques that can be understood and applied by the not-so-lay practitioner. Lastly, any model designed for real world implementation must be available, reliable, and easily maintained. We hope ours meets these criteria.

In this book you will find a run-through example of risk calculations – a Monte Carlo simulation of risk capital – based on data from a quantitative impact study. These computations are too complicated to be performed with short code snippets of a scripting language. Therefore, a prototypical program can be downloaded from www.garrulus.com to re-perform the example. Additional smaller examples are given in the text as R-scripts.

Happy reading, and best of luck negotiating the tricky world of risk calculation and management. May this book serve your purposes well.

I would like to thank many former and present colleagues at Swiss Re, Deutsche Bank, Aon Resolution, Serv and Garrulus Enterprise for valuable discussions over a long span of time and a variety of risk topics. Special thanks go to my family and Markus Stricker.

Claudio Franzetti, Zürich

About the Author

Claudio Franzetti is chief risk officer of Swiss Export Risk Insurance in Zurich and president of Garrulus Enterprise Ltd. Prior to that he held senior positions with Aon Resolution AG, was head of group credit portfolio management and managing director of Deutsche Bank in Frankfurt, head of risk management and controlling of Swiss Re's investment division and senior consultant with Iris AG in Zurich. Before entering the finance industry he started as research engineer with Brown Boveri & Cie., then Asea Brown Boveri, in Baden, in the field of computational fluid dynamics.

Claudio received a masters degree in economics from the University of St. Gall and a master in mechanical engineering from the Swiss Federal Institute of Technology (ETH) in Zurich.

Chapter 1

Introduction to Operational Risk

1.1 Why Regulate Banks?

We pose this question, obviously, for the sake of argument. At this juncture, nothing can endanger the Basel 2 effort: "rien ne va plus, les jeux sont faits." Really? In 2007 American regulators have proposed changes in their version of Basel 2 that will delay its implementation until at least January 2009. Alas, the accord intended as a single worldwide standard is threatened to be different in Europe and America. Posner (2004, 217), an American judge and scholar, says: "As the world's most powerful nation, the United States tends to dominate international organizations, and, when it does not, it ignores them with impunity." Then there is the credit crisis starting in 2008 bringing regulation to the limelight.

The Basel Committee on Banking Supervision (BCBS) stated and re-iterated as objectives (BCBS, 1988, 1) the following:

> Two fundamental objectives lie at the heart of the Committee's work on regulatory convergence. These are, firstly, that the new framework should serve to strengthen the soundness and stability of the inter-national banking system; and secondly that the framework should be in fair and have a high degree of consistency in its application to banks in different countries with a view to diminishing an existing source of competitive inequality among international banks.

Regulation consists of rule-making and enforcement. Economic theory offers two complementary rationales for regulating financial institutions. Altruistic public-benefits theories treat governmental rules as instruments for increasing fairness and efficiency across society as a whole. They assign regulation to governmental entities who search for market failures and correct them. It is taken for granted that governments are well-intentioned to act for the common good (Kane, 1997).

Agency theory recognises that incentive conflicts and coordination prob-lems arise in multi-party relationships and that regulation introduces oppor-tunities to impose rules that enhance the welfare of one sector of society at the expense of another. Agency theories portray regulation as a way to raise the quality of financial services by improving incentives to fulfil contractual obligations in stressful situations. These private-benefits theories count on self-interested parties to spot market failures and correct them by opening more markets (Kane, 1997).

There is also a less favourable explanation for the existence of bank regulation which claims that banks have always been regulated for political reasons. Governments, officials, their friends and bankers either owned these banks or were extended loans at favourable interest rates. In some countries, the state is still directly involved in the banking sector, either through ownership or the provision of state guarantees to certain banks. In a few countries banks are also used to implement social objectives, such as directing credit towards favoured sectors or promoting new enterprises.

Over the past few decades, similarly to insurance, banking has undergone a deregulation accompanied by a re-regulation. There have been relaxations with or abolishments of provisions concerning interest rates, fees and commissions, restrictions on lines of business, ownership, abolishment of reserve requirements, etc.

On the other hand, there has been a strengthening of prudential regulation focused on controls of the capital of banks and an increase of the number and coverage of deposit insurance schemes.

Some arguments for bank regulation hinge upon the role of banks in the payment system. Under conventional payments systems, banks can build up large exposures to one another during a trading day, which are settled at the close of the trading day. The failure of one bank to settle could, it is argued, have a significant domino effect on other healthy banks. As a consequence, admittance to payments systems should be restricted to cautiously regulated institutions.

In the case of banks, market failure arises from banks' difficulty to credibly attest their level of risk to depositors and other lenders. It is argued that, as a result of the absence of regulatory intervention, banks would assume more risk than is prudent, bank failures would be more frequent than is necessary and the financial system as a whole would be unstable. Nonetheless, the last point could not be made in the crisis of 2008.

As with most forms of regulation, the case presumably rests on some kind of market failure, known to economists as so-called externalities. The economist sees negative externalities as a problem to be solved by government regulation or taxation if those harmed by the externality cannot bargain with the injurer in the sense described by Ronald Coase. Regulators would determine the socially optimal amount of externality reduction and require those parties whose activities generate negative externalities to achieve this specified reduction. Alternatively, taxation could achieve the same result by forcing private costs into line with social costs. In our particular case, the externality of bank failures are systemic costs that are not fully borne by the banks themselves. These costs include losses absorbed by governmental deposit insurance, disruptions to other participants in the financial system through spill-over effects and so on. Thus, the regulator's task is to convince the bank to internalise these systemic costs through regulation (which produces the same effects as a tax). Of course the costs will ultimately be sustained by debtors and other stakeholders. Actually, this is not to say that costs are irrelevant for society. The cost of regulation should be lower than the cost of distress and bankruptcy, which include lawyers' fees.

In particular, most deposit insurance schemes and other implicit government measures – such as "too big to fail" now called "systemic" – insulate the depositor from any awareness of their banks' financial condition and, in the absence of other interventions, encourage risk-taking. This is called "moral hazard." The state becomes a creditor to the bank. It is behaving exactly like a bank requiring collateral security for credit.

Benston and Kaufman (1996) posit that if there were no deposit insurance, there would be no appropriate economic role for bank regulation, assuming that the goal is to benefit consumers. This could also be understood by the findings of the following study.

Barth et al. (2006) conclude from their comprehensive global database of bank regulations that the best-intentioned bank regulation does take into account a government's failure to self-regulate. Such failures occur often and their consequences are worse than those that regulation is supposed to address. Their analysis of the relationship between different kinds of regulations and indicators of the state of a country's banking system, include (1) the efficiency of its banks (measured by net interest rate margins and overheads), (2) the extent of corruption, (3) the developmental stage of the system (measured by the amount of credit extended to private firms as a proportion of GDP) and (4) the likelihood of a crisis.

- First, they find that raising capital requirements (Pillar 1) has no discernible impact on the development of a country's banking sector, the sector's efficiency or its likeness to encounter crisis.

- Second, regulatory policies that enhance private-sector monitoring of banks (Pillar 3) tend to develop banking systems, make it more efficient and decrease the likelihood of crisis. However, the more generous deposit insurance is, the more likely is a banking crisis.

- Third, strengthening supervision (Pillar 2) has a neutral or negative impact on banking development, reduces bank efficiency and increases the likelihood of a crisis. This may be because countries with weaker legal systems and political institutions tend to favour stronger supervisors, who in turn are susceptible to corruption. Therefore, the authors worry that the second pillar might do damage in the vast majority of countries.

The question of whether we need regulation, and especially how much, will not be resolved here. Because history takes just one path, it cannot be discerned from experience whether regulation produces real benefit. Regulation must therefore be accepted as a fact of life.

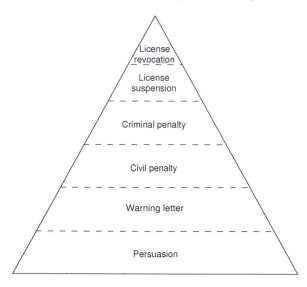

FIGURE 1.1: Regulatory sanction pyramid. (From Grabosky and Braith-waite, 1993, 95. By permission of the Australian Institute of Criminology.)

There is no regulation without potential sanctions. Basel 2 does not pro-vide new and additional sanctions with respect to the measures that the na-tional regulators already have. Sparrow (2000, 40) shows a typical hierarchy of regulatory sanctions. The higher on the pyramid the tougher the penalty.

The foremost intention of sanctions is to massively deter banks from breaching the rules.

1.2   Additional Supervision

A bank is not only subject to banking supervision. There are a handful of additional "supervisions" that are not specific to banks but affect them nonetheless. We cite *inter alia*:

- accounting and audit standards,

- corporate governance,

- religious supervision,

- legal prosecution and the law in general and,

- corporate social responsibility.

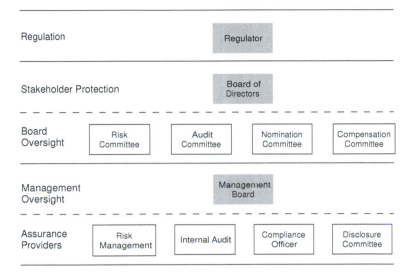

FIGURE 1.2: A corporate governance structure.

With respect to accounting, banks are mainly concerned with the difficulties of the true and fair view on "financial instruments". Both FASB 133 and IFRS 7 are controversial and generally disliked by the financial community. Audits of financial reports lead to an opinion on whether the reports give a true and fair view on the economic resources or obligations of a company. Especially in the controls testing phase of the audit errors and frauds intended to deceive financial statement users can effectively be detected. Although *corporate governance* can be defined in a variety of ways, it generally deals with the mechanisms by which a business enterprise is directed and controlled (EUC, 2002). In addition, it addresses mechanisms by which managers are held responsible for corporate conduct and the bank's performance. Corporate governance is distinct from business management and corporate responsibility. In Europe, corporate governance codes are generally non-binding sets of principles, standards or best practices, issued by a collective body, and referring to the internal governance of corporations. There are over 35 different codes, pertaining to the following issues (among others):

- Shareholder and stakeholder interests,

- supervisory and managerial bodies,

  - board systems,
  - the distinct roles and responsibilities of supervisory and managerial bodies,
  - accountability of supervisory and managerial bodies,
  - transparency and disclosure,

- conflicts of interest,
- the size, composition, independence, selection criteria and procedures of supervisory and managerial bodies,
- the working methods of supervisors and management,
- remuneration of supervisory and managerial bodies,
- the organisation and supervision of internal control systems.

In the US, corporate governance has become important in the wake of the collapses of Enron, WorldCom, Tyco and others. Congress has concocted the so-called "Sarbanes-Oxley Act of 2002" ("SOX"), of which section 404 proves very burdensome. The number of controls that large companies must test and document can top the tens of thousands. J.P. Morgan Chase is reported to have 130 employees working full-time on compliance alone. According to a study by Ernst & Young, one of the major audit firms, half of America's big public companies estimated that they would commit more than 100,000 man-hours to comply with Section 404 in its first year. Yet, section 404 reads innocuously:

> *Rules required.* The Commission shall prescribe rules requiring each annual report required by section 13(a) or 15(d) of the Securities Exchange Act of 1934 to contain an internal control report, which shall (1) state the responsibility of management for establishing and maintaining an adequate internal control structure and procedures for financial reporting; and (2) contain an assessment, as of the end of the most recent fiscal year of the issuer, of the effectiveness of the internal control structure and procedures of the issuer for financial reporting.

> *Internal control evaluation and reporting.* With respect to the internal control assessment required by subsection (a), each registered public accounting firm that prepares or issues the audit report for the issuer shall attest to, and report on, the assessment made by the management of the issuer. An attestation made under this subsection shall be made in accordance with standards for attestation engagements issued or adopted by the Board. Any such attestation shall not be the subject of a separate engagement.

In short, the company must make an assessment and the auditor must make an attestation with a report of the assessment. With this act the quasi-governmental Public Company Accounting Oversight Board (PCAOB) was created, which issued Auditing Standard No. 2. It says that the so-called COSO[1] Integrated Framework is a suitable and available framework for the

---

[1]COSO stands for the Committee of Sponsoring Organizations of the Treadway Commission, which are the American Accounting Association, American Institute of Certified Public Accountants, Financial Executives International, Institute of Management Accountants and the Institute of Internal Auditors.

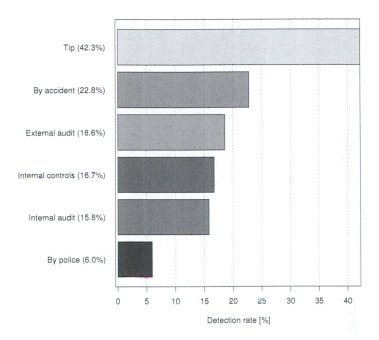

FIGURE 1.3: Fraud detection rate by source, 2008, for frauds greater 1 million USD. (Source: ACFE, 2008, 20. By permission of the Association of Certified Fraud Examiners, Inc.)

purposes of management's assessment. For that reason, the performance and reporting directions in this standard are based on it. It now seems that excessive emphasis is put on the "internal control" structure. Figure 1.3 shows that only eight percent of all known fraud is detected by internal controls. Fraud is more often caught by accident. A report (AeA, 2005) comments: "Investors will not be any better protected from fraud than before implementation of Section 404. The best protection from fraud would be to increase the efforts to support whistler-blower provisions and to increase prosecution efforts. Those provisions will be far more successful at preventing investor losses than Section 404, and they are dramatically more cost effective."

By religious supervision we understand mainly the adherence to Shari'a law, the teachings of the Qur'an and the Sunnah of Islamic banking. Islamic financial institutions have an independent Shari'a Supervisory Board (SSB), recommended by the board of directors and appointed by the shareholders. It consists of at least three jurists specializing in Islamic commercial jurisprudence – referred to as Shari'a scholars – who direct, review and supervise, with rulings or opinions called *fatwa*, the activities of the bank (ex-ante compliance). Similar to the conventional internal audit there is an internal Shari'a compliance unit to ensure that the management implements the rulings (ex-post compliance). Financial instruments, must at the very least conform to the law with regards to Riba, Maysir and Gharar. But operations must also comply, thus extending Shari'a to contracts with suppliers, labour contracts and so forth.

Prosecution, especially like that of the former Attorney General of the State of New York, Eliot Spitzer, has proven very efficient, especially when measured in terms of payments. In 2005 Spitzer is reckoned to have collected 2.8bn USD from several industries for their many unlawful or at least very dubious practices. Given these sums and the driving will behind the prosecution, banking supervision looks quite dim.

TABLE 1.1: Payments in Global Research Analyst Settlement 2002 (Source: New York State Attorney General)

Investment Bank	Retrospective Relief	Independent Research	Investor Education	Total
Salomon Smith Barney	300	75	25	400
Credit Suisse First Boston	150	50	0	200
Merrill Lynch	100	75	25	200
Morgan Stanley	50	75	0	125
Goldman Sachs	50	50	10	110
Bear Stearns	50	25	5	80
Deutsche Bank	50	25	5	80
J.P. Morgan Chase	50	25	5	80
Lehman Brothers	50	25	5	80
UBS Warburg	50	25	5	80
Total	900	450	85	1435

Spitzer sued several investment banks for inflating stock prices, misusing affiliated brokerage firms to give biased investment advice and to sell initial public offerings of stock to preferential customers, viz. corporate executives and directors. In 2002, Spitzer negotiated a settlement of these lawsuits, with federal regulatory bodies, stock exchanges, and the banks and brokerage houses in question.

The terms of the settlement agreement include:

- the insulation of research analysts from investment banking pressure,

- a complete ban on the spinning of Initial Public Offerings (IPOs),

- an obligation to furnish independent research,

- disclosure of analyst recommendations,

- settled enforcement actions involving significant monetary sanctions.

As shown in Table 1.1 the total compensation and fines amounted to USD 1.435bn.

Note the following news release from 11th of April 2006 (reproduced by permission of Thomson Reuters):

> LONDON (AFX) - The Financial Services Authority, the UK's financial markets watchdog, said it has fined Deutsche Bank AG 6.3 mln stg for market misconduct. The fine, the third-biggest ever levied by the FSA, relates to Deutsche's handling of a sale of shares in Swedish truck maker Scania AB, and a separate transaction involving shares of Switzerland's Cytos Biotechnology AG.

Just compare the 6.3m GBP with the figures from Table 1.1 and you get a sense of the regulators.

There is yet another layer of supervision called *social responsibility*. Non-governmental organisations (NGOs) can exert considerable pressure on financial institutions. In 2002, a global coalition of non-governmental organisations including Friends of the Earth, Rainforest Action Network, WWF-UK and the Berne Declaration came together as BankTrack to promote sustainable finance in the commercial sector. In their report (Durbin et al., 2005, 3) they express the opinion that:

> within the banking sector, addressing environmental and social issues is now considered critical to the proper management of transaction, portfolio and reputational risks. The question is no longer whether commercial banks should address the sustainable development aspects of the activities they support, but how they should do it – what substantive standards should they apply? How should they implement them? And how should they assure compliance?

The maximum score of 4 means that "All, or nearly all, of the policy meets or is in line with relevant international standards." They assessed banks' policies in 13 substantive areas such as human and labour rights, indigenous people, climate and energy, biodiversity, etc. The banks chosen have a large project finance activity.

The leading banks in this rating were: ABN Amro and the HSBC Group with a score of 1.31, followed by J.P. Morgan Chase, Rabobank and Citigroup. The laggards in ascending order were: Sumitomo Mitsui with a score of 0, UBS, HBOS, Deutsche Bank, BNP Paribas and Bank of Tokyo-Mitsubishi with 0.08 points.

From the scoring we see that the over-all level is low. Generally US, British and Dutch banks fare quite well, whereas Swiss, German and Japanese banks score poorly.

The Commission of the European Union EUC (2001) has taken up these ideas. It sees *corporate social responsibility* (CSR) as a concept whereby companies incorporate social and environmental concerns in their business operations and in their interaction with their stakeholders on a *voluntary* basis. It is about enterprises deciding to go beyond minimum legal requirements and obligations stemming from collective agreements in order to address societal needs. It seems that there is still room for improvement.

Overall, it is evident that internationally active banks suffer both from over-regulation and overlapping regulation, i.e. similar or identical requirements to satisfy different supervisory requests. BITS (2005, 25), a non-profit US banking consortium, complains about such overlaps:

> The overlaps are a result of a piecemeal approach to the development of examination guidance associated with the implementation of each successive set of regulatory requirements.(...) Unless more coordinated examination guidance is forthcoming to allow banking organizations the necessary flexibility to identify and leverage their internal processes and resources to achieve compliance with the overlaps in these regulations, the existing inefficient and siloed approach will continue to present significant challenges to the banking industry and the regulatory community from both an operational and financial standpoint.

Thus, the onerous compliance burden is not only due to the content of the regulation but also to the unsystematic manner in which they are imposed.

1.3   The Basel Regulatory Effort

The previous rules, known as the "Basel Accord" or Basel 1, were agreed in 1988 and implemented in 1992. They require a bank's capital as defined by the accord to be at least 8% of its risk-weighted assets. The specific weight for each class of asset goes from zero for assets thought to be very safe to 100% for unsecured loans. During the 1990s, banks' risk-management systems became far more sophisticated than required. Moreover, Basel 1 had an unwanted consequence: its weights did not conform to the market's estimate of the bank's risks. Hence, banks played regulatory arbitrage: they disposed of risks for which Basel 1 required more capital than the market deemed appropriate while retaining assets for which the market required more capital than the regulators did. The increase in banks' capital-adequacy ratios thus reflects regulatory arbitrage as well as cautious risk management.

Basel 2 is intended to align regulatory capital requirements with actual risk and to account for improvements in the best banks' practices. The outline has been clear for some time and almost all details have now been sorted out – besides the lessons to be learned from the financial crisis of 2008.

Basel 2 consists of three pillars:

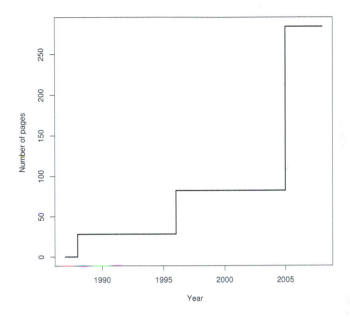

FIGURE 1.4: Number of pages of the Basel Accords with respect to time.

- Pillar 1: Minimum Capital Requirements,

- Pillar 2: Supervisory Review Process,

- Pillar 3: Market Discipline.

Pillar 1 requires banks' capital to be at least 8% of risk-weighted assets, unchanged from the old accord: but the weighting system has been altered. Pillar 2 implies that banks must increase their capital base, if national supervisors think they are too low, even if they are above the minimum capital rate. Pillar 3 stresses the importance of market discipline, and supposes that banks should become more transparent where risk is concerned.

Pillar 1 breaks down regulatory capital into three parts: credit risk, market risk and operational risk. The rules for market risk are identical to Basel 1, amended for this purpose in 1997 (see Figure 1.4). The part for operational risk is completely new.

Banks will now have three options to measure credit risk, each of which will assign more specific weights to their assets than Basel 1 did. The choices are:

- Standardised Approach,

- Internal Ratings-Based Approach,

  – Foundation Approach,

  – Advanced Approach.

Under the simplest "standardised" approach, banks ought to use external data from credit-rating agencies and others. Under the foundation approach, as a general rule, banks will provide their own estimates of default probabilities and rely on supervisory estimates for other risk components, i.e. loss-given-default and exposure at default. Under the advanced approach, banks will provide more of their own estimates of all risk components and their own calculation of the effective maturity.

For operational risk quantification there are also three or more options, namely:

- Basic Indicator Approach,

- Standardised Approach,

  – Alternative Standardised Approach,

- Advanced Measurement Approach.

For many activities, and thus for most banks, the new regime implies massive changes in capital requirements. The new operational-risk item, for instance, will affect banks specialising in areas that involve little lending and have therefore in the past needed little capital to keep supervisors happy. Almost any bank with a lot of retail business can expect its minimum capital to fall.

Second, the majority of American and European banks are comfortably capitalised according to Basel 1. A change in their capital requirements might therefore make no big difference: if they are not constrained by capital, their minimum could go up or down without changing their behaviour. Thirdly, even if a bank's minimum requirements under pillar 1 diminish, its supervisors could still increase the minimum under pillar 2. Fourth, a bank that reduced its capital ratio might generate scepticism in financial markets. So even if Basel 2 relaxes capital constraints, the market might not. Most banks that will have to implement Basel 2 seem content, if not simply resigned. These banks and their regulators, expect it to come into effect more or less on time. But with the crisis of 2008 some haphazard national regulation is on the rise.

In the European Union every bank without exceptions down to the last savings bank will have to comply with Basel 2. In America, only a few large banks will be required to migrate to Basel 2 rules, because the Americans propose that only "big, complex banking organisations" shall adopt the new rules. This encompasses approximately a dozen banks. Other banks are free to change regime if they wish. America's Basel 2 banks, however, will have to use A-IRB, the most advanced assessment method for credit risk. They will also have to apply the most sophisticated of the three methods of measuring operational risk. American banks are also subject to domestic prompt corrective action (PCA) requirements.

There is still much to argue before Basel 2 comes in force. Such a intricate set of rules is almost sure to produce unintended consequences. Implementing Basel 2 will cost the banks an enormous amount of time and resources. Dowd (1998, 222) is especially critical of the process and the outcome:

> One problem is that there is no reason to trust – indeed, every reason to distrust – the process by which capital adequacy regulations are formed. This process is one of negotiation in and around committees, which in itself often leads to compromise decisions that are frequently unsatisfactory and sometimes downright irrational. The pressure to satisfy interested parties also produces excessively long and detailed rulebooks that often reflect the concerns of participating parties rather than real issues on the ground, and these rules are frequently drawn up in almost total disregard of the cost they inflict on regulated institutions.(...) It is therefore naive, to put it mildly, to believe that the combination of this process and these inputs can produce a logically defensible outcome.

This opinion may seem a little harsh, especially as regulation, necessarily has a political aspect, and thus suffers from typical political maladies, i.e., incompetence, lobbying, compromise, bargaining, etc.

1.4 Risk and Capital

1.4.1 How Do We Define Risk?

If we want to manage or discuss risk then we must try to define it. It seems odd that this still needs to be done but our impression is despite much discussion and the undisputed relevance of risk to the banking industry, a solid definition is lacking.

Therefore, let us begin by peeking at "risk" in several dictionaries:

- a factor, thing, element or course involving uncertain danger; a hazard,

- the danger or probability of loss to an insurer,

- the variability of returns from an investment; the chance of non-payment of a debt,

- the possibility of suffering a harmful event,

- uncertainty with respect to loss; also: the degree of probability of such loss (syn: peril),

- a person or thing that is a specified hazard to an insurer "a poor risk for insurance,"

- an insurance hazard from a specified cause or source,

- a source of danger; a possibility of incurring loss or misfortune; (syn: hazard, jeopardy, peril),

- a venture undertaken without regard to possible loss or injury (syn: peril, danger),

- the probability of becoming infected given that exposure to an infectious agent has occurred (syn: risk of infection).

A cursory semantic analysis reveals three clusters of definitions: 1) danger, hazard and peril, 2) loss and harm and 3) possibility, probability and chance. In insurance and medicine the potential subject of harm is also called a risk. Moreover, two additional terms stand out: 1) the negative outcome and 2) its possibility of occurrence with given probabilities. If we take all this together then we can attempt the following definition.

Definition 1.1. We understand by *risk* a negative monetary outcome, stemming from a given source, described by its probability distribution or probability characteristics.   □

Thus, risk will be treated as a so-called *random variable*. By "negative" we also understand a deviation from the expected even in the case where there is no loss. It is not always possible to come up with a distribution. Therefore some reduced features or characterisations, e.g., probability of occurrence, severity, expectation, variance, other moments, is the best one can assess. One may say the same that Saint Augustine said about time: "What then is time? If no one asks me, I know: if I wish to explain it to one that asketh, I know not."[2]

The term "risk" may also be distinguished from uncertainty. In economics and decision-making "risk" carries certain connotations of a certain situation. Knight (1921, 209) made the distinction between risk and uncertainty:

> But Uncertainty must be taken in a sense radically distinct from the familiar notion of Risk, from which it has never been properly separated. The term "risk," as loosely used in everyday speech and in economic discussion, really covers two things which, (...) are categorically different. (...) The essential fact is that "risk" means in some cases a quantity susceptible of measurement, while at other times it is something distinctly not of this character; and there are far-reaching and crucial differences in the bearings of the phenomenon depending on which of the two is really present and operating. (...)

---

[2]*Confessiones* XI, 14: Quid est ergo tempus? si nemo ex me quaerat, scio; si quaerenti explicare velim, nescio.

It will appear that a **measurable** uncertainty, or "risk" proper, as we shall use the term, is so far different from an **unmeasurable** one that it is not in effect an uncertainty at all. We shall accordingly restrict the term "uncertainty" to cases of the non-quantitive [sic!] type.

This view is widely accepted by specialists but, alas, starkly at odds with daily usage of the terms. Ekeland (1993, 141) summarises it in his very readable booklet:

So it is that we find two areas of risk becoming distinct, the realm of the randomness, characterized by probability theory, and the realm of the unknown, in which the only rule is caution. This would no doubt be enough to give a coherent image of human decision, if the border between these areas weren't hazy, and if probability theory hadn't gradually become a universal instrument.

"Randomness" is also called aleatory or stochastic uncertainty while this "un-known" is dubbed epistemic or state-of-knowledge uncertainty. From this it follows that risk is relatively benign in comparison with uncertainty. (Luh-mann, 1991, 1994), a sociologist, tried to make the term "risk" clearer by contrasting it to "danger." Risk involves making a decision and knowing, at the moment the decision is being made, that one may have future cause to re-gret that decision. Thus, risk is the awareness of potential regret at the outset, not the retrospective awareness of danger. Alternatives to the decision taken must be clearly distinguishable in respect of the possibility of loss occurring.

Danger, on the other hand, cannot be influenced by the subject: one is simply exposed to a danger (e.g., from a natural hazard) unless one takes a decision to protect oneself from it. Then the efficacy of the protection becomes the risk. There are situations where the decision that led to a bad outcome can no longer be attributed, though it is evident that without decision the loss would not have materialised. This may be the case when there is a long time period between the decision and the outcome, or when the decision is part of a long and overly complex chain of causal links. If we think of an investment decision (for example, the decision to buy a certain stock) and we suffer a loss because the market has valued it less over time, then we speak of risk. But if there is a systemic crisis, experienced in 2008, we are rather in a "danger" situation which we could call a man-made hazard. Moreover, decisions may have also effects on other, uninvolved people.

Risk is unavoidable because in the modern world we must continuously make decisions: almost nothing happens spontaneously, or at least we perceive the situation as such. Therefore, not making decisions can also become a distinct decision.[3]

---

[3]Yet another sociological view on risk (Beck, 1999, 4): "Risk is the modern approach

1.4.2   What Is Risk Management?

Knowing what risk and what management are is not sufficient to understand the combination of the two. The simplest way to define risk management is to take the description of the Basel Committee (BCBS, 2006, 3) as follows:

Definition 1.2. Risk management is a process to (1) identify, (2) evaluate, (3) monitor and (4) control or mitigate all material risks and to assess the bank's overall capital adequacy in relation to its risk profile.   □

In Section 4.1.1 on page 227 we try to clarify what the complex term "management" means, especially as part of a wider notion of management system.

Identification often starts with categorising and cataloguing – constructing a taxonomy – the known risks and investigating new ones along changes in technology, system, organisation, products and the like. Beyond this potentially reductionist view also emergent risks stemming from the interconnection and interplay of single elements deserves special attention. The Basel initiative has proposes such classifications (see Outline 1 on page 30).

Evaluation can be understood as providing some quantification or at least some metric in order to rank and compare potential negative outcome and to be able to judge them also in economic terms. A good part of this book is concerned with proposing a quantification of operational risk by the construction of a probability distribution.

Monitoring is a process of continuously or regularly assessment of the actual state of risky items. This is needed because the riskiness depends *inter alia* on exogenous, uncontrollable changes of parameters affecting the risky items or situation.

At last, control has at least two distinct descriptions: firstly, control means the process in which a person or group of persons determines what another person will do. Control is thus similar to imparting instructions, commands or orders. The second meaning is used in the sense of meeting some performance standard, which implies (a) there being a defined standard and (b) feedback information to identify variation of actual performance from that required. We assume the latter description as the relevant in our context. Mitigation is the activity of bringing the actual state of risk to the predefined standard. This may be achieved by reducing or avoiding risky business or transferring risk, e.g., by using insurance.

The capital adequacy theme assumes that there is a given risk capacity which must be adequate, i.e., sufficient, in order to warrant the actual risk profile. Because there is only one capital, all risks must be aggregated to make a comparison between actual over-all risk and risk capital useful.

This calculation task is the foundation for the classical understanding of

to foresee and control the future consequences of human action, the various unintended consequences of radicalized modernization."

risk management in a bank. It is the basis for its trans-organisational function. Before the advent of the *value-at-risk* (VaR) methodology (*RiskMetrics*) introduced to the community by J.P. Morgan in 1993, risk management was more or less just the transaction clearance with a right to veto. This changed radically with this new risk measure for a portfolio of assets based on empirical data, statistical methods and finance theory, all made possible by the availability of computer technology. Viewing a bank as a portfolio of assets of different business units made it possible to develop a risk metric for the whole enterprise and simultaneously challenge (with this in-house model) the regulatory conservatism for calculating the appropriate amount of capital. RiskMetrics was a successful attempt to create an industry standard. Under the Basel Accord banks have been permitted to use value-at-risk for market risks since 1996.

Moreover, this metric also improved divisional control by charging activities and transactions with costs for economic capital as defined by VaR and thus requiring a risk-adjusted rate of return while assessing the performance of a business unit in a new way. Capital as a scarce resource could be allocated to business units, typically through an asset and liability committee (ALCO). This capital calculation can be seen as a risk-capital based concept of organisational control. This revolution has also brought about a sub-culture of quantitative experts and risk modellers with a distinct - some would say technocratic – conception of risk management (Power, 2008, 999). Starting from VaR the same spirit was applied to credit risk – CreditMetrics and CreditRiskPlus of Credit Suisse First Boston in 1996 – and as we have seen from this very book, also for operational risk and its loss distribution approach. This view of risk management is strongly contested by other constituencies, especially internal controls and audit. We will come back to this power struggle later on (see Section 4.11).

Besides this procedural description given in the definition there is always the complementary *structural view* of risk management as institution. The risk management organisation of a bank is most often not the decision maker with respect to risk. Its technical activities is to yield the appropriate information, presenting alternatives for decision making and thus supporting and enhancing the management of risk as a responsibility of the line management. Herein the risk management organisation is challenged to communicate its analysis in order to get a judicious response from its target audience. We will fathom this important aspect more fully in Sections 4.3.5 and 4.3.6 starting on page 264.

There should be no ambiguity about the fact that risk management describes processes surrounding choices about risky alternatives.

1.4.3 Capital

For Pillar 1 the quantitative comparison of available capital (also "risk bearing capital") and risk capital is fundamental. The assessment of the available capital as stated in the Basel 1 accord of 1988 (BCBS, 1988) looks at the balance sheet and its accounting rules but goes beyond the legal categorisation of equity to favour of a more economic view. The capital should be freely and readily available for losses. A distinction between so-called *core capital* ("Tier 1") and *supplementary* capital ("Tier 2"). Furthermore, there are deductions and limits on single or groups of items as explained by Table 1.2.

TABLE 1.2: Capital elements of the Accord

Capital Elements	Eligibility
Paid-up share capital/common stock	100%
Disclosed reserves	100%
– Goodwill	100%
Tier 1	100%
Undisclosed reserves	100%
Asset revaluation reserves	100%
General provisions/general loan-loss reserves	100%
Hybrid (debt/equity) capital instruments	100%
Subordinated debt SD	$\min(SD, 50\%$ Tier 1$)$
Tier 2	$\min($Tier 2, Tier 1$)$
Tier 3	
Capital Base	

Disclosed reserves are created or increased by appropriations of earnings or other surplus, e.g., retained profit, general and legal reserves. Undisclosed reserves must be accepted by the regulators as they are not publicly available and may not include hidden reserves on securities held. Provisions ascribed to identified deterioration of specific assets or known liabilities, whether individual or grouped, should be excluded from the general provisions and general loan-loss reserves. However, they are eligible for inclusion in Tier 2 up to to a maximum of 1.25% of risk weighted assets.

In 1996 an amendment to the Capital Accord was issued to provide an explicit capital cushion for the price risks to which banks are exposed, especially those arising from trading activities (BCBS, 1996). It took effect at year-end 1997. This amendment permitted banks to issue short-term subordinated debt (so-called "Tier 3 capital") at their national regulator's discretion to meet a part of their market risks. Eligible capital consists therefore of shareholders' equity and retained earnings (Tier 1 capital), supplementary capital (Tier 2 capital) as defined in the Accord of 1988, and short-term subordinated debt (Tier 3 capital). Tier 3 capital must now have an original maturity of at least two years, is limited to 250% of the bank's Tier 1 capital and must cover market risk, including foreign exchange risk and commodities risk.

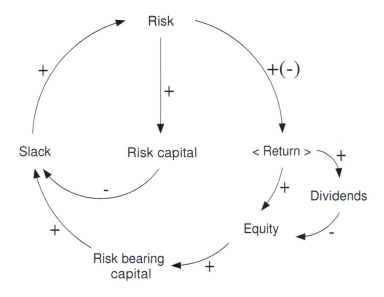

FIGURE 1.5: Influence diagram for risk capital.

On June 26, 2004 the Basel Committee on Banking Supervision published the "International Convergence of Capital Measurement and Capital Standards: a Revised Framework," the new capital adequacy framework ordinarily known as Basel 2. Basel 2 calls yet for another change in the capital requirement specific to the method chosen for credit risk. It reads (BCBS, 2004a, 12):

> Under the internal ratings-based (IRB) approach, the treatment of the 1988 Accord to include general provisions (or general loan-loss reserves) in Tier 2 capital is withdrawn. (...) Banks using the IRB approach for other asset classes must compare (i) the amount of total eligible provisions (...) with (ii) the total expected losses amount as calculated within the IRB approach. (...) Where the total expected loss amount exceeds total eligible provisions, banks must deduct the difference. Deduction must be on the basis of 50% from Tier 1 and 50% from Tier 2. Where the total expected loss amount is less than total eligible provisions (...) banks may recognise the difference in Tier 2 capital up to a maximum of 0.6% of credit risk-weighted assets.

From this historic sketch we see that assessment of available capital is not so simple, and is contingent upon methods chosen. How these rules were dreamt up is a mystery. Figure 1.5 shows how capital requirements act as a circuit breaker. With no minimum requirements the risk could increase indefinitely, as more risk creates the expectation of greater returns, thereby creating more equity, more risk capacity and so on. The capital requirement breaks this otherwise indefinitely growing circuit. The more risk, the more risk capital is needed. This diminishes the slack, i.e., the difference between available and required capital. Smaller slack decreases the growth of risk. In the diagram this implies that in a circle there is an odd number of changes in sign. The capital ratio is calculated using the definition of regulatory capital and risk-weighted assets. The total capital ratio must be no lower than 8%.

$$\text{Capital Ratio} = \frac{\text{Regulatory Capital}}{\text{Total risk-weighted assets}} \geq 8\%. \tag{1.1}$$

Total risk-weighted assets are determined by multiplying the capital requirements for market risk and operational risk by a factor of 12.5 (i.e., the reciprocal of the minimum capital ratio of 8%) and adding the resulting figures to the sum of risk-weighted assets for credit risk.

$$\begin{aligned} \text{Total risk-weighted assets} = {}& 12.5 \times \text{Market Risk} \\ &+ 12.5 \times \text{Operational Risk} \\ &+ \text{risk-weighted assets for credit risk.} \end{aligned} \tag{1.2}$$

Generally, an accounting perspective of risk reveals a historic record rather than a predictive view. The value of a company is more than the shareholders' equity or the above defined "capital base." Therefore, when a bank becomes bankrupt the losses for stakeholders far exceed the capital base. A simpler limit is the *leverage ratio*. American banks are subject also to domestic *prompt corrective action* (PCA) requirements. These were established in order to invigorate the banking system in the early 1990s. A bank to be called "well-capitalised" must have tier-one capital, i.e., mainly shareholders' equity, of at least 5% of its gross assets; below 4%, it is considered "under-capitalised." Taking the reciprocal, the leverage ratio should be below 20. If the required capital under Basel 2 falls below the level to keep equity ratios above 4% then there is an issue. In the wake of the credit crisis regulators and central bankers are prone to favour less sophisticated approaches compared to Basel 2. Nonetheless, regulators allowed broker-dealers, i.e., amongst others the major American investment banks to have even higher leverage ratios (see Figure 1.6) with the well-known detrimental outcomes. Freixas and Parigi (2007) analyse this regime and conclude that in a moral hazard setting, capital requirement regulation may force banks to hold a large fraction of safe assets which may lower their incentives to monitor risky assets.

Example 1.1. In its annual report for the year 2007 Deutsche Bank shows both a regulatory calculation of the capital – alas without capital for operational risk – and an economic compilation (see Table below). The regulatory equivalent is just the regulatory risk position times 8%. Deutsche reports a total regulatory capital of EUR 38.0bn – consisting of 28.3bn Tier 1 and 9.7bn Tier 2 capital.

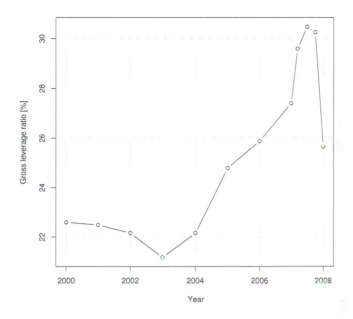

FIGURE 1.6: Gross leverage ratios of the major broker-dealers, based on data for Merrill Lynch, Goldman Sachs, Bear Sterns, Lehman Brothers and Morgan Stanley. (Source: SEC 10-K reports.)

(im mEUR)	Regulatory Equivalent	Economic
Credit Risk	–	8,506
Market Risk	–	3,481
Credit and Market Risk	26,305	11,987
Operational Risk	–	3,974
Business Risk	–	301
Sub-Total	26,305	16,262
Diversification	–	-2,651
Total	26,305	13,611

Without taking the diversification effect into consideration, the sum of the capital components adds up to EUR 16.3bn. Suppose the economic capital for operational risk is added to the regulatory column. Then the *pro forma* total capital is EUR 30.3b, still exceeding the total regulatory capital.

A more economical assessment of the firm is provided by the *franchise value*. This can be defined as the net present value of the future profits that a firm is expected to earn as a going concern. Firms with superior technologies and intellectual capital, good customer relationships, organisation and distribution or scarce input factors like skilled managers create such value. But this value is not reported in the balance sheet and must be derived as an estimate. Franchise value can be defined as the equivalent of the difference between a firm's total market value B_T and its replacement cost C_R, i.e., the expense of rebuilding the firm as is. Neither market value nor replacement cost can be assessed directly. The market value may be estimated as the market value of its equity E_M (shares of stock outstanding times share price) plus the book value of its liabilities L_B. When purchasing an asset at a higher price than its book value, the difference is accounted for in the balance sheet as goodwill. Because this difference is a component of the purchaser's franchise value, we approximate the replacement cost of the assets using the assets' book value A_B minus goodwill G_B. So the franchise value becomes:

$$FV = B_T - C_R \approx E_M + L_B - (A_B - G_B).$$

Valuing companies sustains a whole industry of consultants and there is a plethora of models. As of 30th of October 2004 UBS, the major Swiss bank, has published the figures to be $E_B = 34.5bn$ for the equity at book value while the market value of the equity was $E_M = 95.8bn = 2.8 \times E_B$. Now Demsetz et al. (1996) have shown in a regression exercise that banks with substantial franchise value is correlated with a high degree of protecting the equity. They found that that banks with more franchise value hold more capital and have less asset risk than banks with less franchise value. Although their tendency to hold risky loans is similar to other banks, they maintain better diversified loan portfolios. Thus, franchise value has a disciplinary role.

And what about the Modigliani-Miller proposition? The proposition says that under some quite general conditions the value of the firm is independent of the financial structure (Tirole, 2005, 77). Or put differently, the level of debt V_D – split between different levels of seniority and collateral – has no impact on the total value V of the firm. The simplest way to elucidate this is to assume risk-neutrality, and thus taking the expectation as valuation of two options. The value of the equity V_E is the expectation of the return R minus debt repayment D, if greater than zero. The value of debt is the expected value of the debt repayment $E[D]$ or if R is less then $E[R]$. Thus Thus, V is not a function of D. Decisions concerning the liability structure affect only how the income generated by the firm is shared, but has no effect on the total firm value. This stunning result was produced by Modigliani and Miller (1958, 1963). It is still some sort of puzzle.

1.4.4 Liquidity

The "right" amount of capital alone cannot safeguard a bank from serious trouble. Capital allows banks to cushion unexpected losses and provides financial flexibility to support unexpected asset growth or to sell assets at a loss, if necessary, to meet obligations. Higher levels of capital may provide some reassurance to customers and market participants for meeting liquidity requirements of a bank as the bank may seem better able to sustain its associated obligations or appear more creditworthy. Because the funding is increasingly effected through capital markets and wholesale products, which are more volatile than traditional retail deposits, the funding also becomes more risky. However, since 2005 there have been significant changes in the market and in market conditions. Banks have historically used secured funding sources to fund assets, i.e., securitisations, securities financing transactions and repo financing, and even bonds secured by a pledge of loans. However, in the current stressed market conditions firms are increasingly utilising the assets on their balance sheet to gain liquidity. As a firm increases the assets pledged for the benefit of secured investors, the adequacy of the remaining assets used to meet unsecured creditors' claims may be diminished. The Basel Committee on Banking Supervision offers the following concept of liquidity (BCBS, 2008b, 1):

Definition 1.3. *Liquidity* is the ability of a bank to fund increases in assets and meet obligations as they come due, without incurring unacceptable losses. □

The risks associated with liquidity are at least twofold as liquidity can be produced by borrowing, i.e., funding, or by liquefying assets and thus being exposed to actual market conditions. Therefore the following distinction is made.

Definition 1.4. *Funding liquidity risk* is the risk that the firm will not be able to meet efficiently both expected and unexpected current and future cash flow and collateral needs without affecting either daily operations or the financial condition of the firm. *Market liquidity risk* is the risk that a firm cannot easily offset or eliminate a position at the prevailing market price because of inadequate market depth or market disruption. □

The two aspects are closely related as one way of generating liquidity is selling assets on the market. Not being able to determine the timing of the sale exacerbates the latter. Besides the market there is of course the central bank that has to decide whether to make liquidity available in order to prevent a general problem. Banks are very vulnerable to liquidity risks because one fundamental role of banks is the transformation of size and maturity, as embodied in the collection of short-term deposits and the offering of long-term loans. The interdependencies with other risks is well-summarised in the following quote (BCBS, 2008a, 1).

> The contraction of liquidity in certain structured product and inter-bank markets, as well as an increased probability of off-balance sheet commitments coming onto banks' balance sheets, led to severe funding liquidity strains for some banks and central bank intervention in

some cases. These events emphasised the links between funding and market liquidity risk, the interrelationship of funding liquidity risk and credit risk, and the fact that liquidity is a key determinant of the soundness of the banking sector.

In order to manage this risk, a bank should maintain a cushion of high-quality, unencumbered, liquid assets to be held as insurance against a range of liquidity stresses, e.g., the loss or impairment of available funding sources. As of now, there are no explicit quantified regulatory requirements in the Basel Accord. A very old and unsophisticated method of liquidity management is through simple limits expressed as ratio between certain debt items and liquid assets. Since 1997 the so-called "net capital rule" has been applied by the SEC to broker-dealers, and thus to the now-defunct investment banks. It requires a ratio of no more than 15:1 between indebtedness and liquid assets. In 2004, the SEC decided to allow broker-dealers with assets greater than USD 5 billion to use their own risk models for determining liquidity risk. The then-leading investment banks had ratios of more than twice that. It is argued that these high ratios are responsible for the extinction of these banks, either through absorption by stronger financial institutions (Bear Stearns was acquired by J.P. Morgan Chase), bankruptcy (in the case of Lehman Brothers) and transformation into a Banking Holding Company (as witnessed from Goldman Sachs and Morgan Stanley).

Large capital ratios are good, but before using this associated capital it must become available in the form of liquid assets.

1.4.5 Ruin Theory

Ruin theory links risk as random variable to capital. The classical *Ruin Problem* (Feller, 1957, 313) considers the familiar case of the gambler who wins or loses a dollar with probabilities p and $q = 1-p$, respectively, playing against an infinitely rich adversary who is always accepting to play (our gambler has the privilege of stopping at will). The gambler follows the strategy of playing until he either loses his entire capital S_0 and is thus "ruined" or increases it to S with a gain of $S - S_0$. We are interested in the effect of the initial endowment or capital S_0 on the probability of the gambler's ruin and on the distribution of the duration of the game.

The probability of ruin is given as (Feller, 1957, 314):

$$q_{S_0} = \begin{cases} \dfrac{(\frac{q}{p})^S - (\frac{q}{p})^{S_0}}{(\frac{q}{p})^S - 1} & \text{for} \quad p \neq q \\ 1 - \dfrac{S_0}{S} & \text{for} \quad p = q. \end{cases} \tag{1.3}$$

From the formula we see that the $\partial q_{S_0}/\partial S_0 < 0$, i.e., the probability of ruin is lower the more initial capital is endowed. This is the fundamental rationale for regulators' control of capital levels. Figure 1.7 shows the ruin probability as a function of the relative gain target S/S_0. Obviously, the higher the target,

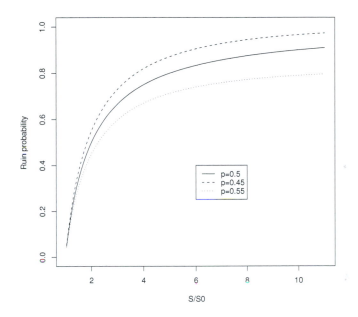

FIGURE 1.7: Gambler's probability of ruin.

the greater the probability of ruin and the lower the probability of winning. Besides the ruin probability, the game's duration relative to the endowment is also of interest. On average, how long – how many tosses – does it take to become ruined? Feller (1957, 317) gives the following formulae for the duration D_{S_0} with initial endowment S_0:

$$
D_{S_0} = \begin{cases} \frac{S_0}{q-p} - \frac{S}{q-p} \times \frac{1-(q/p)^{S_0}}{1-(q/p)^{S}} & \text{for} \quad p \neq q \\ S(S_0 - S) & \text{for} \quad p = q. \end{cases} \tag{1.4}
$$

The analogy to enterprise consists in identifying the capital with the (book) equity of the bank, the wins and losses with either cash flow (for solvency) or income (for bankruptcy) to the bank, and to refer the probabilities to the respective volatilities. The adversary is the market. This approach has been formulated by Lundberg (1909) for insurance companies' provisions, and for corporations by Beaver (1966) and Wilcox (1976).

The above is a *structural model* used for assessing default rates for corporations which pre-dates the Merton model Merton (1974), today's most popular approach. The latter understands equity as a call option on the assets where the option's strike price is the value of the liabilities. This translates neatly into the well-developed theory of option pricing. Simply put, equity E can be described as the asset value A net of liabilities L, but not less than zero, formally $E = \max(A - L, 0)$. Here again, the higher the net asset value the more the option is "in the money" and thus the lower the chance of the option to expire worthless.

Such models are used to determine the default rate for corporations having outstanding loans or bonds in the domain of credit risk.

In these arguments "risk" was understood as the negative outcome resulting from random fluctuations of income or asset value. Many methods for measuring risk have been devised. See Balzer (1994) for an overview. In general, these have been derived from the probability distribution of the negative outcome under analysis that relates the range of probable outcomes to their respective probability of occurring.

From these preliminaries we can deduce the problems we must solve with the Loss Distribution Approach for operational risk: the probabilities of losses due to operational losses over a single year. The corresponding risk capital will be added to market risk and credit risk and compared to the risk bearing capital that has a close link to the equity from above.

On the other hand, regulators aim to improve risk management and are therefore taking "second pillar" measures to curtail efforts to increase capital without improving management.

1.4.6   Risk and Aversion

In 1713 Daniel Bernoulli was asked by his cousin how much he would be willing to pay to play the following game (Bernoulli, 1954):

> Peter tosses a coin and continues to do so until it should land "heads" when it comes to the ground. He agrees to give Paul one Ducat if he gets "heads" on the very first throw, two Ducats if he gets it on the second, four if on the third, eight if on the fourth, and so on, so that with each additional throw the number of Ducats he must pay is doubled.

After n tosses of "tails" in a row the player gets 2^n. With $n = 30$ this is already a very big sum, viz. more than one billion Ducats. Now, most people would offer between 5 and 20 Ducats – but is this fair? On one hand, the probability of winning more than 4 Ducats is just 25%. On the other hand, the potential gain is enormous, because the probability of tossing a very long series of tails before that first toss of heads, is very small but by no means zero. The enormous prize that could be won counterbalances the very small probability of success. Therein lies the paradoxon: If the expected gain is infinite, why is nobody willing to pay a big sum, say 100,000 Ducats, to play the game?

The explanation of this odd behaviour rests on the assumption that 1 Ducat is not always of the same use to its owner. A single Ducat is worth more to a beggar than a billionaire. The utility (called by Bernoulli "moral expectation") of 1 Ducat decreases with increased wealth. If we weight each additional gain x_i by a steadily decreasing utility factor $u_i(x_i)$, the total expected utility U is less than the expected value, consistency with the sums people are willing to pay is reached. Formally stated:

$$U(x) = \sum_{i=1}^{\infty} u_i(x_i)p_i \ll \sum_{i=1}^{\infty} x_i p_i = E(x) \quad \text{with} \quad \partial u/\partial x < 0. \tag{1.5}$$

In 1789 Jeremy Bentham defined "utility" as follows (Bentham, 1907, 44):

> By utility is meant that property in any object, whereby it tends to produce benefit, advantage, pleasure, good, or happiness, (all this in the present case comes to the same thing) or (what comes again to the same thing) to prevent the happening of mischief, pain, evil, or unhappiness to the party whose interest is considered: if that party be the community in general, then the happiness of the community: if a particular individual, then the happiness of that individual.

Nowadays, economist discuss utility much more soberly, understanding it to be a variable whose relative magnitude indicates the direction of preference for some good Hirshleifer (1988, 89). Moreover, they nearly always invoke the so-called rule of *diminishing marginal utility*, that is $u'' \leq 0$. On utility functions see for example Haugen (1993, 132) or Gerber and Pafumi (1998). In this context we are much more concerned with losses than with gains. So the analogy would be that aversion increases with the potential degree of loss. Insurance actuaries' corresponding notion is called *harm function* $h(x_i)$ with the general property $h'' \leq 0$ (Daykin et al., 1994, 190).

Example 1.2. If we assume that the marginal utility $u'(w)$ as a function of wealth w is inversely proportional to w then solving the ordinary differential equation $u' = a/w$ leads to $u(w) = c + a \log w$. If the marginal utility were negatively proportional to the utility, then $u'(w) = -au(w)$ yields $u(w) = ce^{-aw}$. This is the exponential utility function. $\triangle$

In 1961, Daniel Ellsberg[4] (Ellsberg, 1961) conducted a very interesting experiment that clearly elucidates the difference between the so-called *risk situation* and *uncertainty* as introduced by Knight (1921) and discussed previously on page 14.

The game was proposed as follows: the subjects are presented with 2 urns. Urn I contains 100 red and black balls in an *unknown* ratio. Urn II is known to contain exactly 50 red and 50 black balls. The subjects must choose an urn from which to draw and bet on the colour that will be drawn – they will

---

[4]President Nixon's National Security Advisor, Henry Kissinger, called Ellsberg "the most dangerous man in America" – also the title of a documentary film. In 1971 Ellsberg – a senior officer of the Marines – following his conscience published the top-secret so-called Pentagon Papers about the Vietnam war.

receive a \$100 pay-off if the guessed colour is drawn, and \$0 if the other colour is drawn. The subjects must decide which they would rather bet on:

1. a red ball from urn I or a black ball from urn I,

2. a red ball from urn II or a black ball from urn II,

3. a red ball from urn I or a red ball from urn II,

4. a black ball from urn I or a black ball from urn II.

In first and second situations the probabilities are known and these situations are therefore referred to as "risk situations." In the second and third situations the possible outcomes but not their probabilities are known and we speak of "uncertainty." Now, probability theory has established the so-called *principle of indifference* (also called principle of insufficient reason) stating that equal probabilities must be assigned to each of several alternatives, if there is no positive reason for assigning unequal ones. Thus all choices should be equal.

In the first two situations one would expect that the subjects would select each colour with similar frequencies. The experiment confirms this. However, in cases 3 and 4 people uniformly prefer to draw from urn I. It is impossible to determine the subjects' rationale for their choices – do people regard a draw of a particular colour from urn I as more likely? For certain not, because otherwise they would not choose urn I in both cases 3 and 4.

If they choose urn I in case 3, this implies that they believe that urn II has more black balls than red. Yet, if they hold this belief, then they ought to choose urn II in case 4 – but they don't and therein lies the paradox.

The primary conclusion one draws from this very experiment is that people always prefer risk (definite information) over uncertainty (indefinite information). People go with "the devil they know." This is a very, very important finding.

1.5   What Is Operational Risk?

Operational risks are at least as old as business,[5] not to mention banking. Every industry, every activity, is confronted with operational risk because, almost tautologically, operational is equivalent with doing, working, undertaking and industriousness. And therefore operational risks are also involved in all collaborative works such as an enterprise.

Until now, operational risk in banking has been overshadowed by its more

---

[5] Power (2005) argues that "operational risk" is also purposefully constructed to delineate areas of influence, especially high-level calculators like value-at-risk quants and member of the accounting industry.

prominent cousins, credit risk and market risk. However, since the Basel Committee on Banking Supervision (BCBS) has proposed to also introduce capital requirements for operational risk to the Basel Capital Accord, operational risk has gained enormous attention.

Furthermore, increased risk awareness together with stakeholders' demand for greater transparency regarding banks' risks and a capital market that punishes banks that have not properly identified and communicated their operational risk profile, are forcing financial institutions to incorporate operational risk into their risk management systems.

But exactly what are operational risks, or how have they been defined? At an earlier stage of Basel 2 there were just "other risks" until it was determined that this negative definition was too broad. The Basel Committee suggests BCBS (2003b, 120) the following definition :

Definition 1.5. Operational risk[6] is the risk of loss resulting from inadequate or failed internal processes, people and systems, or from external events. □

The four elements, i.e.,

- processes,

- people,

- systems and

- external events,

are called "causes." Undoubtedly, this definition still encompasses a very broad spectrum of risks. It includes legal risk, but excludes explicitly business risks, strategic risk and reputational risk.[7] To explain in more depth the risks involved and to facilitate the collection of loss data across banks, operational risk has been broken down into seven event types (see Outline 1). To keep things simple, we will use the parenthetical names to designate the corresponding event type. Table 1.4 shows major losses classified by events. One initial finding is that in each category there has been at least one loss of more than 100 million USD. The most common and severe event is internal fraud. The data is from Rosengren (2004). Although this definition relies on the categorisation of operational risks based on underlying "causes" (people, processes, systems, and external events) the attribution of an operational loss in any particular case may not be so simple. Losses stemming from damage

---

[6]Up to now, in the U.S.A. another definition according to the FED (2008, 1000.1-4.4) is in vigour: Operational risk arises from the potential that inadequate information systems, operational problems, breaches in internal controls, fraud, or unforeseen catastrophes will result in unexpected losses.

[7]The U.S. Commercial Bank Examination Manual (FED, 2008, 1000.1-4.4) defines reputational risk as: Reputational risk is the potential that negative publicity regarding an institution's business practices, whether true or not, will cause a decline in the customer base, costly litigation or revenue reductions.

1. Internal Fraud ("Internal"),

2. External Fraud ("External"),

3. Employment Practices and Workplace Safety ("Employment"),

4. Clients, Products & Business Services ("Clients"),

5. Damage to Physical Assets ("Physical"),

6. Business Disruption and System Failures ("Disruption"),

7. Execution, Delivery & Process Management ("Execution").

OUTLINE 1: Basel's event types

to physical assets, such as natural disaster, are relatively easier to define than those stemming from internal problems, such as employee fraud or deficiencies in the production process.

In addition, the assignment of a capital charge for operational risk by the New Basel Capital Accord has led to the development of new and competing techniques for identifying, collecting, ranking and measuring operational risk in financial institutions.

Example 1.3. The massive fraud at a French bank lead to a loss of USD 7bn. The board of directors engaged three external experts for a preliminary analysis (Folz et al., 2008). An edited excerpt of the findings reads as follows:

> At Société Générale trading activities are not allowed to take positions on rises or falls in the market (called directional risk) unless they are residual, over a short period, and within strictly defined limits. The author of the fraud departed from his normal arbitrage activities and established genuine "directional" positions in regulated markets, concealing them through fictitious transactions in the opposite direction. The various techniques used consisted primarily of:
>
> - purchases or sales of securities or warrants with a deferred start date;
>
> - futures transactions with a pending counterparty;
>
> - forwards with an internal Group counterparty.
>
> The author of the fraud began taking these unauthorised directional positions, in 2005 and 2006 for small amounts, and from March 2007

TABLE 1.3: Examples of events

Risk Event Types	Examples Include
Internal	Intentional misreporting of positions, employee theft and insider trading on an employee's own account.
External	Robbery, forgery, check kiting and damage from computer hacking.
Employment	Workers' compensation claims, violation of employee health and safety rules, organised labour activities, discrimination claims and general liability.
Clients	Fiduciary breaches, misuse of confidential customer information, improper trading activities on the bank's account, money laundering and sale of unauthorised products.
Physical	Terrorism, vandalism, earthquakes, fires and floods.
Disruption	Hardware and software failures, telecommunication problems and utility outages.
Execution	Data entry errors, collateral management failures, incomplete legal documentation, unapproved access given to client accounts, non-client counterparty misperformance and vendor disputes.

for large amounts. These positions were uncovered between January 18th and 20th 2008. The total loss resulting from these fraudulent positions has been identified and amounts to 4.9 billion EUR, after their unwinding between January 21st and 23rd 2008.

The General Inspection department believes that, as a whole, the controls provided by the support and control functions were carried out in accordance with the procedures, but did not make it possible to identify the fraud before January 18th, 2008. The failure to identify the fraud until that date can be attributed

- firstly to the efficiency and variety of the concealment techniques employed by the fraudster,

- secondly to the fact that operating staff did not systematically carry out more detailed checks

- and finally to the absence of certain controls that were not provided for and which might have identified the fraud.

A Paris court on February 8th, 2008 ordered that Jerôme Kerviel be placed

TABLE 1.4: Some major losses larger than USD 100m by events (Data from Rosengren, 2004)

Internal		
Allied Irish Bank	USD 691m	fraudulent trading
Barings	USD 1bn	fraudulent trading
Daiwa Bank Ltd	USD 1.4bn	fraudulent trading
External		
Republic New York Corp.	USD 611m	fraud
Employment		
Merrill Lynch	USD 250m	gender discrimination
Clients		
Household International	USD 484m	improper lending
Providian Financial Corp.	USD 405m	improper sales and billing
Physical		
Bank of New York	USD 140m	September 11, 2001
Disruption		
Solomon Brothers	USD 303m	change in computer technology
Execution		
Bank of America	USD 225m	systems integration failures
Wells Fargo Bank	USD 150m	failed transaction processing

in custody. He is being investigated on allegations of forgery, breach of trust and illegal computer use. If eventually convicted, he could face a maximum sentence of three years in prison and a fine of EUR 370,000. Kerviel was awarded a bonus of EUR 300,000 for the year 2007 having asked for the double amount.   △

Example 1.4. In the saying: "recessions uncover what auditors do not" the word "recession" could be changed to "financial crisis." From news dispatches one of the biggest frauds ever can be summarised as follows.

Bernard Madoff, a confessed swindler, was arrested and charged on Thursday, December 11th, 2008 with a single count of securities fraud. Besides managing a well-known market-making firm, Madoff also managed a hedge fund that U.S. prosecutors said yielded USD 50 billion of losses, defrauding banks, investment funds, charities and wealthy individuals. The SEC said it appeared that virtually all of the assets of his hedge fund business were missing. The USD 50 billion allegedly lost make the hedge fund one of the biggest frauds in history.[8] Madoff operated a Ponzi scheme, a swindle offering unusually high returns, with early investors paid off with money from later investors. Around the first week of December 2008, Bernard Madoff told a senior employee that hedge fund clients had requested about USD 7 billion of their money back, and that he was struggling to pay them.

U.S. prosecutors said he faces up to 20 years in prison and a fine of up to USD 5 million – actually he was sentenced to 150 years of prison. In January 2009 a U.S. House of Representatives panel started an inquiry into the failure of securities regulators to detect the fraud. An investment professional named Harry Markopolos had begun writing letters to regulators in 1999 alleging that Madoff was conducting a Ponzi scheme. He continued to send similar letters, right through April 2008. Markopolos maintained that his charges were detailed and specific but the SEC ignored them. The criticism, from lawmakers who said the SEC had been conned or was complicit, has raised the odds for a major overhaul of the U.S. financial regulatory system.

One investor has already sued Fairfield Greenwich Group, running some of the largest so-called feeder funds, to try to recoup more than USD 1 billion the investment firm collected and funnelled into funds managed by Bernard Madoff.

Investment advisers sued for investing customer money with Madoff and will try to make claims to their liability insurers. Policies that could come into play include general and professional liability cover, directors and officers or errors and omissions coverage, and even a few home-owners' insurance policies.

Men willingly believe what they wish, as another saying goes. △

The new accord has also defined a two-tiered classification for reporting, shown in Outline 2. So how does operational risk fundamentally differ from market and credit risk? First of all, the fact that operational risks are not controllable either with respect to precision or timeliness is significant. Any market and credit risks assumed by a bank can be almost immediately modified in normal market conditions because there is a market for those risks. In Figure 1.8 we see how significantly a risk profile can be altered by tweaking the amount and degree of risk taken by trading portfolios. For credit an institution may enter into credit derivative contracts, or, less immediately, it may securitise credits.

A second difference concerns *diversification* which works (to a certain extent) with certain risks but not with operational risks. Such instruments are not and will not be available in the near future for operational risks. Yet another argument concerns scale: reducing the amounts invested does not reduce operating risks, for a certain minimal infrastructure and know-how is necessary to conduct certain activities. A last consideration concerns the extent of the risk: operational risk is ubiquitous, whereas market and credit risk are relatively confined. Figure 1.8 reveals the revenue generated in favourable en-

---

[8]When former energy-trading giant Enron filed for bankruptcy in 2001, it had USD 63.4 billion in assets.

1. Corporate Finance

2. Trading & Sales

3. Retail Banking

4. Commercial Banking

5. Payment and Settlement

6. Agency Services

7. Asset Management

8. Retail Brokerage

OUTLINE 2: Basel's line of business classification

vironments, the tendency to increase risk to compensate short-falls in other less profitable activities, the relative linearity between VaR and revenues in a CAPM-like manner and the huge discrepancy in efficiency as measured by revenues divided by value-at-risk. Lehman Brothers have the steepest slope – and highest efficiency – while JP Morgan Chase are the least efficient. But one has to bear in mind that revenues are accumulated over a period while value-at-risk is a point in time measure and thus less reliable. In the light of the sub-prime crisis the measure seems not to be a good predictor.

1.6   Economic Capital for Operational Risk

If we define risk (see Definition 1.1 on page 14) as a distribution of loss, economic capital is its inverse at a given probability.

Definition 1.6. *Economic capital EC* is the loss amount which is not exceeded with a given probability in a specified length of time. It is the inverse of the distribution function of losses. In our context the time horizon will be one year and the probability shall be 99.9%. □

We use Economic Capital interchangeably with Capital at Risk (CaR) and Operational Risk Capital Requirement (ORR). If this amount or more is available as in a bank's own funds then there is a cushion for a firm's risk of unexpected losses at a given confidence level. Thus, the economic capital figure permits management to estimate the amount of own funds that the bank is risking due to operational risks.

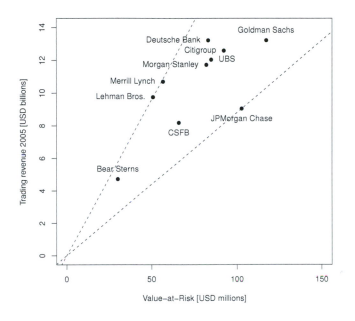

FIGURE 1.8: Value-at-Risk and trading revenues (Data: Boston Consulting Group).

In order to capture a significant proportion of the tail of the credit default loss distribution, the 99.9^{th} percentile unexpected loss level over a one-year time horizon is a suitable definition for credit risk's economic capital. This can be seen in Figure 1.9. For compatibility's sake, the same confidence is applied to operational risks. The increased use of economic capital models requires banks to factor in operational risk. Furthermore, risk management professionals claim that it makes little sense to compute economic capital for credit risk and market risk while omitting operational risk. But non-quants like auditors want to put operational risk in their internal controls framework where quantifications are secondary at best.

The model's output can be used to determine the so called economic capital required to cover losses arising from operational risks. In addition, scenario analysis allows the model to identify the main risk drivers of the economic capital. Thus, the effect of different insurance programmes on the overall required economic capital can be assessed, optimising the insurance programme.

Economic capital as a measure of a firm's risk has several features and benefits: It is a more appropriate measure of the economic risk than that specified under the former regulatory regime; it recognises diversification effects; and it

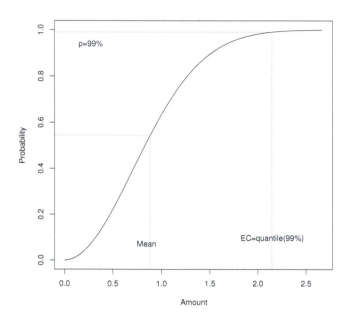

FIGURE 1.9: Economic capital and loss distribution.

allows the integration of different risk types into one common risk measure, i.e., economic capital.

1.7   Operational Risk under Basel 2

The New Basel Capital Accord offers banks a differentiated approach to quantify their exposure to operational risk. According to their specific degree of sophistication, banks can choose between three main approaches (1) the Basic Indicator Approach, (2) the Standardised Approach and (3) the Advanced Measurement Approach. The outline of each of the three approaches is as follows.

1.7.1   Basic Indicator Approach

The Basic Indicator Approach is the simplest method to calculate the capital charge for operational risk. Banks using this plain method are required to hold capital for operational risk equal to a fixed percentage of a single indicator – gross income. Formally the capital charge is described as

$$\text{Required Capital} = \text{AAGI} \times \alpha \qquad (1.6)$$

where AAGI is the average annual gross income over the past three years, and α is a factor set to 15%.

Obviously, this is not a very sophisticated method of calculating capital, as it is not sensitive to differing levels of operational risk in various types of income-generating activities. Because of its simplicity the Basic Indicator Approach is meant to be applicable to any bank. However, under this approach no adjustments to the capital charge with respect to insurance coverage are allowed. Furthermore, more sophisticated, internationally active banks are expected to use one of the more risk sensitive approaches.

1.7.2 Standardised Approach

Under the Standardised Approach the activities in the bank are divided into eight business lines (corporate finance, trading and sales, retail banking, commercial banking, payment and settlement, agency services, asset management and retail brokerage).

Within each business line gross income is used as a proxy for the scale of operations and presumably also the degree of risk. The capital charge is calculated by multiplying the indicator by a factor (beta) set by the Basel Committee for each business line. Formally the capital charge is described as:

$$\text{Required Capital} = \sum_{j=1}^{8} GI_j \times \beta_j, \qquad (1.7)$$

where GI_j is Gross income, β_j are the factors and j is the Line of Business. The total capital charge is the elementary sum of the required capital across

TABLE 1.5: Beta factors by line of business

Lines of Business	Beta Factors
Corporate Finance β_1	18%
Trading and Sales β_2	18%
Retail Banking β_3	12%
Commercial Banking β_4	15%
Payment and Settlement β_5	18%
Agency Services β_6	15%
Asset Management β_7	12%
Retail Brokerage β_8	12%

each business line.

Example 1.5. The Committee has made several investigations into the quantitative impact of the rules. In the data collection exercise of 2002, 89 banks responded on several aspects of operational risk. If we read the given table (see BCBS, 2003a, 39) and divide the figures by 89 in order to create an artificial "average bank" then the capital requirement would be calculated with the weights from Table 1.5 to yield a total of 1.08bn EUR. From

TABLE 1.6: Example of risk charge with the Standard Approach

Line of Business	Gross Income [mEUR]	Risk Capital [mEUR]
Corporate Finance	405.2	72.9
Trading and Sales	1291.5	232.5
Retail Banking	3089.7	370.8
Commercial Banking	1483.2	222.5
Payment and Settlement	512.4	92.2
Agency Services	146.3	22.0
Asset Management	276.8	33.2
Retail Brokerage	280.6	33.7
Total	7485.8	1079.7

these figures we can establish the total percentage of the beta factor to be $0.144 = 1079.7/7485.8$, or 14.4%. The discount with respect to the Basic Indicator approach is a mere 0.6%.   △

At national supervisory discretion a so-called *"Alternative Standardised Approach"* (ASA) can be applied. This uses the standard approach in all but two business lines – retail banking and commercial banking. In these business lines, loans and advances – multiplied by a fixed factor m – replace gross income as the exposure indicator.

The Standardised Approach is designed for banks that have established effective risk management and internal controls, in addition to systems for measurement and validation. This method is more sensitive to risk than the basic approach but still does not recognise insurance coverage or any other risk mitigation techniques.

These two approaches are regarded as "one-size-fits-all" concept, in the sense that these approaches assume that each line of business is exposed to the same degree of risk regardless of a bank's level of risk management and control.

Example 1.6. In the working papers of the Basel Committee on Banking Supervision there was the so-called Internal Measurement Approach (IMA) for calculating the regulatory capital. Although this approach was dropped and included in the final Capital Accord it is worthwhile mentioning because it gives an additional idea on how to devise a bank internal method. It leans on the notion of capital contribution by each combination of loss event i and line of business j and models it as each expected loss leverage by a specific factor γ_{ij}.

$$\text{Required Capital} = \sum_{j=1}^{8}\sum_{i=1}^{7}\gamma_{ij} \times EI_{ij} \times PE_{ij} \times LGE_{ij} \qquad (1.8)$$

$i =$ loss event
$j =$ line of business
$\gamma_{ij} =$ leverage factor
$EI_{ij} =$ exposure indicator
$PE_{ij} =$ probability of loss event
$LGE_{ij} =$ loss given event.

The leverage factor γ can be understood as a fixed relationship between expected loss – as defined by the product of the three terms $EI_{ij} \times PE_{ij} \times LGE_{ij}$ – and the contributory capital, proxied by the "unexpected loss". The exposure indicator is a norming constant in order to derive a monetary unit from the specific exposure, e.g., transaction volume. $\triangle$

Although the Standardised Approach by the differentiation by lines of business is more sensitive to risk than the Basic Indicator Approach, the income reflects not risk but business volume. The assumption that the volume is a good indicator for risk is adventurous to say the least. Furthermore, the loss experience of the institution is not taken into consideration. It is very difficult to envisage how this calculation can be used to manage and control operational risk and its causes in a forward-looking way.

1.7.3 Advanced Measurement Approach

Under the Advanced Measurement Approach (AMA) different methodologies are applicable (provided that the national supervisor has granted permission): Loss Distribution Approach (LDA) or score-card approach, etc. Hence, banks are allowed to use their internal risk models to calculate the appropriate capital cushion for operational risk. Since different banks' methods vary considerably, the Basel Committee has chosen to subject the models to quantitative and qualitative standards rather than force a specific model upon all banks.

The AMA is designed to calculate a bank's underlying risk by using the bank's internal loss data as key input for capital calculation. In the LDA, a bank estimates the frequency and severity distributions of its operational loss events for each business line and loss event combination. Starting from these two distributions, the bank then computes the probability distribution of the aggregated operational losses over a given holding period. The AMA differs in that it allows banks to take into account correlations in operational risk between business lines and event types. This should result in a more precise reflection of operational risk for the bank as a whole.

The AMA can also be realised by the score-card approach. Using different methods, specific risk indicators forming the basis of the score-card are calculated for each business line at regular intervals. The risk capital initially calculated for operational risks (initial capital) – e.g., by means of a loss distribution – is continuously adjusted to the current risk profile and risk control environment by means of score-cards that include a series of indicators used as estimators for special risk types. This involves not only quantitative, but also qualitative, criteria that can provide risk management with insightful information on the assessment and control of potential risks. The data generated are objective and validated by means of historical internal and external loss data. The focus may be with individual business lines or the entire bank. The score-cards are intended to bring a forward-looking component to the capital calculations and to reflect improvements in the risk control environment that reduce both the frequency and severity of losses.

The AMA is intended only for sophisticated banks that have substantial systems in place to control operational risk. The application of this approach depends on the explicit approval of national regulators and requires the compliance with qualified qualitative criteria.

The bank also has to prove that the risk measure reflects a holding period of one year and a confidence level of 99.9%. To ensure that the right "tail" of the operational loss distribution is sufficiently considered, the system also has to be able to survive stress testing to prove it is capable of incorporating low frequency, high impact losses.

The AMA also recognises insurance coverage as a risk mitigation device. However, eligible insurance contracts must meet specific criteria (see The Joint Forum, 2003) and the recognition of insurance mitigation is limited to 20% of the total operational risk capital charge.

1.8   Role of Insurance

We think it worthwhile to cite a definition of the term "insurance" in the interest of elucidating some of the finer points Rejda (1995, 21).

Definition 1.7. *Insurance* is the pooling of fortuitous losses by transfer of such risks to insurer, who agree to indemnify insured for such losses, to provide other pecuniary benefits on their occurrence or to render services connected with the risk. Fortuitous means unforeseen, unexpected and resulting from chance. Indemnification means restoring the insured to his or her approximate financial position prior to the occurrence of the loss. Loss includes harm.   □

Insurance is thus not equivalent to hedging. Hedging financial instruments pay on occurrence of a stated event, not necessarily tied to a loss of the hedger. In most such financial contracts the triggering event is objectively defined, whereas in insurance contracts, it is often open to debate whether the event is covered or not.

Banks can mitigate the risk they are facing either by hedging financial transactions, purchasing insurance, or finally, by avoiding specific transactions. With respect to operational risk, several measures can be devised to mitigate such losses. For example, damages from natural disasters can be insured. In addition, precautions can be taken against system failures (e.g., faulty installation of backup facilities). However, other risks, such as internal fraud, can hardly be mitigated through process-related measures.

Risk transfer or risk mitigation techniques such as insurance covers are not designed to control risk. Rather, they help a firm either to avoid altogether or actively manage the probability of loss and lessen the amount of damage experienced when a loss event occurs. Thus insurance is a form of *contingent financing*.

In order to decide whether it is worthwhile to transfer operational risks to an insurer, a bank needs to identify the relative benefits of the insurance, and assess whether they overbalance the associated costs. The optimal amount of operational risk insurance can be determined by comparing the marginal benefit of insurance coverage against its marginal cost. In short, a bank will only buy insurance to mitigate its operational losses if the cost of insurance is less than the cost of using its own internal capital or the rate at which the bank can raise new capital in the market. Thus, insurance replaces capital as a cushion against unexpected losses.

In theory the price of insurance must equal the expected value of loss during the policy period plus the administrative costs and profit for the insurance company. The administrative costs and profit may, as illustrated in the Figure 1.10, actually increase the cost paid by the bank, (higher mean) so that it would be cheaper for the bank to cover such losses out of its own capital. The fact that a bank is willing to pay a premium higher than the expected value of loss demonstrates that banks find the reduction in uncertainty valuable in and of itself. Yet, it is important to remember that risk could be less costly to the insurance company than to the bank because of the insurance company's diversification, improved risk management and pooling of risks. Therefore, both the standard deviation (and variance) and the mean of the loss distribution function would be reduced. The loss distribution of the insured will have more mass in the vicinity of the expected loss and less mass in the tail.

In addition, insurance can enhance shareholder value by smoothing a bank's cash flows and preventing financial catastrophes due the occurrence of extreme events.

The insurance industry offers a broad range of classical insurance products. The non-life part of insurance consists of property and liability (or casualty) insurance. Within property insurance there is a third line of products called *special lines* (see Table 1.7).

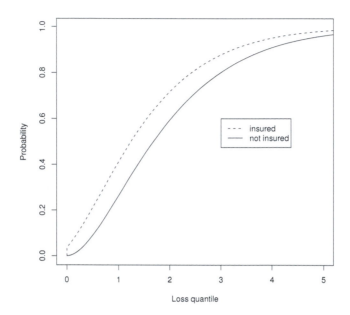

FIGURE 1.10: Influence of mitigation on the LDA (stylized).

Casualty insurance is defined as that which remedies the effects of harm done to a person's legal rights. Such wrongs include crimes, breaches of contract or torts. Torts must fall into one of the three following categories: intentional torts, negligence or absolute liabilities. The last term, "absolute liabilities" means that liability is assumed even if fault cannot be proven. The classic example is a company's liability for occupational injuries. Table 1.7 lists the usual kinds of casualty insurance type.

Of the above-listed insurance products, some reduce (not necessarily intentionally) risks incurred from operational losses. Furthermore, multi-peril basket insurance products, such as all-risks insurance and organisational liability insurance, are also being offered to the banking sector.

Although other potential risk financing alternatives, such as mutual self-insurance pools, securitisation, finite risk plans are also available, the alternative risk transfer market is still embryonic. Thus, for the near future, traditional insurance is likely to be the only effective and cost-efficient instrument for banks to mitigate operational risks.

The Joint Forum (2003, 23) notes that:

> The current market for various types of operational risk transfer is not a mature market. While there are longstanding insurance markets for many specific elements of operational risk (e.g., theft, prop-

TABLE 1.7: Property and casualty synopsis

Property Insurance	Special Lines
Fire Insurance	Bankers Blanket Bond
Extended Coverage	Fidelity/Crime/Computer Crime
Increased Cost of Working	Burglary/Theft/Robbery
All Risks (Property)	Kidnap and Ransom
Construction/Erection All Risk	Unauthorised Trading
Glass Insurance	
Electronic Insurance	
Transit Insurance	
Insurance Valuable Items	
Travel/Baggage Insurance	
Fine Art Insurance	
Business Interruption Insurance	

Casualty Insurance
Casualty and Accident Insurance
General Liability
Professional Indemnity (Error and Omission)
Directors & Officers Liability
Property Owner's Liability Insurance
Environmental Impairment Liability
Workers Compensation
Employers Practice Liability
Personal Accident Insurance
Motor Insurance

erty losses resulting from external events such as fires, etc.), this is primarily piecemeal. (...) In the current market, there is little appetite in the insurance industry to develop innovative new products to more broadly cover operational risks, and any products would likely be prohibitively expensive for protection buyers. Because the insurance market is cyclical, however, when the current hard market ends insurers may attempt to develop innovative new products – either insurance or alternative risk transfer instruments – to facilitate operational risk transfer.

The financial crisis of 2008 will not help to induce insurance companies to new products in this field. The nationalisation of the AIG, American Insurance Group, the then biggest insurance company world-wide has evidenced the close intertwining of the participants in the financial industry. It could well be that in the aftermath of the crisis many industry participants will focus on their core business that they should understand best.

As of writing we cannot find any concrete offer of description or an operational risk insurance on the market. The fundamental differences between an insurance policy and a financial guarantee is hard to be overcome. The hype for this kind of mitigation was during the consultative phase of the Basel 2 Accord, i.e., around 2003.

1.9   Regulation after the Crisis

It is a fact that at the time of this writing we are in a deep crisis which started as a sub-prime lending problem and then grew into a financial system crisis and credit crunch, and that is now developing into a severe recession. While popular opinion believes that bankers' excessive greed combined with poor risk management that overemphasized quantitative models are to blame, the analysis of causes – circumstances and coincidences included – goes far beyond these culprits.

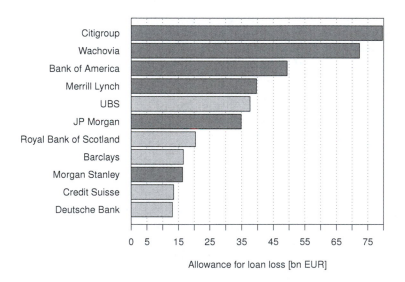

FIGURE 1.11: Major banks' allowance for loan losses from 3rd quarter 2007 to 2nd quarter 2009.

In order to give an idea of the severity of the crises we show in Figure 1.11 the allowances for loan losses for seven quarters starting in 2007 effected by the major banks. They total 1133.4bn EUR worldwide, 762.9bn in the USA and 341.3bn in Europe. The governmental commitments, consisting of capital injections, debt guarantees, asset purchases and asset guarantees – excluding injections to Freddie Mac and Fannie Mae and the USD 700bn of TARP commitments – are depicted in Figure 1.12. For the shown ten countries they total EUR 4994bn, a fantastic figure.

The macro-economy was characterised by an enormous savings glut or excessive liquidity that had piled up in the preceding decades. Moreover, soaring oil prices increased liquidity further. Naturally, investors went in search of higher yields as the interest rates were very low. Especially troublesome was the speed of accumulation. The flow did not go to or stay in emerging markets, exceeding their investment opportunities, but rather went into developed markets, especially real estate in the U.S.A. Simultaneously, securitisation of low-quality assets together with inadequate regulatory oversight created a situation where the leading banks basically converged into a single highly correlated risk pool. The banks actually ignored their own business model of securitisation – shed risks off the economic balance sheet – by choosing *not* to simply transfer risk.

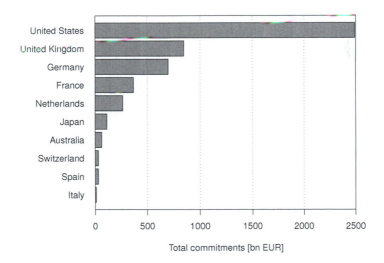

FIGURE 1.12: Governmental total commitments to the financial sector as June 2009 (Data: Panetta et al., 2009, 13).

As we see in Figure 1.13 banks transferred risky assets to a special purpose vehicle and financed them with loans or commercial papers riding the yield-curve. Recourse provisions meant that although the risks still loomed large, the bank's balance was alleviated and could be further leveraged. It must be inferred that the banks thought – helped by overly-optimistic ratings of the assets – that the recourse was so remote that it needed not to be taken into consideration. The credit default swap (CDS) market exploded in volume and only an exiguous number of credit insurers covered a myriad of credit issues. A downgrade of such an insurer meant the downgrade of thousands of credits. Figures on volumes were hard to assess. Following the bankruptcy of Lehman Brothers around USD 400bn of CDS were presented for settlement but only USD 6bn had to be paid, demonstrating the lack of transparency in these markets.

The response included massive liquidity injections as no bank trusted another, bail-outs and nationalisation of ailing banks and insurances, state guarantees, increase of deposit insurance, transfer of toxic assets to governmental funds, increase of funding to the International Monetary Fund (IMF) and changes in accounting rules for valuating assets in a lacklustre market. Then the crisis spread to the real economy, hitting the auto industry in particular and the economy in general. What are the proposed changes in regulation

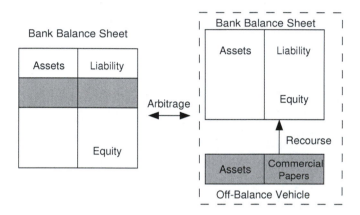

FIGURE 1.13: Off-balance sheet vehicle.

for the financial markets as a result of this crisis? Although the stated ambition is still to guarantee the stability to of the financial system, the proposal nonetheless targets individual institutions and not the "financial system." The Financial Stability Forum, chaired by Mario Draghi, actual Governor of the Italian central bank, proposes the following (FSF, 2009):

- Strengthened prudential oversight of capital, liquidity and risk management

  - capital requirements,
  - liquidity management,
  - supervisory oversight of risk management, including of off-balance sheet entities,
  - operational infrastructure for over-the-counter derivatives;

- enhancing transparency and valuation

- risk disclosures by market participants,
- accounting and disclosure standards for off-balance sheet entities,
- valuation,
- transparency in securitisation processes and markets;

- changes in the role and uses of credit ratings

  - quality of the rating process,
  - differentiated ratings and expanded information on structured products,
  - credit rating agency assessment of underlying data quality,
  - uses of ratings by investors and regulators;

- strengthening the authorities' responsiveness to risks

  - translating risk analysis into action,
  - improving information exchange and cooperation among authorities,
  - enhancing international bodies' policy work;

- robust arrangements for dealing with stress in the financial system

  - central bank operations,
  - arrangements for dealing with weak banks.

At first glance it seems like "just more of the same." The systemic aspect requiring a holistic approach is watered down when it designates financial institutions as "of systemic relevance" – interpreted as "too big to fail" and thus having a governmental guarantee ensuing the notorious moral hazard. What makes sense is to make arbitrage almost impossible, i.e., a given risk may not be transformed or repackaged in such a way that the attached regulatory capital is altered. This should be applied both to individual institutions and the financial system as a whole. For example, the charge for the risk of a loan granted from a bank cannot be lowered by moving it into another account, or by putting it into a special purpose vehicle under control of the bank or by transferring it to another country or selling it to an insurance company. This implies higher trans-national coordination and homogenisation and an increase in scope including all participants in the financial risk-bearing business.

As we have said about the Basel 2 effort, it is a political process involving even more players jockeying for influence. Only the banks are being apologetic, just pressing the accounting standard setters for change. In the menu above we cannot discern any special mention of operational risk.

Chapter 2

The Problem Context

2.1 General Remarks

Modelling always means reducing the object under investigation to its main relevant features. It is a process of optimisation that gauges the complexity and tractability of the object in question using an agreed-upon methodology. In this instance we must considering mathematical models whose "main features" are represented by logical and quantitative relationships.

Models are useful because they (Fishman, 1978):

- enable investigators to organize theoretical beliefs and empirical observations about a system and to deduce the logical implications of this organisation,

- lead to improved Operational Risk Management system understanding,

- bring into perspective the need for detail and relevance and

- provide a framework for testing the desirability of Operational Risk Management modifications.

It is evident that several models can co-exist. We present just one model that:

- complies with the restrictions of the Basel 2 framework,

- is based on long-standing actuarial methodology and,

- tries to balance available data and data modelling.

There is an intimate relationship between data and model which is analogous to the logic of reasoning (see Figure 2.1). The structure consists of inductive and deductive inferences. In the first case, we infer from the specific to the general by drawing conclusions from a set of premises, such as experience, evidence and data. The conclusions contain more information than the premises and may be disproved by further evidence. In the case of deductive inferences, the reverse is true: we move from the general to the specific by drawing conclusions that necessarily follow from the general set of premises. In our context, induction is associated with Statistics while deduction pertains to Probability

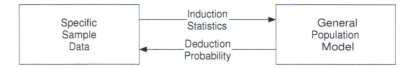

FIGURE 2.1: Data and model.

theory. As our model is based upon given and accumulated loss data, it may be that future data will necessitate a revision of our model. Therefore the LDA with its actual premises must be considered a temporary phenomenon.

2.2   The Data Problem

The data available for modelling losses with a distribution approach is bound by quantity and quality. For most financial institutions the paucity of data can only be overcome with the use of industry data. Even the Basel regulations take this into account.

Because these and other problems have been known for many years in the actuarial community, there are methods in place to cope with these issues. Specifically, we see the following issues in the context of operational risk data modelling.

As a first phase the more sophisticated banks built new loss databases from scratch or enhanced pre-existing ones to satisfy the Basel requirements. Each loss must now be classified with both an Event and a Line of Business identifier. Otherwise it is quite self-evident what kind of data should be collected. In Figure 2.2 on page 52 you see a so-called *entity-relation diagram* often used for database design. The boxes are "entities" which contain items that can be thought of as the header of a corresponding table. The bold typed key-words are primary keys for uniquely identifying the entries, while the italicized items point to related entries in other tables.

Because there are three levels per event, one would assign a loss to the lowest level, which is linked to the next level above and so on to the top level. See Appendix 5.3 on page 351 for the detailed list of event definitions. Furthermore, like an event, a loss belongs to an organisational unit that in turn is mapped by the entity "Organisation" to the Basel categories. Thus the regulatory requirements are fulfilled. Additionally, the loss may be covered by an insurance policy leading to a claims payment. Lastly, we must deal with indices. In order to scale and put the loss in perspective with previous losses or losses from other external industry participants, it is important to collect potential indicators, e.g., number of employees, turnover, balance sheet. These are primarily related through the time-stamp to the loss.

The collection of data and its entry into a database requires some attention as the categorisation may not always be evident. In order to guarantee inter-subjectivity of the data collection, i.e., the fact that two different individuals engaged in data collection must yield the same result, formal prescriptions and manuals must be issued to workers in the field.

A decision also has to be made on how to handle non-monetary losses and "near misses" (or rather near hits). These are difficult to evaluate, but can provide important information. Additional thoughts are required on how to treat opportunity costs or loss of profit resulting from operational risks.

Furthermore, operational losses frequently have some sort of life cycle. They are not confined to a single point in time, but develop over time and only gradually become apparent. The estimation of the ultimate loss may also change due to new information. Finally, compensations paid under insurance contracts or lawsuits affect the loss amount. But it often takes relatively long until the ultimate loss amount is ascertained. As a result, loss databases should be appropriately flexible in order to take account of such changes and their historic evolution.

2.2.1 Scaling

Longitudinal (over time) and cross-sectional (over industry) data need to be appropriately scaled because conditions may have changed. One example is the effect that the value of money, the structure of the institution, the number of employees or customers, etc. may have on the frequency of occurrence or the amount lost. Some of these indicators also vary among institutions and in the industry. It may also be that some data need to be truncated because a certain threshold cannot be exceeded.

With its simple indicator approaches (see page 36) the Basel 2 regulations have implicitly assume that *gross income* – or a sliding mean of it – is the most important basis for comparisons. In addition, the simple indicator approach assumes linearity. With the Alternative Standardised Approach (see page 38) a second index, viz. loans and advances (balance sheet sums) are given.

2.2.2 Mixing

Pooling or simultaneously considering data from different sources, such as external and internal data, raises compatibility and quality issues. There may be systematic differences in the manner of collection, for example. Data may have been stored with different deductibles, net or gross of mitigants, etc. Statistical tests will reveal the homogeneity of the sample.

2.2.3 Paucity

Because there are few large operational losses and that serious data collection has just begun, the data base is small and requires robust statistical meth-

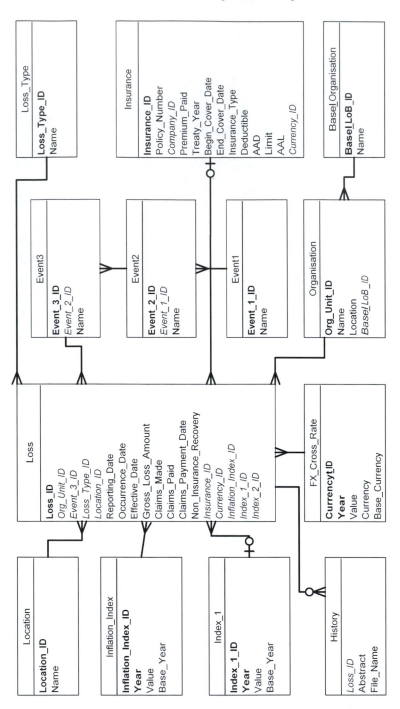

FIGURE 2.2: Database entity-relation diagram.

ods. The publicly-known large losses are so extensively discussed and analysed that it is almost impossible to attribute them to a single cause (Event). Moreover, the question of how to infer the frequency of these "Publicly Known Matters" (PKM) from a more or less unknown sample needs careful consideration. The scarcity of data requires the modeller to refrain from trying to use too many elements with their own distributions. Too high a granularity will reduce the quality of the analysis because of small sample sizes.

2.2.4   External Loss Data

External loss data, i.e., on operational losses experienced by other banks, are collected by several data consortia and, additionally, there are a few commercial providers. Consortia allow their members to exchange loss data in a standardised, anonymous and quality-controlled form. At present, the major data consortia are GOLD (Global Operational Loss Database) in Great Britain and ORX (Operational Risk data eXchange association) in Switzerland. The first was established on the initiative of the British Bankers' Association in the year 2000. The latter was set up one year later and as of writing has some twenty members. An example of a national initiative is DIPO (Database Italiano delle Perdite Operative) in Italy, a consortium founded by the Italian bank association ABI (Associazione Bancaria Italiana) in the year 2000. The use of external data is most problematic with respect to their classification and scaling. Moreover, a loss that can be easily assumed by one bank may be life-threatening for another. But, as suitable data are only available to a limited extent, pragmatic solutions are needed.

2.3   The Dependency Problem

The regulators want to lump together all risk capital linked to the single Lines of Business and Event combinations unless well founded correlations can be produced. This would result in an unrealistic large number which would even depend on the number of combinations used. In order to circumvent this simplification of the situation, modelling is required. The simplest approach is to introduce a so-called copulae where the "true" joint probability function – almost always unavailable or difficult to estimate – is replaced by the marginal distributions plus a decoupled function providing the correlation of the variables. The easiest way to account for dependencies of operational risk losses is by the use of copulae.

2.4   The Insurance Problem

Regulators must be persuaded not only to take insurance into account but to do so in a way appropriate to the problem at hand. All too often the impression arises that modelling of insurance is not a pressing matter and can therefore be dismissed with some simple allowances or "hair cuts." Because insurance products are highly non-linear in their effect constructing a loss distribution of gross losses and then trying to incorporate insurance – often through premiums or "limits" – will not work satisfactorily, if at all. Insurance must be incorporated into the net loss generation before the final loss distribution is calculated. This is only feasible when using Monte Carlo simulations.

2.5   The Mapping Problem

In the context of operational risk there are several overlapping classifications. Because Basel regulators have defined Events and Lines of Business there is a need for a mapping between (1) the real organisation and the Lines of Business and (2) between insured perils/risks and Events.

Classifications are always arbitrary aside from a single exception: the taxonomy of living creatures. The latter is a strict hierarchy: there are no creatures belonging to more than one sub-class of class (Dawkins, 1988, 259). But financial classification systems may be chosen for convenience, utility or clarity. Therefore, the Basel's class system must be judged on these grounds.

Events and Lines of Business can be defined in such a way that all occurrences interpreted as instances of operational risk are classified into only one class. For all practical purposes, the insured-risks classification has grown over the centuries to provide ease of communication. Nonetheless, we must remember that "insured risks" hardly constitutes an ideal basis for classification.

As an example think of a fraud: How to classify a deed perpetrated by an outsider with the collusion of an employee? Is this an external or an internal fraud, or both?

Generally, there are the following cases (see Figure 2.3): (a) an element does not belong to a class, (b) only partially, (c) partially but probably completely or finally (d) only to one class completely.

Suppose the classes are more or less a partition. Now for the mapping there are several cases (see Figure 2.3): (A) One class of A is identical with a class from B, (B) one class of A is completely within a class of B, (C) one class of B is completely within a class of A and (D) a class from A is only partially within a class of B.

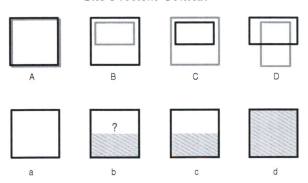

FIGURE 2.3: Relations between classes and elements.

From this one sees immediately that besides the trivial case I mentioned the mapping is not bi-directionally unambiguous. Therefore one has to decide on the direction of the mapping, i.e., "from insurance to Events" or "from Events to insurance." It is evident that mapping from the granular (specific) to the coarse (general) is easier and thus preferable.

Because the Basel regulations require the operational risk quantification to be posted per Event and Lines of Business, the direction is already defined with respect to insurance risks.

The Accord defines principles for business line mapping (BCBS, 2005, 255):

> All activities must be mapped into the eight level 1 business lines in a mutually exclusive and jointly exhaustive manner.
>
> Any banking or non-banking activity which cannot be readily mapped into the business line framework, but which represents an ancillary function to an activity included in the framework, must be allocated to the business line it supports. If more than one business line is supported through the ancillary activity, an objective mapping criteria must be used.

The mapping of typical banking activities to the given Lines of Business proposed by the Accord are summarised in Table 2.1. The more granular level 2 is not shown here.

Special attention must be given to *legal risk*. Indeed, legal risk is included in Basel's definition of operational risk as proposed on page 29. The Accord says in a footnote (BCBS, 2005, 140): "Legal risk includes, but is not limited to, exposure to fines, penalties, or punitive damages resulting from supervisory actions, as well as private settlements." Clearly, this is not a definition. In another document titled "Customer due diligence for banks" (BCBS, 2001a, 4) the Banking Committee on Banking Supervision has given the following definition:

TABLE 2.1: Mapping of lines of business

Level 1	Activity Groups
Corporate Finance	Mergers and Acquisitions, Underwriting, Privatisations, Securitisation, Research, Debt (Government, High Yield) Equity, Syndications, IPO, Secondary Private Placements.
Trading & Sales	Fixed income, equity, foreign exchanges, commodities, credit, funding, own position securities, lending and repos, brokerage, debt, prime brokerage.
Retail Banking	Retail lending and deposits, banking services, trust and estates. Private Banking, private lending and deposits, banking services, trust and estates, investment advice. Card Services Merchant/commercial/corporate cards, private labels and retail.
Commercial Banking	Project finance, real estate, export finance, trade finance, factoring, leasing, lending, guarantees, bills of exchange.
Payment and Settlement	External Clients Payments and collections, funds transfer, clearing and settlement.
Agency Services	Custody Escrow, depository receipts, securities lending (customers), corporate actions. Corporate Agency Issuer and paying agents. Corporate Trust.
Asset Management	Pooled, segregated, retail, institutional, closed, open, private equity.
Retail Brokerage	Execution and full service.

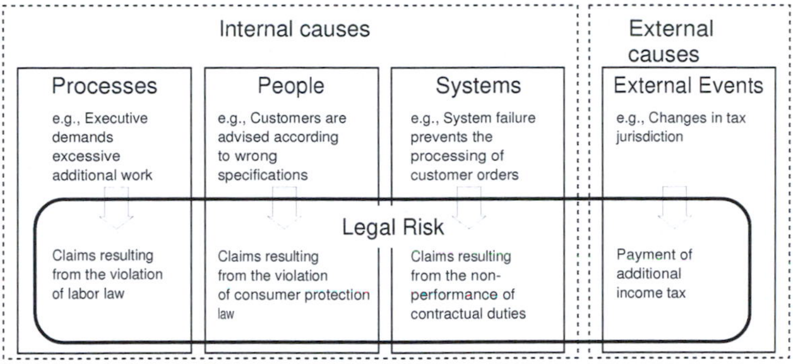

FIGURE 2.4: Legal risk as an intersection. (From OeNB and FMA, 2006, 74. By permission of the Oesterreichische Nationalbank.)

> Legal risk is the possibility that lawsuits, adverse judgements or contracts that turn out to be unenforceable can disrupt or adversely affect the operations or condition of a bank.

Yet another definition, this one from the International Bar Association Working Party on Legal Risk is (McCormick, 2004, 13):

> Legal risk is the risk of loss to an institution which is primarily caused by:

- a defective transaction; or

- a claim (including a defence to a claim or a counter-claim) being made or some other event occurring which results in a liability for the institution or other loss (for example, as a result of the termination of a contract) or;

- failing to take appropriate measures to protect assets (for example, intellectual property) owned by the institution; or

- change in law.

The latter definition extends the scope to exogenous causes like change in law. One could also mention changes in legal market practice and documentation, changes in the application of laws and regulations, significant change in advice received from external lawyer on a substantial point or erosion of rules concerning lawyer-client privileges.

As legal risk is part of operational risk, but is not defined and not an event class, a mapping is necessary if one wants to understand legal risk as an internal class.

Figure 2.4 tries to show how legal risk intersects with the official Basel classification. It appears as some kind of derived risk, because it is due to a previously occurring event that leads to some legal activity. Many banks mention legal risk in their annual reports as a major threat to the organisation.

2.6   The Management Problem

Suppose we have a satisfactory model for the quantification of operational risk. The risk capital is a function of both the model and the relevant parameters. These pertain to probability distributions. Now the most urgent question is how to relate the input to management action. In a first step it is useful to understand the sensitivities of the result from the parameters, because this is a first indication of how changes would have the biggest impact on the result. In order to achieve some real risk reduction, which shows up also in the risk quantity, we need to know how these parameters relate to management action. If we increase the number of risk controllers or simplify risk policies, re-design some critical procedures or extended staff education and so forth, how do these initiatives translate into capital reduction and to what extent? Are there external factors like increased competition in Mergers and Acquisitions (M&A) activities that have an effect on operational risk? Thus, one major issue is the link between both management action and external circumstances and the risk capital amount. This link may be achieved eventually with a general model which could be substantiated as index values.

A second problem is the time dimension. While some actions may have an impact within a short time others will need longer to take effect. The quantification based on historic data will change slowly because of the sparse supply of data. In addition, new data will be averaged with past data and thus further slow the correlation of the data to the changes.

We see this issue as the biggest challenge with the quantitative approaches. Quantification is neither a sufficient, nor even a necessary precondition for management.

2.7   Strategic Risks of a Bank

Strategic risk is not covered by the Basel 2 accord. It could not be otherwise, as free enterprise is the very core of our market system and regulation would impede progress. Therefore it should not be an issue in this book. Nonetheless, we want to spend some time on this topic because any failure of future business plans may pose a much bigger threat to the financial system than operational or other risks a financial institution faces.

Drucker (1999, 43) defines strategy as the set of assumptions as to what its business is, what its objectives are, how it defines results, who its customers are, what the customers value and pay for. He calls this the company's Theory of Business. One important socio-economic fact is the declining birth rate in developed countries, which makes financial services for an affluent, ageing population one of the most important business lines for banks. There is and will be an increased demand for products that guarantee some financial security in the decades ahead. Drucker (1999, 55) says that in the 1990s banks were slow to pick up on this trend and over-invested in the dwindling corporate business sector rather than in the expanding mutual and pension funds sector.

Other strategic threats come from new entrants, especially other institutions that are able to leverage information efficiently over large populations, e.g., telephone companies, big retail stores and peer-to-peer exchange facilitators. As an example, the nascent peer-to-peer lending boom operates on the principle that people needing to borrow money and those able to lend it are better off dealing (almost) directly with each other rather than using a bank or another middleman. Currently, two companies, Zopa and Prosper, are demanding a fee for facilitating such person-to-person lending while providing safeguards and services to both parties. Such institutions may have the potential to expand into mortgages and acquire a share of the banks' retail business. Investment banking, for example, could lose its most profitable M&A business as small teams establish boutiques. There is enough available capital seeking an investment, enough reputable and capable consultants, and sufficiently sophisticated IT systems are available. Many argue that only exorbitant salaries and bonuses keep these M&A specialists from spinning off.

In short, the above examples demonstrate that long-term goals and their pursuit must be the ultimate purpose of a bank. Risk management and the Basel initiative can only help to maintain and direct that purpose.

2.8 AMA Standards

The Basel Committee on Banking Supervision does not recommend a specific approach for operational risk. However, it requires the application of a detailed set of *quantitative* standards to internally-generated operational risk measures for the purpose of calculating the regulatory minimum capital charge. In addition, banks that intend to apply for the AMA have to comply with a second set of equally detailed *quantitative* standards.

The comprehensive and flexible model presented here is intended to fulfil all requirements set by the Basel Committee on Banking Supervision.

Given the continuing evolution of analytical approaches for operational risk, the Committee does not specify the modelling approach or distributional assumptions used to generate the operational risk measure for regulatory capital purposes. However, a bank must be able to demonstrate that its approach captures potentially severe "tail" loss events. Whichever approach is used, a bank must demonstrate that its operational risk measure is comparable to that of the internal ratings-based approach for credit risk, (i.e., comparable to a one year holding period and a 99.9th percentile confidence interval). An important

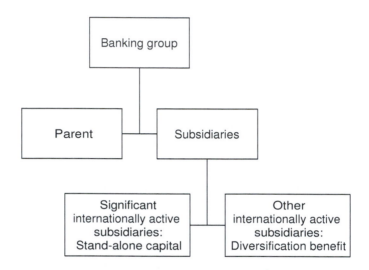

FIGURE 2.5: Home-host structure.

issue is the allocation of capital within banking groups consisting of several substantial subsidiaries abroad (see Figure 2.5). In a BCBS principle (BCBS, 2004b, 3) it is argued that experience demonstrates that capital is usually not freely transferable within a banking group, especially during times of stress. Therefore, each *significant internationally active* banking subsidiary within the group must be adequately capitalised on a stand-alone basis, while others need an allocated portion of the group-wide AMA capital requirement. In calculating stand-alone AMA capital requirements, subsidiaries may incorporate a judicious estimate of the diversification benefits of its own operations, but may not account for group-wide diversification benefits. The fitness of such allocation methodology will be reviewed with special consideration given to the development stage of risk-sensitive allocation methodologies and the extent to which it reflects the operational risk in the legal entities and across the banking group. This "hybrid" approach assumes frictionless co-operation between several national regulators that has the potential to be problematic.

2.9   The Knowledge Problem

Operational risks may be analysed from different points of view: auditors, lawyers, insurance experts, process specialists, strategists, software engineers, financial experts and last but not least, mathematicians, actuaries and statisticians all belong to different "communities," each with its own language, mental models and so forth. Quantification rests on mathematics, and therefore on mathematical models. Successful management requires a broad knowledge base, to say the least. But while such knowledge may be necessary, it alone may not suffice. As Pfeffer and Sutton (2000) depict nicely, there is a wide-spread knowing-doing gap in companies which fail to turn knowledge into action and profit.

The genuine knowledge problem with the Advanced Measurement approach has to do with the model language, i.e., mathematics, and very abstract terms, e.g., probability. We think it useful to take a closer look at some underlying concepts. From our perspective, it turns out that almost everything is vaguer and more debatable than previously thought and that many terms are more a matter of convention than a real necessity. On the other hand, as the flood of companions proves, it is vital to simplify complicated ideas so that they may be acted upon efficiently. The knowledge about the simplifications need also be kept in mind.

2.10   Probability, Causality and Other Primitives

Not surprisingly, probability, causality and correlation are three very important and fundamental terms used in this book. Problems related to ascertaining or eliciting probabilities all derive from them. For example, probability is fundamental to the loss distribution approach as the distribution is actually a probability distribution. Causality in turn is a concept basic to both *explaining* and *predicting* events in general and losses specifically. Correlation is somewhat akin to causality but rooted in probability theory. By looking at these basic concepts, we will understand why it is perfectly legitimate to use the LDA and what it takes to systematically learn from the past with respect to preventing operational losses from arising in the future. Furthermore, we will spend some time examining how to best estimate probabilities.

Taleb (2005, 290) notes that "many readers in the technical profession, say engineering, exhibited some difficulty seeing the connection between probability and belief and the importance of scepticism in risk management." This chapter may be of help for this purpose too.

2.10.1 Probability

We start with the definition of the mathematical term "probability" by means of axioms, i.e., propositions that can be neither proved nor disproved are assumed to be evidently true. These axioms are named after Kolmogorov.

Definition 2.1. A *probability (measure)* denoted $Pr(A)$ is a function mapping a set of events A of the sample space Ω to the real numbers satisfying the following axioms:

1. $Pr(\Omega) = 1$,

2. $Pr(A) \geq 0$ and

3. $Pr(A_1 \cup A_2 \cup A_3 \ldots) = Pr(A_1) + Pr(A_2) + Pr(A_3) + \ldots$ for A_i a finite or infinite collection of pairwise mutually disjoint events of Ω.

$\square$

A *sample space* of an experiment, i.e., measurement or observation, denoted by Ω is the set of all possible outcomes. Each element of a sample space is called an *element* of Ω or a sample point. An *event* in the probability sense is any collection of outcomes contained in the sample space. A *simple* event consists of exactly one element and a *compound* event consists of more than one element. Ω may be finite, countable, discrete or continuous.

It is important to be aware that the theory of probability does not require an interpretation of the term "probability." The axioms tell us how to develop new probabilities from given probabilities without the need for interpretation. The reverse, however, is also true: when unable to define the term to everyone's satisfaction, resorting to axioms provides a handy way out.

But to *ascertain* or measure probabilities – to relate the abstraction to reality – we need a prescription, an interpretation and thus a theory about the nature of the probabilities. The following is mainly due to Hájek (2003).

Jaynes (2007), in a very insightful book, derives the above mentioned axioms from the Pólyia-Cox desiderata of rationality and consistency. The first consists of (1) a relationship between degree of plausibility and real numbers and (2) qualitative correspondence with common sense. The latter claims that rules for manipulating probabilities should be consistent.

2.10.1.1 Classical Probability

The classical interpretation of Laplace assigns probabilities in the absence of any evidence, or in the presence of equally balanced evidence. In such circumstances – especially common in many games of chance – probability is shared equally among all possible outcomes, so that the classical *a priori* probability of an event A is simply a fraction of the total number of possible outcomes[1]

$$Pr(A) = \frac{N_A}{N} \qquad (2.1)$$

with N the number of possible outcomes, N_A the number of favourable outcomes and A an event.

There are some questions concerning this formulation. N and N_A are finite integers. Consequently, irrational-valued probabilities such as $1/\sqrt{2}$ are automatically eliminated. Thus quantum mechanics cannot be accommodated (Hájek, 2003). Moreover, the definition is circular, because it assumes equal probability of the outcomes, which we set out to define. The *principle of insufficient reason* says that in the absence of any prior knowledge, we must assume that all events are equally probable. This principle was also coined *principle of indifference* by John Maynard Keynes. But, if a dice is loaded, i.e., if not all faces are equally probable, then this definition cannot be applied. (Actually, the previous sentence is flawed because it is not compatible with the definition.) Laplace's theory is confined to finite sample spaces. When the spaces are infinite, the spirit of the classical theory may be preserved by appealing to the information-theoretic principle of maximum entropy (Papoulis, 1991, 533; Shannon, 1948), a generalisation of the principle of indifference.

Example 2.1. The principle of indifference can lead to a paradox named after Joseph Bertrand. A factory produces cubes with side-length between 0 and 1 unit of length. What is the probability that a randomly chosen cube has side-length between 0 and 1/2? The tempting answer is 1/2, as we imagine a process of production in which side-length is uniformly distributed. But the question could be restated and thus complicated: A factory produces cubes with face-area between 0 and 1; what is the probability that a randomly chosen cube has face-area between 0 and 1/4? Now the tempting answer is 1/4, as we imagine a process of production in which face-area is uniformly distributed. But we cannot allow the same event to have two different probabilities. And what about a volume of 1/8? Now the tempting answer is, not surprisingly, 1/8. It therefore becomes evident that the paradox arises because the principle can be used in incompatible ways. △

2.10.1.2 Logical Probability

Logical theories of probability maintain that probabilities can be determined *a priori* by an examination of the space of all possibilities. However, they generalise the classical view in two ways: the possibilities may be assigned unequal weights, and probabilities can be computed irrespective of any

---

[1]Laplace (1825, 7): "La théorie des hasards consiste à réduire tous les évènements du même genre à un certain nombre de cas également possibles, c'est-à-dire, tels que nous soyons également indécis sur leur existence; et à déterminer le nombre de cas favorables à l'évènement dont on cherche la probabilité. Le rapport de ce nombre à celui de tous les cas possibles, est la mesure de cette probabilité qui n'est ainsi qu'une fraction dont le numérateur est le nombre des cas favorables, et dont le dénominateur est le nombre de tous les cas possibles."

evidence. Indeed, the logical interpretation, seeks to encapsulate in full generality the degree of support or confirmation that a piece of evidence confers upon a given hypothesis. The most thorough study of logical probability was undertaken by Rudolf Carnap (1950). He draws a sharp distinction between two sorts of probability: frequency-probability and confirmation-probability. For the latter, he thought that a measurement of the overlap of two propositions' ranges could be worked out by purely logical methods, independent of statistical observations (Passmore, 1994, 419). Some features of this inductive logic are by no means clear.

2.10.1.3   Frequency Interpretations

Gamblers, actuaries and statisticians often identify relative frequency and probability. Thus, the probability of "heads" on a certain coin is the number of heads in a given sequence of tosses of the coin, divided by the total number, i.e.,

$$Pr(A) = \frac{N_A}{N} \qquad (2.2)$$

with N the number of *actual* outcomes, N_A the number of favourable outcomes of an event A. We see that Eq.(2.2) is identical with Eq.(2.1), but the interpretation is different. Here too each outcome has the same probability. This definition was proposed by John Venn (1876). Problems arise, however, when the number of trials is limited to zero or one: The faces of a dice never tossed have no probabilities, a coin tossed an odd number of times must be biased for heads or tails and if tossed only once gives the probabilities 1 and 0 for either side.

The response to such criticism led frequentists to consider infinite reference classes, defining probability as the limit of the relative frequencies,

$$Pr(A) = \lim_{N \to \infty} \frac{N_A}{N}. \qquad (2.3)$$

But now there are other problems. Generally, the first k outcomes are compatible with any limit or even with a non-existing limit. The same outcomes could be reordered and thus lead to yet another relative frequency. If an infinite sequence does not exist, frequentists must imagine *hypothetical* infinite extensions of an actual sequence of trials.

Another insurmountable difficulty is posed by the single case. A surgeon asserts: "The probability that John dies within a year is 20%." In the frequentist view this is another way of saying that John is member of a group of patients who on average die within a year in 20 percent of cases. Richard von Mises (1936, 1939) accepts this consequence, insisting that the notion of probability only makes sense relative to a collective. In particular, he regards single case probabilities as nonsense.

Let us then see how the frequentist interpretations fare according to the axioms. Finite relative frequencies of course satisfy finite additivity. In a finite reference class, only a finite number of events can occur, so only a finite number of events can have positive relative frequency.

Limiting relative frequencies violates countable additivity and thus does not provide an admissible interpretation of Kolmogorov's axioms. On the other hand, science is very interested in finite frequencies, and indeed, working with them makes up much of statistics' business!

2.10.1.4 Propensity Interpretations

Probability is thought of as a physical propensity, or property, or disposition or tendency of a given type of physical situation to yield an outcome of a certain kind, or to yield a long run relative frequency of such an outcome. This view was prompted by the craving for making sense of single-case probability ascriptions such as "the probability that this radium atom decays in 1600 years is 1/2." Indeed, Karl Popper (1959b) advances his propensity theory based on such quantum mechanical probabilities (Hájek, 2003). This interpretation seams rather obscure.

2.10.1.5 Subjective Probability

Subjective Bayesians advertise probability as a degree of belief, confidence or credence of suitable agents. For Keynes (1921, 5) the probability concerns propositions, i.e., statements that can be either true or false, and not "events" or "occurrences." Thus he states (Keynes, 1921, 4):

> Let our premises consist of any set of propositions h, and our conclusion consist of any set of propositions a, then, if a knowledge of h justifies a rational belief in a of degree α, we say that there is a *probability-relation* of degree α between a and h.

We would denote this relation by $Pr(a \mid h) = \alpha$ that reads from left to right: "the probability of a given h to be true," or from right to left: "the probability that h leads to or implies a." This belief measure also obeys the basic axioms of probability calculus. Let us identify e with "evidence" and H with a hypothesis. The heart of this interpretation of probability lies in the famous *Bayesian inversion formula*:

$$Pr(H \mid e) = \frac{Pr(e \mid H)Pr(H)}{Pr(e)}. \tag{2.4}$$

This formula states the belief we accord a hypothesis H upon obtaining evidence e which can be computed by multiplying our previous belief $Pr(H)$ by the likelihood $Pr(e \mid H)$ the e will materialise if H is true. $Pr(e)$ serves as normalisation term. In this view $Pr(A)$ is a short-hand for $Pr(A \mid K)$ where K is our tacit body of knowledge.

Now, what is a suitable agent? As mentioned, various psychological studies show that people commonly violate the usual probability calculus in spectacular ways (see Section 2.10.3 on page 74). An agent is required to be logically consistent. These subjectivists argue that this implies that the agent obeys the axioms of probability. Subjective probabilities are traditionally analysed in terms of betting. It can easily be proven that if your subjective probabilities violate the probability calculus, you are susceptible to systematic loss.

For example, suppose that you violate the additivity axiom by assigning $Pr(A \cup B) < Pr(A) + Pr(B)$, where A and B are mutually exclusive. This is equivalent to $Pr(A) + Pr(B) - Pr(A \cup B) > 0$, which means that both selling a bet on A and B individually and buying a bet on $(A \cup B)$ yields a sure profit at no risk. Actually the betting must be taken *cum grano salis*, i.e., the probability is the betting price you regard as fair, whether or not you actually enter into such a bet.

But if utility is not be a linear function of such sums, then the size of the prize will make a difference to the putative probability: winning a dollar means more to a pauper than it does to a billionaire, and this may be reflected in their betting behaviours in ways that have nothing to do with genuine probability assignments. Better, then, to let the prizes be measured in utilities: after all, utility is infinitely divisible, and utility is a linear function of utility.

Subjective probabilities cannot tell the difference between uncertainty due to lack of information and uncertainty that no possible increase of knowledge could eliminate. Actually, Bayesians must assume that all uncertainty is *epistemic*, i.e., arises from not knowing. However, we can try to fill this gap by consulting more knowledgeable people. We consult our doctors on medical matters, our meteorologist on weather forecast, and so on. There is more on this topic in Section 2.10.3.

In conclusion, there is no universally accepted interpretation of probability. The main opposing views are still the finite frequency approach and the subjectivist interpretation. A casual look into a philosophy book like Passmore (1994) shows to what degree this problem has confounded various scientists from disparate disciplines. In short, the whole edifice of probability is firmly rooted in the air. Nonetheless, many other fields must take recourse to probability, and in the following section we will examine some of their approaches. The three interpretations have different domains of applicability (see Figure 2.6). Subjective probabilities cover everything in the world that people have opinions on as long as these opinions follow the aforementioned axioms of probability. Thus even events that have never happened may have a probability. Probabilities as limits of frequencies can only be used when sufficient amount of data is available for apparently similar elements of a reference class. The propensity interpretation has an even more limited domain, because it applies only to items whose causal structure, construction or design is known (Gigerenzer, 2008, 150).

Jaynes (2007, xxiii) sees the frequentist definition as a limiting case or special case of the subjective probability:

> The traditional "frequentist" methods which use only sampling distributions are usable and useful in many particularly simple, idealized

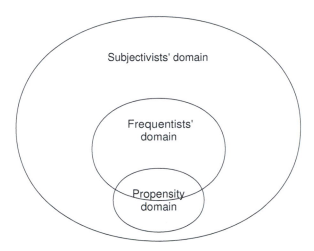

FIGURE 2.6: Probability domains.

problems; however, they represent the most proscribed special cases of probability theory, because they presuppose conditions (independent repetitions of a "random experiment" but no relevant prior information) that are hardly met in real problems.

For risk management, the objective frequentist approach is often deplored for its lack of data, and the subjective interpretation is seen as arbitrariness of experts' opinion.

2.10.2 Causality and Correlation

Causality is one of the most prominent terms in philosophy and science – as a glimpse into a book like Mittelstrass (2004) confirms immediately. The term is used in physics and sociology, in history, jurisprudence, medicine and sciences. While cause and effect pairs are a fundamental way to interpret our reality, nobody seriously doubts that there are reasons why a particular effect comes about. We have little doubt that if the dog swallows poison and dies the next day, the poison is the cause of death. This common attribution of the relationship of cause and effect to a phenomenon precedes all scientific analysis. Objects of sense are categorised roughly into those that act and those that are acted upon. No conscious reflection seems to enter into the judgement that partitions natural things into causes and effects. But when we ask ourselves precisely what we mean when we say, that A is cause and B effect, that A causes B, or that B is the result of A, we raise the question of causality. Whatever answer we give, it is the statement of our conception of causation. It is our judgement as to the actual relationship between A and B.

It will probably become clear that much more is involved than we had thought at first sight.

In science, the distinction is sometimes made between the *principle of causality* and the *law of causality*. The first states that nothing happens without a cause (*nihil fit sine causa*). The latter says that the same causes lead to the same effect, or in a strict reading, if A causes B, then A must always be followed by B. In this sense, smoking does not cause cancer. After Aristotle with his four causes, i.e., material, formal, efficient and final; and its scholastic interpretation by Saint Thomas Aquinas and after Hume and Kant, legions of scientists considered this matter. The concatenation of causes and effects, while assuming that nothing is caused by itself, stretches back to a first cause – or God – as theologians like Saint Thomas Aquinas were convinced.

David Hume (1978, 77, 87) posited that three sufficient conditions must hold for causality:

1. antecedence of cause with respect to effect,

2. spatial proximity of cause and effect and

3. a constant *empirical* connection.

He held that causes and effects are not knowable, but are habits of our mind to make sense of the observation that A often occurs together with or slightly before B. All we can observe are correlations from which we make inductive inferences. Immanuel Kant replaced experience with *reason* in the sense that the connection must be intelligible *a priori*. Thus he escaped the circularity of relying on experience which in turn is shaped by cause and effect. The contrast between induction and deduction as a means for explaining things arises naturally from the cause and effect scheme. At the turn of the last century some influential physicists turned philosophers, like Ernst Mach, were intent upon denying that science explains "why" things happen as they do, or in short that "why" is "how." Banishing causality to the realm of metaphysics, explanations are *descriptions* of regularities and probabilities. Modern physics with relativity theory, quantum mechanics and chaos theory has made things even worse by questioning simultaneity, determinism and so forth. And induction method of acquiring knowledge, supposing that one can move from a series of singular existential statements to a universal statement, has been fiercely attacked by Karl Popper's empirical falsifiability (Popper, 1959a).

In short, causality has shifted from determinism to probability. In this framework causality can be formalised. An event $B(t_0)$ is the cause of an occurrence $A(t_1)$, if:

1. $t_0 < t_1$ and

2. $Pr(B(t_0)) > 0$ and

3. $Pr(A(t_1) \mid B(t_0)) > Pr(A(t_1))$.

Note: $Pr(A)$ stands for the probability that event A occurs, while $Pr(A \mid B)$ is the *conditional probability* of A occurring given that the event B has occurred.

Moreover, $B(t_0)$ would be the *direct cause*, if all possible intermittent events $C(t)$ with $t_0 < t < t_1$ had no effect in the sense of $Pr(A \mid B(t_0), C(t)) = Pr(A(t_1) \mid B(t_0))$ with $Pr(B(t_0), C(t)) > 0$. Finally, a *determinate cause* is given with $Pr(A(t_1) \mid B(t_0)) = 1$. Now, this does not help those interested in a real-world problem that cannot be tested by experiment.

Looking at cause-and-effect atoms is a little bit too simplistic. A first generalisation would be causal chains, i.e., concatenations of cause-and-effect links. More realistically, one has to acknowledge that events happen in a given environment of many circumstances and conditions of which some may be necessary or sufficient but not relevant.

John Mackie (1974) states that it is very rarely possible to state a single cause of an effect. Any event might include a number of factors that play a role in its cause. When an event Z happens regularly after the occurrence of some other factors, say A, B, and C, it is the conjunction (union) of those factors $ABC := \{A, B, C\}$ that he calls a "minimal sufficient condition" for Z. So, the conjunction of the events of taking a lethal dose of poison, failing to drink an antidote and not having the stomach pumped constitutes a minimal sufficient condition for death (Stempsey, 1998).

Any minimal sufficient condition ABC may be enough to bring about Z, but it is usually not necessary. Other conjunctions of factors, e.g., DEF or $PQRS$, would also suffice. The disjunction (intersection) of all the conjunctions that represent minimal sufficient conditions, would be both a necessary and sufficient condition to bring about Z. This disjunction $\{ABC \vee DEF \vee PQRS\}$ is what Mackie calls the "full cause." No one factor alone, not any conjunction of just two, is sufficient to bring about Z. A, B, and C are individually insufficient, but they are non-redundant parts which jointly constitute a sufficient condition ABC. These factors are called "INUS" factors, i.e., "insufficient but non-redundant part of an unnecessary but sufficient condition." The concept of INUS corresponds to the logic of "conditional counter-factuals," i.e., to asking whether in the absence of any component of the causal complex the effect would have developed anyway.

In design and industrial production many so-called root cause analysis tools are used. The key aim is to identify the root causes of problems or events. It is often asked why a fault arises ("cause") and how it comes about ("mode"). There are at least five different "schools" categorised by their problem solving field, e.g., based on safety (accident and health), production (quality control for industrial manufacturing), process (production and business processes), failure (engineering and maintenance) and systems (more general). They sail under the names of "Causal factor tree analysis," "Fault Tree Analysis," "Failure Modes and Effect Analysis" FMEA and the like. There are inductive (bottom-up) and deductive (top-down) approaches. One such method, Apollo Root Cause Analysis ARCA (Gano, 2008), uses the causal model as depicted in Figure 2.7. This model is a simplification of the above-mentioned. The "action

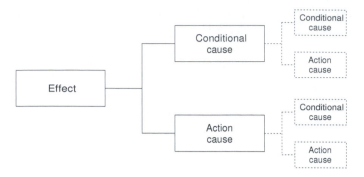

FIGURE 2.7: A causal model for analysis. (From Gano, 2008, 47. With permission.)

cause" is a sort of a trigger, e.g., the striking of a match, while the "conditional cause" is the context, e.g., the presence of oxygen and a flammable material. The effect is the burning.

In jurisprudence *legal cause* must be demonstrated in order to hold a defendant liable for a crime or a tort. It must be proven that causality, or a "sufficient causal link" relates the defendant's actions to the criminal event or damage in question. This has led, prominently in Germany's civil code of 1900 (Traeger, 1904), to the *theory of the adequate causality* mainly in private law where the cause is the event that from experience leads to the effect. In addition within criminal law the *conditio sine qua non*, also known as *theory of the equivalence of conditions*, is sometimes applied. This is the contra-factual argument that the cause is each condition that leads to the effect and which cannot rationally be discarded. Again, experience and reason emerge. For responsibility there must be an element of either intention or negligence where someone fails to take reasonable precautions under the circumstances. If an event was unforeseeable or very unlikely to result in an accident there may plainly be no one to blame. In this framework not every effect, i.e., loss or harm, has a cause.

According to Fumerton and Kress (2001) the law recognises the need to distinguish legally relevant causal factors from irrelevant ones. The concept of tortuous conduct serves to narrow the set of all causal factors to the tortuous causal factors. In addition, the concept of proximate cause narrows this class to the legally responsible tortuous factors. As a result, the notion of proximate cause (*causa proxima*) plays a prominent role in judicial decisions. They write (Fumerton and Kress, 2001, 87), that a proximate cause is typically defined as a cause that produces injury in a natural, direct, and continuous sequence without any supervening causes that interrupt the causal chain. Alternatively, a proximate cause is defined as a tortuous cause that leads to a reasonably foreseeable injury to a reasonably foreseeable plaintiff. Sometimes the relation between an action by the defendant and the injury to the plaintiff is simply too weak to warrant a judgement of legal responsibility and damages. They give a funny illustration quoted below (by permission of Duke University School of Law).

> For example, suppose a defendant recklessly endangers nearby persons and property by throwing a stone, which hits a bird, causing the bird to lose consciousness, and fall into a smokestack. The bird catches fire, regains consciousness, flies out, and expires over a nearby fireworks factory. It falls into the main storeroom – still flaming – and ignites several cases of fireworks, which explode. The vibrations from the explosion cause a flowerpot to fall off a windowsill ten blocks away and land on the plaintiff's head, causing injury. Under the foreseeability test – and perhaps the superseding cause test – for proximate causation, the defendant would be excused from liability even though the defendant's reckless behaviour was an actual cause of the plaintiff's injury, because it was not foreseeable that throwing the stone would result in the flowerpot hitting the plaintiff on the head.

And in medicine? In 1882, as the germ theory of disease – postulating proximate, singular microbial causes of a disease – arose, Jakob Henle and his pupil Robert Koch formulated postulates (or rules) from which the inference could be made that a specific living organism caused a particular disease. In simplified form there were three:

1. the organism is always found with the disease,

2. the organism is not found with any other disease and,

3. the organism, cultured from one with the disease through several generations, produces the disease.

In a rather simplified way, causation involves the relationship between at least two entities, an agent and a disease. Both can be easy to define and identify, or, on the other hand, they can be "fuzzy sets" with blurred boundaries. Smallpox is a good example of the first, while tuberculosis represents the latter (Vineis, 2003, 83).

German medical scientists soon saw that the postulates do not allow for multiple causes nor do they account for causal relations not susceptible to experimentation. They established a 20th century consensus about multifactorial causality. In the age of chronic diseases Austin Bradford Hill advanced standards that can be used to decide when an association might be causal. He called his nine criteria "viewpoints."

There may be several causal "pathways" to a disease, i.e different combinations, each sufficient to cause the disease. For instance, what causes tuberculosis? A particular micro-organism, immune deficiency, malnutrition, crowded living conditions, silicosis, and in addition some genetic factors. In any one case two of these conditions – the mycobacterium and immune deficiency for example – may cause tuberculosis. In another case three of these causes – the mycobacterium, silicosis, and malnutrition – may contribute. In short, one cause, the mycobacterium, is necessary; various combinations that include this bacterium may suffice.

Thus, a first but not completely satisfactory answer to the problem of causality has been to define it by a *set of rules*.

It has long been thought unlikely that a complete explanation of causality will ever be forthcoming (except, perhaps, in natural sciences). The real world may be just too complex and too full of confounders to allow for confident inferences.

Weber (1991) investigates causality and history. Kern (2004) gives a cultural account of causality and its historical and cultural context, relying heavily on literature as evidence of belief.

Now we must turn to *correlation*, because it is so often confused with causality. In statistics the relation between two entities (such as cause and effect or agent and disease) are analysed in terms of association, contingency or correlation. We may use association as the over-arching term and the others as models for measuring the association, or they may be treated synonymously. On page 68 we have identified, in a probabilistic setting, causality with conditional probability. Association is the degree of statistical relationship between two or more events. Events are said to be associated when they occur more or less frequently together than one would expect by chance. Events are said not to have an association when the independent variable has no apparent effect on the dependent variable. The independent variable X is a characteristic that is measured in an experiment or study and that is assumed to have an effect on the outcome, i.e., the dependent variable Y. Therefore, it is assumed on whatever grounds that X is the cause and Y is the effect because the data gathered does not give any information on which variable actually is the cause. For the confirmation of a causal hypothesis the association is a necessary but not sufficient prerequisite, and for linear models it is not even a necessary condition (Bortz, 1999, 457). The measurement does not tell whether:

1. X is the cause of Y,

2. Y is the cause of X,

3. X and Y are mutually causing each other or

4. X and Y depend on a third variable Z (or further variables).

The third variable Z is known as *confounder* or *lurking variable* that may have been omitted from the analysis. For example, there is among school children an association between shoe size and vocabulary. However, learning words does not cause feet to grow, and swollen feet do not make children more verbally able. In this case, the third variable is easy to spot, viz. age. In more specific examples, the driving variable may be harder to identify.[2]

---

[2] Actually, the effect can be "partialled out" by forming the residuals $x_j^* = x_j - \hat{x}_j$ and

There are *asymmetric* measures of association, e.g., conditional probabilities and the Goodman-Kruskal lambda. The asymmetry means that one cannot interchange dependent and independent variables without changing the metric. The classical correlation measures as defined in Section 3.6.2 starting on page 153 are typically symmetric, i.e., $Corr(X, Y) = Corr(Y, X)$. In Section 3.6.7.3 on page 175 we will show an example for lambda.

Another classification consists of three classes:

1. sub-group association,

2. proportionate reduction in error measures (PRE-measures) and

3. deviation from stochastic (statistic) independence measures.

The first class is represented by conditional probabilities and derived measures. For example, one compares $Pr(X \mid Y < a_1)$, $Pr(X \mid a_1 \le Y < a_2)$ and so forth – the sub-groups – and analyses whether these values are deviating.

The PRE-measures are not so ambitious as to aspire to infer causality but humbly want to indicate whether two variables X and Y as observed as $\{x_j, y_j\}$ contain *redundant information*. Thus the aim is rather predictive than explanatory. Lambda and the several correlation measures introduced in Section 3.6.2 belong to this category. With $PE(\cdot)$ the *prediction error*, it takes the general form

$$\zeta_{XY} = \frac{PE(X) - PE(X \mid Y)}{PE(X)}. \tag{2.5}$$

The PRE-measures compare the errors that we could have made, if we had little or no information about the independent variable, to the errors that we could have made, were such information in our possession.

Example 2.2. We have observed the data $\{x_j, y_j\}$ for $j = 1, \ldots, N$. The estimation error of X alone is $PE(X) = \sum_{i=1}^{N} (x_j - \bar{x})^2$. The conditional prediction error follows from a linear regression:

$$PE(X \mid Y) = \sum_{i=1}^{N} (x_j - \hat{y}_j)^2 \tag{2.6}$$

$$= \sum_{i=1}^{N} (x_j - a - bx_j)^2. \tag{2.7}$$

With the settings $S_{xx} = PE(X)$ and $S_{xy} = \sum_{i=1}^{N} (x_i - \bar{x})(y_i - \bar{y})$, the fact that the least square estimate for the slope is $b = S_{xy}/S_{xx}$ and further simple transformations the PRE-measure turns out to be

$$\zeta_{xy} = \frac{S_{xy}}{S_{xx}} \equiv r^2. \tag{2.8}$$

This is identical with the so-called *coefficient of determination r^2* (see Kokoska

---

$y_j^* = y_j - \hat{y}_j$, where $\hat{x}_j$ and $\hat{y}_j$ are estimates from linear regressions on Z. The association between $\hat{x}_j$ and $\hat{y}_j$ is removed from the influence of the third variable.

and Nevison, 1989, 31) and is an empirical covariance divided by the empirical variance. The correlation in this context is

$$r = \frac{S_{xy}}{\sqrt{S_{xx}S_{yy}}}, \tag{2.9}$$

thus very similar to the PRE-measure. $\triangle$

There exist a variety of techniques called "causal modelling" which can also be classified in the same manner. Bayesian networks on the one hand represents the conditional probability approach while so-called *structural equation models* SEM or linear structural equation systems are grounded in correlations. Both approaches start with a graphic representing the structure as causal hypothesis. In the first domain they are called influence diagrams while in the latter they are called path diagrams. Again, the causality is defined *a priori*, for these models are confirming hypothesis rather than explaining relations.

In short, correlation cannot reveal anything about causality, and most especially cannot "prove" it, because correlation is a symmetric measure and there may be confounders making it spurious. Nonetheless, there are a handful of so-called "causal" modelling techniques. The more pragmatic view concentrates on the information gain derived from associated variables – and this is not to be disdained at all.

2.10.3   Elicitation and Expert Knowledge

Most of this section is a summary of a very good article by Baddeley et al. (2004), which in turn draws much from Daniel Kahneman and Amos Tversky. The latter two authors are amongst the most cited authors in economics: their so-called *prospect theory* is a psychological extension or substitute for rational utility, and Kahneman is a Nobel laureate. Bernstein (1996, 270) in his "Against the Gods" dedicates more than a chapter to these two men. Although this seminal work was debated at the time, later research seems to have refuted some of these hypothesis but has been ignored by the literature on expert elicitation (Kynn, 2008).

Experts' opinions are formed despite a paucity of data, and are usually based on anterior experience. In such situations humans use heuristics, i.e., rules of thumb, to aid analysis and interpretation of data. As a result, future judgements of subjective probabilities are biased by both prior experience and the heuristics employed.

2.10.3.1   Individual Bias

At least two main types of individual bias can be distinguished:

- motivational bias and

- cognitive bias.

Motivational biases may be induced by the interests and circumstances of the expert. Her job or reputation may depend on her assessment and she wants to be perceived as knowledgeable. This may lead to overconfidence. These motivational biases can be reduced because they are often under rational control. Bazerman and Neale (1992, 61) list three need-based illusions leading to overconfidence in decision-making. The so-called illusion of superiority is based on an unrealistic view of the self, that is, one's perception of oneself as being more capable, honest, intelligent, etc. than average. Individuals subject to this illusion attribute successes to their skills and failures to external factors. They share no credit and assume no blame. The illusion of optimism means that people underestimate the likelihood of experiencing bad outcomes or events and overestimate the chances of good outcomes. Thirdly, some people are subject to the illusion of control, where they believe they have more control over outcomes than they actually do.

Cognitive biases are more problematic because they emerge from incorrect processing of information. They are typically the result of using heuristics derived from experience to make quick decisions in uncertain or ambiguous situations. They are used because a further assessment of available information is difficult or time consuming, or when information is sparse.

At least four types of heuristics that bring about cognitive bias are usually employed:

- availability,

- anchoring and adjustment,

- representativeness and

- control.

Availability is the rule of assessing an event's probability by the ease with which instances of the event are brought to mind. Sometimes it works quite well. But the mind might stock more dramatic events better than everyday situations. For example, most people think a shark attack or an airplane crash is a more probable cause of death than a bee sting or a car crash. Similarly, risk managers may be likelier to recall a spectacular case of fraud (like Barings) than more frequent, less publicized and less damaging frauds. Thus, people are swayed by information that is vivid, recent or well-publicised.

Anchoring and adjustment is a heuristic that begins with a tentative estimate of a probability called an anchor. In a further step it is revised or adjusted in the light of new information (Tversky and Kahneman, 1974). The assessment clearly revolves around the initial estimate. For example, in deciding about an appropriate remuneration to demand in the face of an uncertain economic environment, workers will tend to anchor their pretensions around their existing wage.

The *control heuristic* highlights people's tendency to act as though they can influence a situation over which they have no control. As an example, people often think that the outcome of a football game depends on their watching.

The *representativeness heuristic* is where people use the similarity between two events, one with known probability and one unknown, to estimate the probability of one from the other (Tversky and Kahneman, 1982).

Example 2.3. Tversky and Kahneman (1974) give the following experimental test example: "Linda is 31 years old, single, outspoken, and very bright. She majored in philosophy. As a student, she was deeply concerned with issues of discrimination and social justice, and also participated in anti-nuclear demonstrations."

- *A1*: is Linda a bank teller?

- *A2*: is Linda a feminist bank teller?

In probabilistic terms test persons find answer *A2* more realistic and grant it a higher probability than *A1*. Now, knowing some statistics, it is obvious that $Pr(A1) \geq Pr(A2)$, as *A2* pertains to a much smaller collective than *A1*.

But there is some more research into this fallacy. Gigerenzer (2007, 95) argues that the problem is wrongly posed and thus missing the interesting point. The difficulty arises with the term "probable," which is not understood in a logical-statistical sense but rather paraphrased with "possible," "conceivable," "plausible," "reasonable" or "typical." Then the answers become also more plausible.   △

Well-known biases induced by the representativeness heuristic include the gambler's fallacy and base-rate neglect.

The *gambler's fallacy* is the belief that when a series of runs have all had the same outcome, e.g., "heads" in coin tossing, then the opposite outcome is more likely to occur in the following runs, since random fluctuations seem more representative of the sample space. People have a misconception of randomness. Watzlawick (1977, 68) gives the example of a sequence $2, 5, 8, 11, \ldots$, where it seems obvious that the rule is to augment the next figure with 3 and the series $4, 1, 5, 9, 2, 6, 5, 3, \ldots$. The latter seems random, unless one is aware that the figures might stem from $\pi = 3.1415926535 \ldots$ and its number starting at the third place. Randomness is perceived as lack of regularity which in turn is based on a lack of knowledge of the underlying generating mechanism. By the way, in 2005 Google raised some 4.2 billion USD through a secondary stock offer issuing 14,159,265 shares. Here we go again.

Base-rate neglect is the bias in making inferences about probability ignoring the background frequencies of events. For example, if the probability of any given woman having cancer is known to be 1/10,000, but a test on 10,000 women gives 100 positive results, people will tend to overestimate the probability that any one of them testing positive actually has cancer, rather than considering the possibility of false positives. In general, poor probability estimates occur because humans are often forced to make quick judgements based on poor information. It has been found that human reason often ignores the base rate even when the information is readily available. This has generated important results for the social sciences and economics, among other fields. This bias is often mentioned jointly with Bayes' rule. Other findings about erroneous estimations of probability include:

- people are insensitive to sample size and draw strong inferences from small number of cases,

- they think chance will "correct" a series of "rare" events,

- they deny chance as a factor causing extreme outcomes.

A strange combination of the gambler's fallacy and base rate neglect is called *probability matching*. Here, an action from given alternatives is chosen in proportion to the probabilities of the range of outcomes. An example by Andrew Lo, cited by Baddeley et al. (2004, 11) is from World War II.

Example 2.4. Bomber pilots were allowed to carry either a flak jacket or a parachute, but not both because of weight limitations. They knew that the probability of getting strafed by enemy guns – requiring a flak jacket for protection – was three times that of being shot down – requiring a parachute for survival. However, pilots were observed to take flak jackets three times out of four, and parachutes on the fourth time. Alas, this is not the optimal strategy and the best use of information. A better survival strategy would have been to take a flak jacket all the time, because to reach an optimal result one has to safeguard against the event with the highest probability – in this case, being strafed by enemy fire. △

Other cognitive biases reflect *emotional responses*. For example, in elicitation the experts involved have individually been overconfident about their knowledge. Several experts undergoing the same elicitation procedure often produce barely- or not even overlapping estimates of elicited parameter values. Even groups of experts are observed to show overconfidence in their consolidated results. Overconfidence is especially a problem for extreme probabilities (close to 0% and 100%) which people tend to find hard to assess.

As Ekeland (1993, 142) notes, in areas like space flight or nuclear safety, the experts' probability estimates are at odds with public opinion and, alas, experience. For example, NASA estimated the probability of an accident occurring to the space shuttle to be 1 in 100,000, while the carrier rocket exploded on the twenty-fifth launch. This gap needs not to be attributed to shady politicians or fanatical engineers. An article of Henrion and Fischhoff (1986) investigated the measurement of the speed of light over a time-span of approximately one hundred years. In physical experiments the researchers themselves estimate the margin of error of the measurement due to the experimental setting. It came out that for half of the experiments the "true" value of 1985 was not even within the interval suggested by the measurements and their error bands. This confirms the thesis that experts – like NASA – are in all good faith overconfident.[3]

Baddeley et al. (2004) conclude that of all of these biases, the most prevalent may be overconfidence and base-rate neglect.

Griffiths et al. (2006) assert that as a result of the great attention given to the work of Kahneman, Tversky and their colleagues, cognitive judgements under uncertainty are often characterised as the result of error-prone rules of thumb, insensitive to prior probabilities. They maintain:

> This view of cognition, based on laboratory studies, appears starkly at odds with the near-optimality of other human capacities, and with people's ability to make smart predictions from sparse data in the real world.

Their survey results on conditional situations suggest that everyday cognitive judgements follow the same optimal statistical principles as perception and memory, and reveal a close correspondence between people's implicit probabilistic models and the real world statistics.

2.10.3.2 Group Bias

Up to here we have assumed that experts are atomistic agents. In reality, experts congregate and deliberate and, in so doing, generate and reinforce more complex forms of bias stemming from group interactions. Difficulties in attaining probabilistic knowledge by individual experts are compounded because mistakes and misjudgements may be conveyed to other experts. This process is made more complex by a source of individual bias that emerges from anchoring effects: one's judgements are "anchored" to another's. This implies that expert knowledge will not necessarily evolve along a predetermined path but instead may exhibit path dependency and persistence.

This is paradigm anchoring: experts' beliefs are anchored to the predominant views. Such paradigm-based research, however, restricts thinking within certain boundaries and this may lead to convergence of expert opinion towards prevailing hypotheses and approaches. Experts will be unwilling to overthrow an old paradigm without a new paradigm with which to replace it, and it may take time for new paradigms to gain acceptability.[4] In normal times, the anomalies that do not fit with the current paradigm are discarded – a cognitive dissonance. Once a theoretical paradigm has become established, alternative approaches are refused. Then, paradigms will shift only when evidence and anomalies accumulate to such an extent that a scientific crisis develops. When

---

[3]The probability given by NASA could also be interpreted as propensity (see Section 2.10.1.4 at page 65), not taking into account human errors.

[4]Max Planck said: "A new scientific truth does not triumph by convincing its opponents and making them see the light, but rather because its opponents eventually die, and a new generation grows up that is familiar with it."

this happens, a scientific revolution is precipitated. This view stems from Thomas Kuhn 1996 and was first published in 1961.

Example 2.5. As a depiction we show some consecutive draws of a so-called Pólya urn (Sigmund, 1995, 77). We start with an urn containing one red and one black marble. After each draw we replace the marble we just picked and add an extra marble of the same colour. So, drawing a red ball will increase the probability for a red ball in the next draw. This process drifts to some stable proportion between red and black marbles. But the final levels are completely randomly distributed between 0 and 1. The analogy to talking themselves into a rut is obvious. See Figure 2.8 for ten paths and 1000 runs.   △

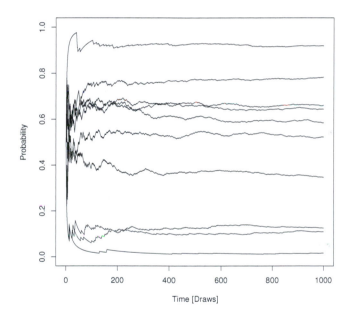

FIGURE 2.8: Ten paths of the Pólya urn.

Rational economic agents may have an incentive to follow the crowd. Thus, *herding* results as a response to individuals' perceptions of their own ignorance. If an individual has reason to believe that other agents' judgements are based upon better information, this is nothing less than a rational approach. Other people's judgements become a given data-set and will be incorporated into the insecure person's prior information set. Thus, their posterior judgments may exhibit herding tendencies. Valid results will only be achieved if the herd is following a path of increasing common knowledge.

The concept of herding on financial markets has a long-standing tradition.

John Maynard Keynes with a view on reputation stated (Keynes, 1936, 158): "Worldly wisdom teaches that it is better for reputation to fail conventionally than to succeed unconventionally." In order to formulate short term expectations of the market, participants may try to guess what the not-necessarily rational expectations of others may be. Keynes (1936, 156) writes:

> (...) professional investment may be likened to those newspaper competitions in which the competitors have to pick out the six prettiest faces from a hundred photographs, the prize being awarded to the competitor whose choice most nearly corresponds to the average preferences of the competitors as a whole; so that each competitor has to pick, not those faces which he himself finds prettiest, but those which he thinks likeliest to catch the fancy of the other competitors, all of whom are looking at the problem from the same point of view. It is not a case of choosing those which, to the best of one's judgement, are really the prettiest, nor even those which average opinion genuinely thinks the prettiest. We have reached the third degree where we devote our intelligences to anticipating what average opinion expects the average opinion to be.

This approach leads also to a – probably unstable – biased opinion. For later literature on this topic see Shiller (2001). For interests in bubbles Mackay (1995) is still a good choice.

2.10.3.3   Elicitation Theory

The core of *elicitation theory* is how to best interrogate experts or laypeople in order to obtain accurate information about a subject in question. There is no universally accepted protocol for probability elicitation in literature, but a range of different approaches have been adopted by scientists. There are three common assessment protocols: the Stanford/SRI protocol, Morgan and Henrion's protocol, and the Wallsten/EPA protocol (Morgan and Henrion, 1990). The two first protocols include five principal phases:

1. motivating the experts with the aims of the elicitation process,

2. structuring the uncertain quantities in an unambiguous way,

3. conditioning the expert's judgement to avoid cognitive biases,

4. encoding the probability distributions and

5. verifying the consistency of the derived distributions.

Assume that we want to appraise probabilities based on the estimates of several experts, each of whom has slightly different background knowledge. To reconcile these differences we can either combine the individual assessments into a single one, or we can request a consensus opinion. How we go about such a reconciliation of opinion leads either to *mathematical aggregation methods* or *behavioural* approaches. Clemen and Winkler (1999) give a good overview of the methods of combining expert estimates to both a point estimate and a full distribution.

The mathematical aggregation methods are mainly based on Bayesian techniques assuming implicitly that nothing is gained when experts share knowledge. The second approach can be jeopardised by group interaction problems, e.g., the dominance of one expert or the pressure for conformity. Perhaps the best known results from the behavioural group-judgment literature is *group polarisation*, the tendency of groups to take more extreme positions than would individual members (Clemen and Winkler, 1999, 196). It remains unclear which method produces more accurate final probability estimates.

Prior information in operational risk management is often a partial opinion or judgment. When prior information is augmented with new evidence, it is necessary to be aware of the heuristics and biases discussed so that their extension to future accepted knowledge can be limited.

In this section we have scratched a little bit on the surface of fundamental concepts for operational risk management. The aim was to show firstly that the complaint about the lack of data and the conclusion that quantifying is not possible may be seen in a different light as soon as probability is interpreted. Furthermore, the two main AMA methods, i.e., loss distribution approach and score-card approach may be seen as representative of either view. In Chapter 4 we will see that the loss distribution approach must be augmented with subjective estimates in order to make it fruitful for managing operational risks. Similarly, "causal modelling" is an often heard approach to operational risks. However, causality is hard to establish in a complex organisation. Albeit useful, some care need be applied to such models.

For the pleasure of argument let us go one step further and ask: "What is scientific?" The U.S. Supreme Court, in *Daubert v. Merrell Dow Pharmaceuticals Inc.*, had to define science in terms of its methods. It reached the following conclusions concerning scientific methods (Goodstein, 2000, 82):

1. the theoretical underpinnings of the methods must yield testable predictions by means of which the theory could be falsified,

2. the methods should preferably be published in a peer-reviewed journal,

3. there should be a known rate of error that can be used in evaluating the results,

4. the methods should be generally accepted within the relevant scientific community.

From this list we see that while Popper's view is in part included, other points show that sciences is also a convention. The daring idea that convention – human decision – lies at the root both of necessary truths and much of empirical science is due to Henry Poincaré.

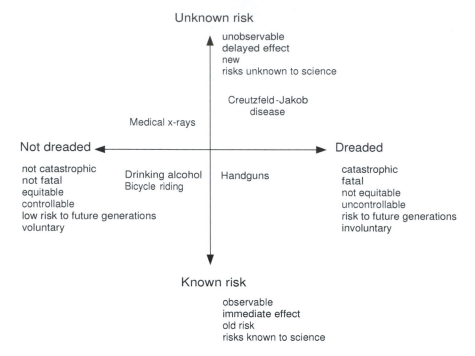

FIGURE 2.9: Main dimension of risk perception factors. (From Hyer and Covello, 2005, 113. By permission of the World Health Organization.)

2.10.4 Risk Perception and Risk Judgements

Perception is the process of taking intellectual cognizance of something, of apprehending by the mind and being convinced of something by direct intuition. Between the presentation of the risk to its representation in the brain, past experience and other factors affect perception and judgement. Many mental short-cuts, known as heuristics, have been presented in the Section 2.10.3, where the two distinct categories of motivational and cognitive biases were discussed. There are some well-known risk-specific factors predisposing any emotional reaction to risk information. In Section 2.10.3.1, we have already mentioned several heuristics or mental short-cuts that lead to biased judgements.

In addition, research as mentioned by Hyer and Covello (2005, 110) indicates that risk perception is a function of two sets of factors (see Figure 2.9), i.e., those reflecting the degree to which

- the risk is *unknown* and

- the risk is *dreaded*.

Unknown risks include those that are highly uncertain, not well understood, unobservable, new or with delayed consequences. Dread, on the other hand, summarises the perception of catastrophic potential, which is involuntary, uncontrollable, affecting future generations and possessing an unfair degree of risk. There can be strong emotions involved such as outrage, fear and anxiety. Table 2.2 gives a short summary, in Appendix 1 on page 349 there is a more extensive compilation from Covello and Allen (1988) and Covello et al. (1988, 54).

Even objectively presented risk may be received subjectively, i.e., it is perceived in the context of individual, social, cultural and other constraints.

In addition to the individual-psychological determinants of risk perception there is a sociological theory called Cultural Theory of Risk aiming at understanding why different people and different social groups fear different risks (Douglas and Wildavsky, 1982). This theory explains why, for example, white men feel themselves less subject to risk than anybody else (Kahan et al., 2007). The foundation of that theory is a two-dimensional scheme for worldviews or "ways-of-life" with the axis "group" and "grid." "Group" runs from individualist to solidarist, and "grid" covers hierarchist to egalitarian. White men are both individualists and hierachists. One proposition of the theory is that individuals gravitate toward perceptions of risk that advance the way of life they adhere to. That is, they are subconsciously disposed to notice and credit risk claims that cohere with their way of life and to ignore and dismiss risk claims that threaten it. Empirical evidence for this theory seems to be rather meager.

TABLE 2.2: Factors influencing risk perception. (From SAMHSA, 2002)

Risks perceived to ...	are more accepted than risks perceived ...
be voluntary	as being imposed
be under an individual's control	to be controlled by others
have clear benefits	to have little or no benefit
be distributed fairly	to be unfairly distributed
be natural	to be man-made
be statistical	to be catastrophic
be generated by a trusted source	to be generated by an untrusted source
be familiar	to be exotic
affect adults	to affect children

This whole chapter was dedicated to the very foundations of those concepts that are needed or used for the risk quantification and thus also for the Advanced Measurement Approach discussed here. We wanted to show that – because these concepts are debatable – multiple methods may be applied for our purposes.

Chapter 3

The Modelling Approach

In this chapter we wish to describe a complete model that complies with the requirements of the Advanced Measurement Approach and to illustrate it with an example based on the Quantitative Impact Study conducted by the banking supervisors. The language of the description is necessarily mathematical. We try to make the description self-contained, though this assumes that many facts are already known to most casual readers. The main themes are data modelling, simulation, risk measurement and results. Data modelling embraces the data model, an array of probability distributions to which the data is fitted with the pertinent techniques. Special emphasis is given to correlations, or associations, as these are crucial for a reasonable regulatory capital assessment. Risk mitigation is also treated, especially the modelling of standard insurance products.

Simulation is chosen for several reasons. Alternative convolution models imply independence of the random loss variables and thus do not allow for correlations. In addition, insurance as a non-linear function cannot readily be adapted to other procedures. Thus, Monte Carlo simulation is a method of "last resort." Here we concentrate on parametric functions and do not consider empirical sampling – sometimes called historical simulation.

3.1 Simulation and the Monte Carlo Method

3.1.1 General Concepts

A bank is a *system* because it consists of many interrelated and interacting elements, viz. individuals, tools, machines, assets, etc., related by some common law, principle or end.

Simulation is a numerical technique for conducting experiments chiefly on a computer, which involve certain types of mathematical and logical models describing the behaviour of some system or some of its parts, possibly over an extended period of time.

A *model* can be defined as an abstraction of a real system. It is a simplification that should nonetheless represent the features of interest closely so as to still validly represent the system. Skilful modelling is a navigation between the Scylla of oversimplification and the Charybdis of impenetrable detail. A scientific model can be used for prediction and control. A model must not be misunderstood for a method. A *method* is a means of getting from a given initial state to a determined terminal state or result. While with different methods the result must be the same, this is not true for models. With different model the results will generally differ.

Simulation in the quantitative field – where sequences of events are not of primary interest – is often contrasted with analytical methods where the system or problem is described in terms of a set of mathematical equations. Very occasionally, closed form solutions may be the result; otherwise numerical methods must be employed. On the other hand simulation – which is also a numerical method – can also be applied where analytical methods work, e.g., for solving a system of equations.

When simulations are restricted to problems where time and ensuing serial correlations are of less importance and where the many observations are independent and the system's response is rather simple, e.g., a risk capital figure, then generally we speak of the *Monte Carlo method* (Law and Kelton, 1991, 113; Rubinstein, 1981, 12). However, any simulation involving random numbers is understood as "Monte Carlo." The method and its name were apparently suggested by John von Neumann and Stanislaw M. Ulam in a study examining the feasibility of the atomic bomb using simulations of nuclear fission. The first publication about Monte Carlo seems to be Metropolis and Ulam (1949). Nicholas Metropolis (1987, 127) also claims to have invented the name. Segrè (1980) reports that Enrico Fermi, a physics Nobel laureate, used the method some fifteen years earlier but not under this heading. All these men worked together on the Manhattan project.

When a model consists only of fixed, as opposed to random, relations and variables, it is called a *deterministic* model. Otherwise, it is a *stochastic* model. We intend to use simulation for exploring how banks behave when faced with loss due to operational risks. This choice is based on the difficulty or impossibility of deriving analytical solutions to correlated and highly nonlinear elements like insurance and on the fact that a great number of elements like loss generators, events and lines of business are involved. A complex model is necessary for while the discrete elements are fairly simple, the number and the quality of the relationships between them pose an analytical challenge.

3.1.2   Monte Carlo Simulation

The oldest conscious simulation is the so-called Buffon's needle experiment for estimating the value of π. It was presented to the Royal Academy of Sciences in Paris in 1773.

Example 3.1. Buffon's Needle refers to a simple Monte Carlo method *ante definitionem*. The experiment is as follows: Take a needle of length L. Take also a table-top with parallel lines drawn on it, equally spaced with distance L between them. (Although this is not a requirement it simplifies the example.)

Now, if you drop the needle on the table, you will find that one of two states occurs: The needle crosses or touches one of the lines, or the needle does not cross a line. The simulation consists in dropping the needle over and over and recording the results. In short, we want to keep track of both the total number N of times that the needle is randomly dropped on the table and the number C of times that it crosses or touches a line. With successive attempts, one sees that the number $2N/C$ approaches the value of π with $N \to \infty$ rather slowly.
△

The above Example 3.1 is a rare representative of a physical model solving a mathematical problem by simulation.

The Monte Carlo method is often used to compute integrals that lack regularity or are of high dimension of domain where quadrature formulae are less efficient.

Example 3.2. Each integral can be expressed as an expected value. We want to estimate the simple one-dimensional integral:

$$I = \int_a^b g(x)dx. \tag{3.1}$$

By choosing any probability function $f_X(x)$ such that $f_X(x) > 0$ when $g(x) \geq 0$ we can write

$$I = \int_a^b \frac{g(x)}{f_X(x)} f_X(x)dx \tag{3.2}$$

and therefore

$$I = E_{f_X}\left(\frac{g(x)}{f_X(x)}\right). \tag{3.3}$$

Most simply we choose $f_X(x)$ to be uniformly distributed according to $U(b,a)$. Then we can write for the expectation:

$$E(g(X)) =) \frac{I}{b-a}, \tag{3.4}$$

and from there

$$I = (b-a)E(g(X)). \tag{3.5}$$

By applying the sample mean estimator we can calculate

$$\hat{I} = (b-a)\frac{1}{N}\sum_{i=1}^N g(X_i) \tag{3.6}$$

by drawing random numbers X_i from $U(b,a)$, evaluating $g(X_i)$ and sampling.

By choosing $f_X(x)$ judiciously you can reduce the variance of the estimator dramatically.   △

A third example shows how to solve a deterministic problem by recasting it into a probabilistic setting and then applying the Monte Carlo method. This is taken from (Demidovic and Maron, 1981, 722).

Example 3.3. We want to solve the following system of linear equations $\mathbf{x} = \mathbf{Ax} - \mathbf{b}$, $\mathbf{A} = [a_{ij}]$, with a MC method:

$$x_1 = 0.1x_1 + 0.2x_2 + 0.7$$
$$x_2 = 0.2x_1 - 0.3x_2 + 1.1.$$

Thus, one can assume the solution to be $x_1 = 1$ and $x_2 = 1$.

The less obvious approach is to define a so-called Markov transition matrix $\mathbf{P}$ of dimension three because for each unknown there are two states plus the final state on the absorbing border, say Γ. The matrix looks like:

$$\mathbf{P} = \begin{pmatrix} p_{11} & p_{12} & p_{13} \\ p_{11} & p_{11} & p_{11} \\ 0 & 0 & 1 \end{pmatrix}. \tag{3.7}$$

The index of the row designates the actual state ("from") and the index of the column indicates the next state ("to"). The interpretation of p_{ij} is the probability of making a transition from state S_i to S_j. The last row says that when you are on the border you will remain there with probability 1. To define these probabilities one may choose factors v_{ij} in order to ensure that $a_{ij} = p_{ij}v_{ij}$, $p_{ij} > 0$ for $a_{ij} \neq 0$ and $p_{ij} \geq 0$ otherwise for $i = 1, 2$ and $j = 1, 2$. The elements of the last column are given from $p_{i3} = 1 - \sum_{j=1}^{2} p_{ij}$ for $i = 1, 2, 3$.

Now the trajectory of state transitions starting in a given state S_{i_0} and ending eventually at $S_{i_{m+1}} = \Gamma$ is called a finite discrete *Markov chain*. And now comes the most astonishing fact, proved in Demidovic and Maron (1981, 718): if we associate with a random trajectory

$$T_i = \{S_{i_0}, S_{i_1}, \ldots, S_{i_m}, S_{i_{m+1}} = \Gamma\} \tag{3.8}$$

a (random) variable

$$X_i(T_i) = b_{i_0} + v_{i_0,i_1}b_{i_1} + v_{i_0,i_1}v_{i_1,i_2}b_{i_2} + \ldots + v_{i_0,i_1}\cdots v_{i_{m-1},i_m}b_{i_m}, \tag{3.9}$$

then the expectation $E(X_i)$ equals the sought after unknown x_i.

For executing the simulation we choose $v_{11} = 1$, $v_{12} = 1$, $v_{21} = 1$ and $v_{22} = -1$, complying with the above mentioned prescriptions and accordingly we set the Markovian transition matrix at:

$$\mathbf{P} = \begin{pmatrix} 0.1 & 0.2 & 0.7 \\ 0.2 & 0.3 & 0.5 \\ 0 & 0 & 1 \end{pmatrix}. \tag{3.10}$$

To implement the calculation of x_1 we need to draw a random variate u from a uniform distribution $U(0,1)$ and reckon the transition according to the following rules: for being actually in the state S_1 if $0 \leq u < 0.1$ then $S_1 \to S_1$, if $0.1 \leq u < 0.3$ then $S_1 \to S_2$ and if $0.3 \leq u < 1$ then $S_1 \to \Gamma$. Analogous rules apply for S_2: if $0 \leq u < 0.2$ then $S_2 \to S_1$, if $0.2 \leq u < 0.5$ then $S_2 \to S_2$ and if $0.5 \leq u < 1$ then $S_2 \to \Gamma$. The unknown x_1 is calculated from a sample whose trajectories always start in S_1, respectively x_2 from starting in S_2.

In Table 3.1 we give the estimates depending on the number of draws. We can recognise the convergence of the procedure to the correct values. In Appendix 5 is an implementation of the algorithm together with a sample of trajectories T_1 and values X_i. The fact that we only come close to an

TABLE 3.1: Results of the simulation

Number of draws [power of 10]	x_1	x_2
1	0.840	1.060
2	1.028	0.968
3	1.006	1.001
4	1.010	0.988
5	1.002	0.994
6	0.999	0.999

acceptable solution after one million draws should serve as a warning against inadequate simulations. △

Various conclusions may be drawn from the above example. Firstly, we can confirm that the MC method is a generic tool capable of solving a vast array of problems. Therefore, it has also been called the "method of last resort." (The slightly negative connotation stems from the fact that a closed form solution is amenable to further analysis. Sensitivities to model parameters are particularly easy to derive, meaning that the model can be analysed to a greater extent.)

Secondly, the result of greatest interest to us is a statistical measure, i.e., the expectation of a sample of outcomes. Thus, our confidence rests on the sample size and on the speed of convergence. It is therefore crucial to estimate the confidence interval of this figure. Conversely, one may try to infer the number of simulation runs necessary to achieve this precision by working backward from a predetermined number. Thirdly, the Example 3.3 shows that for many MC simulations that the convergence is slow. A simulation with 10,000 draws produces $\hat{x}_1 = 1.010$ and $\hat{x}_2 = 0.988$, still 1% from the exact solution. No wonder that such effort has been expended to develop spatial techniques to reduce the variance of results. In the case of the Loss Distribution Approach to interest we must be aware that Regulatory Capital is in the 99.9% quantile, which is chiefly determined by very few in the tail. On average, 1000 draws are required to create an outcome beyond the threshold.

3.1.3   Loss Distribution Approach and Simulation

The Advanced Measurement Approach (AMA) allows for the so-called Loss Distribution Approach.

Definition 3.1. A *random variable* is a function that maps a set or events of the set of all possible events – the so-called sample space – onto a set of real values. They are represented by capital letters, for example X.   □

In this context we are interested by losses in monetary units. The *distribution* is used synonymously with *cumulative distribution function* denoted by $G(x) = Pr(X < x)$. It is the probability that the random variable X, i.e., the loss, is less than a given value x. The Risk Capital is read off the distribution at the specified confidence level, i.e.,

$$EC_\alpha = x_\alpha = G^{-1}(\alpha). \tag{3.11}$$

In essence, the Loss Distribution Approach consists in finding the probability distribution of the losses due to operational events. This function needs to be constructed given the available information including the requirements of the Basel Committee.

There are two main reasons why this function $G(x)$ can only be calculated with the Monte Carlo method: firstly, such losses can stem from a very different and numerous causes and circumstances. This is modelled by thinking of the different sources of loss as different probability distributions $F_i(x)$. The related random variables must be summed. The "summation" of random variables is not evident, especially when the single losses show some dependency, i.e are neither fully dependent nor independent. What is $G(X)$ when $X = X_1 + X_2$ and X_1, respectively X_2, is distributed according to $F_1(X)$, respectively $F_2(X_2)$? Secondly, mitigation through insurance changes gross losses to net losses in a highly non-linear fashion. This involves a cascade of min(.) and max(.) functions that make an accurate non-numerical treatment impossible.

The method of choice is the Monte Carlo simulation, though its typical problems arise. Fortunately, the ever increasing power of computing machines alleviates many issues.

Simplifying a little, we must cope with N simulation draws representing N independent and identically distributed random values X_i, $i = 1, 2, \ldots, N$ from which the so-called *sample cumulative distribution function* $\hat{G}_N(x)$ will be constructed as

$$\hat{G}_N(x) = \frac{1}{N} \sum_{i=1}^{N} J_{\{X_i \leq x\}} \qquad (3.12)$$

where $J_{\{X \leq x\}}$ is an indicator variable with values:

$$J_{\{X \leq x\}} = \begin{cases} 1 & \text{if } X \leq x \\ 0 & \text{if } X > x. \end{cases}$$

The economic capital or risk capital for operational risk is then estimated as

$$\widehat{EC}_\alpha = \hat{G}_N^{-1}(\alpha). \qquad (3.13)$$

In practical terms this means to order the N outcomes x_i to give $x_{1:N} < x_{2:N} < \ldots < x_{k:N} < \ldots < x_{N:N}$ and read off the value that is indexed by the related confidence value (see Eq.(3.170) on page 189 for details). Of course this is the value-at-risk style of quantification where risk is measured as quantile of a loss distribution. We will see late on that there are alternative risk measures based on the same distribution.

3.2 General Model Structure

The main task of the LDA is to generate a cumulated density function of the total loss due to operational events $G(f(X_1 + X_2 + \ldots + X_n))$, where the X_j represent the random (net) losses from sub-sets of the operational events and $f(.)$ represents additional risk mitigation.

From this stylized representation we can understand the general structure of the model. Firstly, we have to decide what and how many sub-sets are useful. Secondly, we have to assess the distributions belonging to these losses; and lastly, we must model the mitigation.

The regulators have developed a two-dimensional grid of objectives: "Events" and "Lines of Business." Loss records must be collected under these two aspects. The events are organized on several levels, ranging from seven on the top level to a high number on level 3. There are also eight business lines, for a total of 56 combinations.

We believe that such fundamental problems are not best solved by naively calibrating 56 Lines of Business/Events combinations. A mathematical fact called *Central Limit Theorem* – actually there are many different formulations – leads to certain assumptions for a normal distribution as a limit of the number of random variables. And 56 is too high a number for permissible random variates having finite variance and third moments.[1] In short, less is

---

[1] If n random variables X_i are independent, with expected value μ_i, finite variance σ_i^2 and finite third moments, then the summation statistic $T = \sum_{i=1}^{n} a_i X_i$ is asymptotically normally distributed with mean $\mu = \sum_{i=1}^{n} a_i \mu_i$ and variance $\sigma^2 = \sum_{i=1}^{n} a_i \sigma_i$.

more: A sensible lower number of distributions representing better represents the structure of losses and leads to greater confidence of its parameters. In

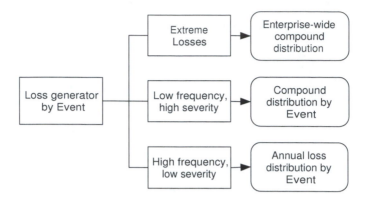

FIGURE 3.1: Loss generator modelling.

principle, the structure is very simple, as depicted in the Figure 3.1. But as is usually the case, the devil is in the details.

One has to decide in which dimension the loss generation is carried out. One may either begin with Events and work from there to Lines of Business, or being with Lines of Business and work back to Events.

Let us put Events in the driver's seat. Obviously, this involves a trade-off. Risk managers are more interested in the risk-taking units and especially the effectiveness of those units' management actions. On the other hand, Lines of Business embody all types of operational risk and are thus heterogeneous from a data point of view. Because data must be gathered with both attributes attached, this dichotomy has no influence on the data collection.

The supervisory text (BCBS, 2005, 149) places some restriction on the data to be gathered:

> To assist in supervisory validation, a bank must be able to map its historical internal losses into the supervisory categories defined in Annexes 6 [Events] and 7 [Lines of Business].

On the other hand, we believe that the insured risks have a high affinity to the Events, while there is no evidence of such a relationship with the Lines of Business. The scarcity of data furthermore dictates the use of the highest level of classification. Therefore, we use the 7 Events as loss-generating sources.

3.2.1   Modelling Losses

From a regulation perspective the loss generation must fulfil the following quantitative requirement (BCBS, 2005, 147):

A bank's risk measurement system must be sufficiently granular to capture the major drivers of operational risk affecting the shape of the tail of the loss estimates.

The way we propose to model losses addresses this regulatory requirement. The loss generation is done through two sources per Event. There is an aggregate loss generator for the attritional High Frequency/Low Severity (HFLS) mode of losses. Because these losses are almost guaranteed, they will be drawn from one specific distribution. This is the standard actuarial approach.

The Events' Low Frequency/High Severity (LFHS) losses are modelled by compounding a frequency distribution and a severity distribution. The number of annual losses and amount of losses are taken as independent. This is also a standard approach in actuarial science.

The Quantitative Impact Study (BCBS, 2003a, 10) in its Table 6 (see Table 3.7 on page 147) shows strikingly the two regimes of LFHS and HFLS, demonstrating why the separation of the two and the generic treatment of HFLS as an aggregate are warranted.

There is an extra "Event" or loss-generating source: catastrophic losses that are characterized by their rareness and severity. We think it irrelevant to which Event category they may belong. Attributing extreme losses to any of the seven Events would just distort the data and pose additional problems for the estimation of the tail.

Catastrophic losses also form a compound frequency/severity distribution. Although this proposal is a dated actuarial approach something similar has been much advertised by Taleb (2007) under the heading of "Black Swan." Taleb and Martin (2007, 188) make the following analogy:

> There are two classes of risk. The first is the risk of volatility, or fluctuations – think of Italy: in spite of the volatility of the political system, with close to 60 postwar governments, one can consider the country as extremely stable politically. The other is a completely different animal: the risk of a large, severe and abrupt shock to the system. Think of many of the kingdoms of the Middle East: where countries exhibit no political volatility, but are exposed to the risk of a major upheaval.

Dependencies (or correlations) are modelled through the frequencies of the compound distribution. The model uses simple copulae that require an estimate of the correlations. As long as they are consistently derived from the same data pool they will comply with all applicable restrictions.

The allocation to Lines of Business is carried out in a tabular form with simple weights adding up to 100%. These weights' derivation is left to the user. This does not preclude sophisticated allocation mechanisms like marginal contributions to the total risk, etc. In essence the allocation is top-down.

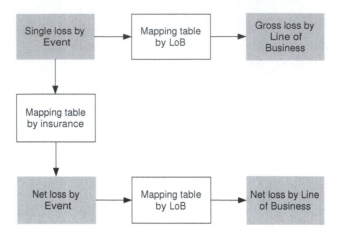

FIGURE 3.2: Model overview.

The second mapping, in addition to the allocation to Lines of Business, concerns the effect of insurance. After drawing the (compound) losses from the loss generators of the Events, we allocate each loss according to the mapping table to the potential insurance policies. The allotment is done randomly. The number of insurance policies can be defined by the user. Typically, a policy incorporates a deductible (franchise) and a limit on each claim. Policies may also stipulate annual aggregate limits.

Finally, there is the possibility to apply general Operational Risk provisions to the annual aggregate net loss. Where the institution has a captive insurance company its potential Stop-Loss retrocession may be taken into account through a similar mechanism.

3.3   Data Requirements

Our approach assumes that the user has the input needed for the specification of the model. In essence, the data consists of distributional parameters, mapping quantities, correlations, insurance parameters and so on.

The simulation model has an upstream influence on the data modelling and manipulation. When we prescribe what distributions may be used the user must be in a position to generate the parameters in question. Nonetheless, we regard such data modelling as a craft, if not an outright art. With regard to data, the regulators require the incorporation of external data as a prerequisite for using the most advanced approach (BCBS, 2005, 149):

A bank's operational risk measurement system must use relevant external data (either public data and/or pooled industry data), especially when there is reason to believe that the bank is exposed to infrequent, yet potentially severe, losses. These external data should include data on actual loss amounts, information on the scale of business operations where the event occurred, information on the causes and circumstances of the loss events or other information that would help in assessing the relevance of the loss event for other banks.

The use of external data needs to take into consideration its relevance for the institution.

3.3.1 External Data Usage

In the quote above the mandatory use of external data is not well defined. From actuarial experience we could infer several usages:

- derive from the more copious data the family of distribution and then fit the internal to this distribution,

- mix the data by mixing the (empirical or fitted) distributions and

- specific to our loss modelling use the "Extreme" event distribution for external and the other events for internal data.

With the second approach several methods are possible. The simplest is a two-point mixture of the internal and external loss distributions, i.e., $G_X(x)$ and $H_X(x)$ to yield

$$F_X(x) = \gamma G_X(x) + (1 - \gamma)H_X(x) \tag{3.14}$$

with $0 \leq \gamma \leq 1$.

In 1967 the so-called *Credibility theory* (Klugman et al., 1998, 385) was introduced into the actuarial world by Hans Bühlmann (Straub, 1988). Although the theory was originally concerned with rate making and pricing, it also addresses how to estimate losses for historically rare occurrences. The theory allows users to quantify how much weight should be placed on the empirical data and how much on other (even prospective) information. This leads naturally to estimating parameters and thus also to Bayesian methods.

So-called *Bayesian estimator* is another approach (Klugman et al., 1998, 107; Rice, 1995, 571). This method is also appropriate to transfer expert know how or score-card findings into parameter of loss distributions (Clemen and Winkler, 1999, 187).

The *prior distribution* (which may be even an improper distribution) is a probability distribution $\pi(\theta)$ of possible parameter values representing the user's belief that various values of θ are the true values.

The *model distribution* (or likelihood function) $f_{X|\Theta}(x|\theta)$ is the probability function of the collected data, given a particular value for the parameter θ. The *posterior distribution* $\pi_{\Theta|X}(\theta|x)$ is the conditional probability distribution of the parameters given the observed data. The *predictive distribution* $f_{Y|\mathbf{x}}(y|\mathbf{x})$ is the conditional distribution of a new observation given the collected data:

$$\text{posterior Distribution} = \frac{\text{Model Distribution} \times \text{prior Distribution}}{\text{normalizing Constant}}, \quad (3.15)$$

more formal

$$\pi_{\Theta|X}(\theta \mid x) = \frac{f_{X|\Theta}(x \mid \theta)\pi(\theta)}{\int f_{X|\Theta}(x \mid \theta)\pi(\theta)d\theta}. \quad (3.16)$$

This Eq.(3.16) is actually the *Bayes' theorem* for probability densities.[2]

$$f_{Y|\mathbf{x}}(y|\mathbf{x}) = \int f_{Y|\Theta}(y|\theta)\pi_{\Theta|X}(\theta|x)d\theta. \quad (3.18)$$

The Bayesian approach has the advantage that it is applicable both to historical data and expert estimates (score-cards) as it is a "behavioural" approach.

Depending on the models chosen, the calculation can be straightforward or more complex.

3.4   Data Modelling and Distributions

The axiomatic definition of the probability has been given on page 62. Both relative-frequency and degree-of-belief probabilities satisfy the mathematical theory of probability, i.e., the above-mentioned so-called Kolmogorov axioms.

The large uncertainties typically encountered in risk analysis make such distributions an indispensable part of the analysis. The probabilities in both the aleatory and the epistemic models are fundamentally the same and should be interpreted as degrees of belief. For rare events that have not yet materialised only the second interpretation makes sense. For communication purposes we think it important to keep this distinction in mind.

---

[2]Better known is the Bayes' theorem for discrete probabilities (Kokoska and Nevison, 1989, 12): For $A_1, A_2, \ldots, A_n$ a mutually exclusive and exhaustive set of events with $Pr(A_k) > 0$ and B an event for which $Pr(B) > 0$ then for $k = 1, \ldots, n$

$$Pr(A_k \mid B) = \frac{Pr(A_k \cap B)}{Pr(B)} = \frac{Pr(B \mid A_k)Pr(A_k)}{\sum\limits_{j=1}^{k} Pr(B \mid A_j)Pr(A_j)}. \quad (3.17)$$

3.4.1 Distributions

Because the size N of samples of operational risks, especially beyond substantial losses, is often small, it is usually difficult to discriminate among the many model candidates for the distributions on the basis of available data alone. Therefore, it makes sense to augment the information base by selecting models that have a long tradition with proven reliability in property and liability insurance. The following distributions have been chosen with this in mind.

Although many textbooks contain descriptions of distributions, we will give a brief summary of those functions that are often used in actuarial science and in the context of operational risk. The fact is that these distributions are not so common and we have not found one single and widely used source that employs all of the following functions. We have collected the information from Law and Kelton (1991), Evans et al. (1993), Kokoska and Nevison (1989), Hogg and Klugman (1984), Bury (1999) and Mack (2002).

Definition 3.2. A *discrete distribution* for a discrete random variable X is a function $F(A)$ which equals the probability that $X = A$ for a set of distinct possible values of A and 0 otherwise. A synonym is *probability mass function* "pmf." □

Definition 3.3. A *continuous distribution* for a continuous random variable X is a function $F(x)$ which equals the probability that $X < x$ for any x, i.e., $F(x) = Pr(X < x)$. It is also called *cumulative density function* "cdf." □

Statistical distribution functions are used as models for the empirically observed data because the empirical distribution lacks mass in the right tail (too few big losses) and in addition the biggest observed loss would bound the losses from above (although bigger losses are conceivable.)

Definition 3.4. The *empirical distribution* – more precisely the empirical cumulated distribution function $F(x)$ – of a sample $x_1, \ldots, x_n$ of size n is given by

$$F(x) = \frac{1}{n} \sum_{i=1}^{n} J_{\{x_i < x\}} \tag{3.19}$$

with the indicator function

$$J_{\{x_i < x\}} = \begin{cases} 1 & \text{if} \quad x_i < x \\ 0 & \text{if} \quad x_i \geq x. \end{cases}$$

□

The ecdf is a typical step function. We have already introduced it as *sample cumulative distribution function* on page 91.

Definition 3.5. A *compound distribution* refers to a random variable $Y = \sum_{j=1}^{N} X_j$ where N is a discrete random variable and the X_j are also randomly distributed. $\qquad\square$

In our context N represents the number of losses occurring (frequency) and X_j represents the amount ("severity") of the single loss. The standard modelling in insurance assumes that losses are independent and identically distributed and loss sizes are independent of the number of occurrences. Single losses from compound generators are available and thus amenable to insurance. The distribution $F(Y)$ can be calculated numerically with different methods depending on the frequency distribution, i.e., recursion, transformation and simulation techniques.

Because we assume operational risk losses stem from one of two different generators, i.e., the (annual) *aggregate losses* for high frequency and low severity (HFLS) and *compound losses* for low frequency and high severity (LFHS), two distinct groups of pertinent distributions or distribution families are permitted, or, in fact, encouraged. Compounding means the joint draw of the number of losses N and then of N amounts, one for each loss N. Aggregates represent the base or "attritional" losses, i.e., not split into frequency and severity.

For the 8th event, *catastrophic loss* generator, the same group applies as for Low Frequency and High Severity (LFHS).

The compound distribution is the result of the convolution produced with (1) a discrete frequency distribution and (2) a severity distribution that is either continuous or given as table.

All the distributions besides those for frequencies presented here have a maximum of three parameters. They are called *location a*, *scale b* and *shape c*. Location is a characteristic point on the abscissa; it can be the starting point or the mean, for example. The scale is often associated with the standard deviation or a measure that stretches the independent values. The shape parameter obviously has to do with the shape of the distribution, but is not present in every distribution.

3.4.2   Annual Aggregate Loss Distributions

In the property and casualty insurance business there has been much research into the best distributions (Mack, 2002, 69; Hogg and Klugman, 1984). For aggregate annual losses a number of distributions are considered standard. These are the gamma, inverse-Gaussian and the lognormal distribution.

Now let us describe the characteristics of these three distributions. We either present the function or the density with its inverse, the maximum likelihood estimators (where there is simple enough a representation) and mean and variance.

Gamma

Distribution:	No closed form	
Density:	$f(x) = \frac{1}{\Gamma(c)} b^{-c} x^{c-1} \exp\{-x/b\}$	
Range:	$x \geq 0,\ b > 0,\ c > 0$	
MLE:	$\hat{c}\hat{b} = \frac{1}{n} \sum_{j=1}^{n} x_j$	
	$\log \hat{b} + \Psi(\hat{c}) = \frac{1}{n} \sum_{j=1}^{n} \log x_j$	
Mean:	$\mu = cb$	
Variance:	$\sigma^2 = cb^2$	

The following definitions hold: $\Gamma(z) = \int_0^\infty t^{z-1} \exp\{-t\}\, dt$ is the gamma function, $\Psi(z) = \frac{\Gamma'(z)}{\Gamma(z)}$ is the so-called digamma function. Hints for numerical calculation can be found in Law and Kelton (1991, 332). The gamma distribution has a so-called *reproductive property*: if k independent gamma-distributed variables X_i with shape parameter c_i and a common scale parameter b are summed as $T = a \sum_{i=1}^{k} X_i$ the T is also gamma-distributed with scale ab and shape $\sum_{i=1}^{k} c_i$. Thus it is easy to calculate the convolution. This property makes the gamma distribution a plausible candidate for aggregate losses.

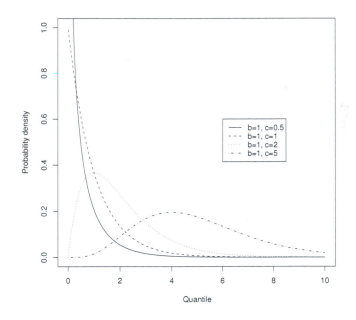

FIGURE 3.3: Gamma distribution densities.

Lognormal

Distribution:	No closed form
Inverse:	no analytical expression
Density:	$f(x) = \frac{1}{x\sqrt{2\pi b^2}} \exp\left\{\frac{-(\log x - a)^2}{2b^2}\right\}$
Range:	$x \geq 0,\ b > 0,\ c > 0$
MLE:	$\hat{b} = \frac{1}{n} \sum_{j=1}^{n} \log x_j$
	$\hat{c} = \left[\frac{1}{n} \sum_{j=1}^{n} (\log x_j - \hat{b})^2\right]^{1/2}$
Mean:	$\mu = b \exp\left\{0.5c^2\right\}$
Variance:	$\sigma^2 = b^2 \exp\left\{c^2\right\}(\exp\left\{c^2\right\} - 1)$

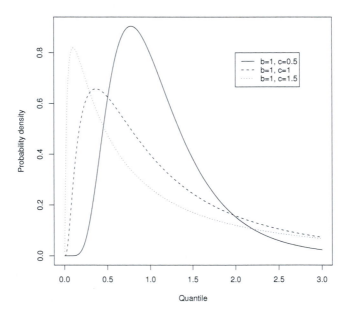

FIGURE 3.4: Lognormal distribution densities.

The random variable X is lognormally distributed, if $Y = \log(X)$ is normally distributed. What the addition is to the normal distribution is the multiplication to the lognormal. For example, if X_1 and X_2 are both lognormally distributed then the product $X_1 X_2$ is also lognormally distributed. But what kind of model underlies this distribution? A process $\{X_j\}$ is said to obey the *law of proportionate effect* if it is represented by $X_j = X_{j-1}(1 - \epsilon_j)$ where $\{\epsilon_j\}$

is a set of mutually independent, identically distributed random variables independent of X_j. The formula above leads to $X_n = X_0 \prod_{j=1}^{n}(1 + \epsilon_j)$. With small absolute values of ϵ_j taking logarithms and approximating $\log(1 + \epsilon_j)$ by ϵ_j, it follows $\log(X_n) = \log(X_0) + \sum_{j=1}^{n} \epsilon_j$. The central limit theorem has $\log(X_n)$ asymptotically normally distributed (Crow and Shimizu, 1988, 5).

Inverse Gaussian (Wald)

Distribution:	$F(x) = \Phi\left(\sqrt{bx/a^2} - \sqrt{b/x}\right) + e^{2b/a}\Phi\left(-\sqrt{bx/a^2} - \sqrt{b/x}\right)$
Inverse:	no analytical expression
Density:	$f(x) = \left[\frac{b}{2\pi x^3}\right]^{1/2} \exp\left\{\frac{-b(x-a)^2}{2a^2 x}\right\}$
Range:	$x > 0,\ a > 0,\ b > 0$
MLE:	$\hat{a} = \frac{1}{n}\sum_{i=1}^{n} x_i$
	$\hat{c} = \dfrac{n}{\sum_{i=1}^{n}\left[x_i^{-1} - (\hat{a})^{-1}\right]}$
Mean:	$\mu = a$
Variance:	$\sigma^2 = a^3/c$

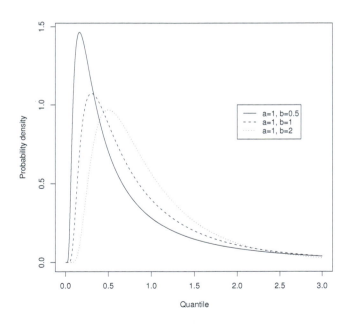

FIGURE 3.5: Inverse Gauss distribution densities.

An annual loss S consists of many single losses $X_1, X_2, \ldots, X_n$ or $S = X_1 + X_2 + \ldots + X_n$. Now the arguments for these distributions: Firstly, these distributions are well-suited to model a distribution for a single loss and secondly, the convolution of the so-distributed losses is a distribution of the same family. The distribution of a single loss will have its mass concentrated at zero.

This distribution was first derived by Schrödinger in the context of physics. It describes the first passage time of a particle moving according to Brownian motion with positive drift hitting a given barrier. If the random walk were not too simple an assumption for stock prices, these distributions could be used for barrier options and similar derivative features like knock-in and knock-out.

The probability distributions for the aggregate losses must describe the total loss around the mean. In Figure 3.6 we see how similar the densities are for a sensible set of identical mean and dispersion. The gamma density can hardly be distinguished from the inverse Gaussian with this specific set of parameters.

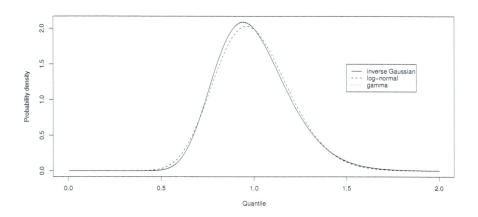

FIGURE 3.6: Comparison of the three densities with mean 1 and coefficient of variation 0.2.

3.4.3   Compound Distributions

Compounding means the use of two distributions, one for frequency and one for severity in order to generate the total or aggregate loss.

A *compound distribution* is the probability distribution of the (doubly stochastic) aggregate loss, where both the number of losses and the loss amount are mutually independent random variables (see Definition 3.5 on page 98).

Each class contains distributions that are particularly popular due to certain characteristic properties. The frequency is modelled by the Poisson and the negative binomial distribution. The latter can be thought of as a Poisson distribution whose parameter is weighted by a gamma-distributed mixing variable in order to account for the uncertainty of the correct parameter (Daykin et al., 1994, 48).

The total loss is modelled from a frequency distribution ("how often does such a loss occur") and a severity distribution ("how bad is the loss when it occurs"). The construction of the relevant loss distribution leads to the so-called convolution that we outline further on.

The number of occurrences K is random; so are the losses Z_j with $j = 1, \ldots, K$. The total loss S is the sum of the single losses, i.e.,

$$S = \sum_{j=1}^{K} Z_j. \tag{3.20}$$

The question that we need answered is: How do we determine the distribution of the total loss? For reasons of simplicity we start with two losses Z_1 and Z_2.

The inequality $Z_1 + Z_2 \leq x$ is equivalent to $Z_1 \leq y$ and $Z_2 \leq x - y$. Because we assume that Z_1 and Z_2 are independent yet possess the same distribution $F(z)$, the probability of the sum is the integral of the product of the probabilities of the summands:

$$\Pr[Z_1 + Z_2 \leq x] = \int_{-\infty}^{\infty} \Pr[Z_1 \leq x - t] \cdot \Pr[t < Z_2 \leq t + dt]. \tag{3.21}$$

With the distribution $F(z) \equiv \Pr(Z \leq z)$ it is

$$\Pr[Z_1 + Z_2 \leq x] = \int_{-\infty}^{\infty} F(x - t) \cdot dF(t)$$

$$\equiv F * F(x)$$

$$\equiv F^{2*}(x)$$

The operator $*$ thus defined is called *convolution*. The convolution is both associative and commutative. This distribution describes the severity of two losses. It is contingent upon the fact that two losses occur. The probability that this will happen is $\Pr(K = 2)$. Independence of number of occurrences and the severity of losses motivates the multiplication of these probabilities for the total loss probability:

$$G_2(x) = \Pr(K = 2) \cdot \Pr(Z_1 + Z_2 \leq x). \tag{3.22}$$

From this we can generalise the formula:

$$\Pr(Z_1 + Z_2 + \ldots + Z_k \le x) = \int_{-\infty}^{\infty} F^{(k-1)*}(x - t) \cdot dF(t)$$

$$= F^{(k-1)*} * F(x))$$

$$\equiv F^{k*}(x)$$

$$G(x) = \Pr(X \le x) = \sum_{k=0}^{\infty} p_k \cdot \Pr[\sum_{j=1}^{k} Z_j \le x] \qquad (3.23)$$

$$G(x) = \sum_{k=0}^{\infty} p_k \cdot F^{k*}(x). \qquad (3.24)$$

If the cdf $F(z)$ can be differentiated we can formulate the following result for the densities:

$$g(x) = \sum_{k=0}^{\infty} p_k \cdot f^{k*}(x) \qquad (3.25)$$

for $x > 0$, respectively $G(x) = g(x) = p_0$ with $x = 0$. The convolution of the densities is obviously

$$f^{k*}(x) = \int_{-\infty}^{\infty} f^{(k-1)*}(x - t) \cdot f(t)dt. \qquad (3.26)$$

Another way to perform this convolution is to describe it as multiplication in Fourier-transformed variables. Back-transforming the product yields the convolved result. In this so-called frequency space the convolution becomes a multiplication which is easier to perform. Now the Fourier-transformed of a probability density is the so-called *characteristic function* (Papoulis, 1991, 157).

3.4.3.1　Frequency Distributions

Poisson

Distribution:	$F(x) = \sum_{i=0}^{x} \lambda_i \exp\{-\lambda\}/i!$
Inverse:	no analytical expression
Mass density:	$p(x) = \lambda^x \exp\{-\lambda\}/x!$
Range:	$x \in \{0, 1, \ldots\}$
MLE:	$\hat{\lambda} = \frac{1}{n} \sum_{j=1}^{n} x_j$
Mean:	$\mu = \lambda$
Variance:	$\sigma^2 = \lambda$

The parameter λ is called *intensity*. This distribution is a discrete probability distribution derived by Siméon Denis Poisson in 1838 as a limit function of the Binomial distribution. The accordingly distributed discrete variate N counts, among other things, the number of occurrences (sometimes called "arrivals") that take place during a given time interval.

In 1898 Ladislaus von Bortkiewicz published a work commonly called "Das Gesetz der kleinen Zahlen" (The Law of Small Numbers) (von Bortkiewicz, 1898). Therein he noted that events with low frequency in a large population followed a Poisson distribution even when the probabilities of the events varied. Gumbel (1968) writes: "A striking example was the number of soldiers killed by horse kicks per year per Prussian army corps. Fourteen corps were examined, each for twenty years. For over half the corps-year combinations there were no deaths from horse kicks; for the other combinations the number of deaths ranged up to four. Presumably the risk of lethal horse kicks varied over years and corps, yet the over-all distribution was remarkably well fitted by a Poisson distribution." The intensity parameter λ equals 190/280 or

TABLE 3.2: Deaths by horse kicks

Number of deaths	Frequency	Relative Frequency	Poisson prediction
0	144	0.51429	0.49659
1	91	0.32500	0.34761
2	32	0.11429	0.12166
3	11	0.03929	0.02839
4	2	0.00714	0.00497

0.679. On average, fewer than one soldier died in this manner per corps-year.

Negative Binomial

Distribution:	$F(x) = \sum_{i=1}^{x} \binom{n+i-1}{i} p^n (1-p)^i$
Inverse:	no analytical expression
Mass density:	$p(x) = \binom{n+x-1}{x} p^n (1-p)^x$
Range:	x and $n \in \{0, 1, \ldots\}$, $0 < p < 1$
MLE:	numerical (Law and Kelton, 1991, 348)
Mean:	$\mu = n(1-p)/p$
Variance:	$\sigma^2 = n(1-p)/p^2$

Some authors call the above distribution the *Pascal distribution* and designate with negative binomial the mass that allows for non-integer parameter n. The probability mass is $\Gamma(n+x)p^n(1-p)^x/\Gamma(n)x!$ (Evans et al., 1993, 110).

The Negative Binomial distribution is also called *Pólya distribution*. The latter derived the distribution by combining the Poisson with a Gamma distribution (see Example 3.15 on page 138 for the derivation). In this way the Negative Binomial discrete distribution can be understood as a kind of Poisson distribution with uncertainty of the value of the parameter. Therefore, this function has one parameter more than the latter. The Poisson and the negative binomial distribution belong with the binomial and the geometric distributions to the so-called (a,b,0)-class distributions. These can be recursively represented:

$$p_k = p_{k-1}(a + b/k) \quad \text{with} \quad k = 1, 2, \ldots \tag{3.27}$$

The parameter a and b are given as

- Poisson with $a = 0$, $b = \lambda$,

- Negative binomial with $a = p$ and $b = (n+1)p/(1-p)$.

This important property is used in the numerical convolution. Panjer (1981) demonstrates that the following result holds for the above mentioned number distributions (Daykin et al., 1994, 120):

$$g(x) = p_1 f(x) + \int_0^z (a + by/x)f(y)g(x-y)dy \quad \text{with} \quad x > 0. \tag{3.28}$$

If the claims amount (severity) distribution is discrete or discretised on the positive integers $x = i \cdot h$, $i = 1, 2, \ldots$, the total claims can be represented according to

$$g_i = g(i \cdot h) = \sum_{j=1}^{\max i, r} (a + bj/i)f_i \cdot g_{i-j} \quad \text{with} \quad i = 1, 2, \ldots \tag{3.29}$$

r is the index of the last mass point of f_i and

$$g_0 = p_0 \quad \text{for} \quad x = 0. \tag{3.30}$$

The cumulated probability is simply

$$G(x) = G(i \cdot h) = \sum_{j=0}^{i} g_j. \tag{3.31}$$

The starting value $g_0 = p_0$ is

$$g_0 = \begin{cases} e^{-\lambda} & \text{in the Poisson case} \\ \left(1 + \frac{aN}{a+b}\right)^{-\frac{a+b}{a}} & \text{in the Negative Binomial case,} \end{cases} \tag{3.32}$$

where a and b are defined according to the above.

So why should we use Monte Carlo simulation when we can compute the compound distributions by this algorithm? Quite simply, because it is not possible to incorporate insurance into these formulae. If we model insurance trivially then we can do without MC.

Example 3.4. Let us assume a Pareto distributed loss with parameters location $a = 1$ and shape $c = 1.1$. This loss severity is combined with either a Poisson distributed frequency with parameter $\lambda = 10$ or alternatively, with a negative binomial distribution of identical mean with parameters size $n = 10$ and probability $p = 0.5$. We take a R-function called **panjer** of a package **actuar**.

```
library(actuar)
dPareto<-function(x,a,c){if (x>a) {c/a*(a/x)^(c+1)} else {0}}
for (i in 0:50) {s[i]<-dPareto(i*0.1,1,1.1)}
s<-s/sum(s)

a<-panjer(fx=s,freq.dist="negative binomial",par=list(size=10,prob=.5))
b<-panjer(fx=s,freq.dist="poisson",par=list(lambda=10))

plot(b,type="l",lty=1,xlim=c(0,600),xlab="Quantile",
     ylab="Probability density")
lines(a,lty=2)
legend(325,.004,legend=c("Poisson frequency",
    "Negative binomial frequency"),lty=c(1,2))
```

Figure 3.7 shows the result of the compounding. The mean of both distributions is the same, but the negative binomial distribution leads to greater mass in the upper tail and less mass in the lower region. This is what we expect as the variance in the second case must be greater than in the first. For a better discretisation approximation see Daykin et al. (1994, 123). △

Yet another interesting feature of these two discrete distributions is their behaviour when compounded if severity is not considered from ground-up but in excess of a threshold. The number of exceedances N_D is $\sum_{i=1}^{N} J_{\{X_j>D\}}$ where $J_{\{X_j>D\}}$ is the notorious indicator variable. The probability of losses in exceedance shall be $q = Pr(X > D)$. Now it can be shown that the following relation between the distributions of N and N_D holds:

$$N \sim Poisson(\lambda) \longrightarrow N_D \sim Poisson(q\lambda)$$
$$N \sim negBin(n, p) \longrightarrow N_D \sim negBin(n, \tfrac{p}{p+q(1-p)}) \tag{3.33}$$

This proposition is very useful in calculating the statistical effect(s) of insurance.

3.4.3.2 Severity Distributions

We foresee the distributions for extreme values as most relevant to the modelling of the LFHS representation.

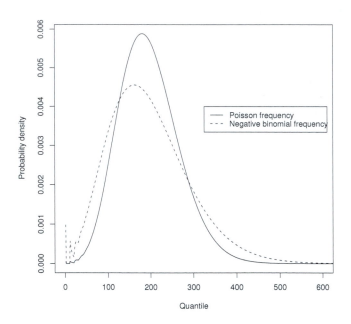

FIGURE 3.7: Compounding frequency and severity.

Besides this distribution family for severity we must also consider the lognormal distribution to be valuable in certain circumstances.

Weibull (EV III)

Distribution:	$F(x) = 1 - \exp\left\{-(x/b)^c\right\}$
Inverse:	$x_p = b[\log(1/(1-p))]$
Density:	$f(x) = (cx^{c-1}/b^c)\exp\left\{-(x/b)^c\right\}$
Range:	$0 \leq x < \infty,\ b > 0,\ c > 0$
MLE:	$\dfrac{\sum\limits_{j=1}^{n} x_j^{\hat{c}}\log x_j}{\sum\limits_{j=1}^{n} x_j^{\hat{c}}} - \dfrac{1}{\hat{c}} = \dfrac{1}{n}\sum\limits_{j=1}^{n} x_j^{\hat{c}}$
	$\hat{b} = \left(\dfrac{1}{n}\sum\limits_{j=1}^{n} x_j^{\hat{c}}\right)^{1/\hat{c}}$
	(numerical solution see Law and Kelton, 1991, 334.)
Mean:	$\mu = b\Gamma(1+1/c)$
Variance:	$\sigma^2 = b^2[\Gamma(1+2/c) - \Gamma(1+1/c)^2]$

The Weibull, Fréchet and the Gumbel distributions belong to the Extreme Value distributions. There exists a generalisation such that the three become special cases. These three distributions share the so-called *reproduction property*. Let us denote with $X_{1:n}$ the first order and with $X_{n:n}$ the n-th order statistics, i.e., the smallest and the largest realisation of a sample of size n. In this case, the property states that if X is distributed from either Gumbel or Fréchet, then $X_{n:n}$ is also from Gumbel or Fréchet. This holds true for the Weibull distribution as well, but with respect to the least order $X_{1:n}$ (Bury, 1999, 23).

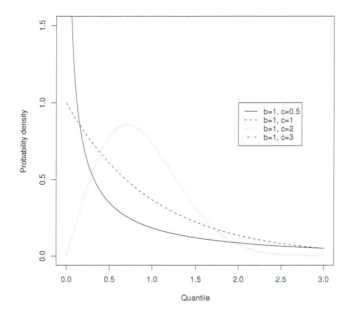

FIGURE 3.8: Weibull distribution densities.

Fréchet (EV II)

Distribution:	$F(x) = \exp\left\{-(\frac{x-a}{b})^{-c}\right\}$	
Inverse:	$x_p = a + b\left[\log(1/p)\right]^{-\frac{1}{c}}$	
Density:	$f(x) = \frac{c}{b}(\frac{x-a}{b})^{-c-1}\exp\left\{-(\frac{x-a}{b})^{c}\right\}$	
Range:	$x \geq a,\ a \geq 0,\ b > 0,\ c > 0$	
MLE:	numerical estimation	
Mean:	$\mu = a + b\Gamma\left(1 - 1/c\right)$ for $c > 1$	
Variance:	$\sigma^2 = b^2\Gamma\left(1 - 2/c\right) - \mu^2$ for $c > 2$	

The Fréchet cumulative distribution function is the only well-defined limiting function for the maxima of random variables. Thus, the Fréchet cdf is

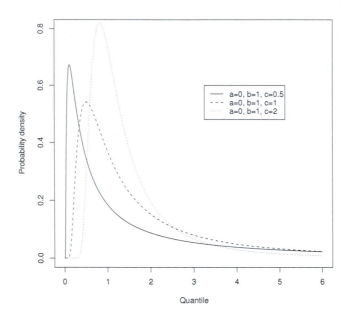

FIGURE 3.9: Fréchet distribution densities.

well suited to characterise variables of large features. As such, it is important for modelling the stochastic behaviour of large losses.

Kotz and Nadarajah (2000) list over fifty applications for the Fréchet distribution ranging from earthquakes through floods, horse racing, rainfall, queues in supermarkets, sea currents, wind speeds and track race records.

Gumbel (EV I)

Distribution:	$F(x) = \exp\left\{-\exp\left\{-(x-a)/b\right\}\right\}$
Inverse:	$x_p = a - b\log[\log(1/p)]$
Density:	$(1/b)\exp\left\{-(-x-a)/b\right\}\exp\left\{-\exp[-(x-a)/b]\right\}$
Range:	$-\infty < x < \infty,\ b > 0$
MLE:	$\hat{a} = -\hat{b}\log\left[\dfrac{1}{n}\sum\limits_{j=1}^{n}\exp\left\{\dfrac{-x_j}{b}\right\}\right]$
	$\hat{b} = \dfrac{1}{n}\sum\limits_{j=1}^{n}x_j - \dfrac{\sum\limits_{j=1}^{n}x_j\exp\left\{\dfrac{-x_j}{b}\right\}}{\sum\limits_{j=1}^{n}\exp\left\{\dfrac{-x_j}{b}\right\}}$
Mean:	$\mu = a - \Gamma'(1) = a + 0.57721$
Variance:	$\sigma^2 = b^2\pi^2/6$

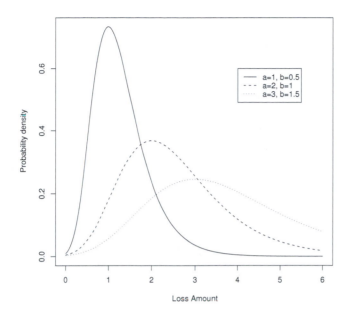

FIGURE 3.10: Gumbel distribution densities.

The Gumbel distribution is sometimes called the double exponential distribution due to its functional form. Because the transformations $Z = c\log((X-a)/b)$ and $Z = -c\log(-(X/b))$ lead the Fréchet and the Weibull to the Gumbel, it is also called the Extreme Value distribution *tout court*.

Generalised Extreme Value (GEV)

Distribution:	$F(x) = \exp\left\{-\left[1 - c\frac{x-a}{b}\right]^{1/c}\right\}$
Inverse:	$x_p = a + c\{1 - (-\log p)^c\}/c$
Density:	$f(x) = (1/b)\exp\left\{-\left[1 + c\frac{x-a}{b}\right]^{1/c}\right\}\left[1 + c\frac{x-a}{b}\right]^{1/c-1}$
Range:	$a \le x < \infty,\ b > 0$
MLE:	numerical estimation
Mean:	$\mu = a + \frac{b}{c}\left[\Gamma(1-c) - 1\right]$
Variance:	$\sigma^2 = \frac{b^2}{c^2}\left[\Gamma(1-2c) - \Gamma^2(1-c)\right]$

The GEV distribution embodies the three aforementioned distributions. If we name the shape parameter of the GEV c_{GEV} then the GEV corresponds to:

- the Fréchet distribution with $c_{\text{GEV}} = c_{\text{Fréchet}}^{-1} > 0$,

- the Gumbel distribution with $c_{\mathrm{GEV}} \to 0$ and

- the Weibull distribution with $c_{\mathrm{GEV}} = -c_{\mathrm{Weibull}}^{-1} < 0$.

The second assertion follows from the well-known fact that $\lim_{n\to\infty}(1 + 1/n)^n = e$.

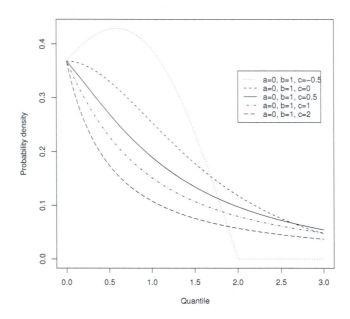

FIGURE 3.11: Generalised Extreme Value distribution densities.

Exponential

Distribution:	$F(x) = 1 - \exp\{-x/b\}$
Inverse:	$x_p = -b\log(1-p)$
Density:	$f(x) = (1/b)\exp -x/b$
Range:	$0 \le x < \infty,\ b > 0$
MLE:	$\hat{b} = \frac{1}{n}\sum_{j=1}^{n} x_j$
Mean:	$\mu = b$
Variance:	$\sigma^2 = b^2$

The exponential distribution can be modelled as the waiting times W between two events whose occurrences $X(t)$ are Poisson distributed with parameter νt. The following holds: $Pr(W \le t) = 1 - Pr(W > t) = 1 - Pr(X(t) = 0) = 1 - exp(-\nu t) = F(t)$.

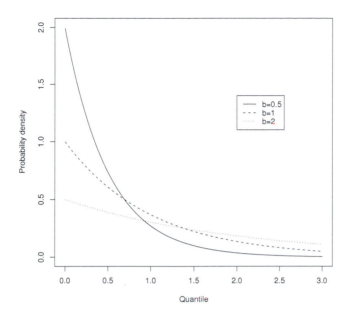

FIGURE 3.12: Exponential distribution densities.

Pareto

Distribution:	$F(x) = 1 - [a/x]^c$
Inverse:	$x_p = a(1-p)^{-1/c}$
Density:	$f(x) = ca^c/x^{c+1}$
Range:	$a \leq x < \infty,\ a > 0,\ c > 0$
MLE:	$\hat{a} = \min(x_j)$

$$\hat{c} = \left[\frac{1}{n} \sum_{j=1}^{n} \log(x_j/\hat{a}) \right]^{-1}$$

Mean:	$\mu = ca/(c-1)$ for $c > 1$
Variance:	$\sigma^2 = ca^2/[(c-1)^2(c-2)]$ for $c > 2$

There are three kinds of Pareto distributions. The above is of of the first kind and has two parameters. The Pareto is of the second kind and has $F(x) = 1 - [a/(a+x)]^c$ as its distribution. It is also called *Lomax distribution*. The third is $F(x) = 1 - [K_2 \exp\{bx\}/(a+x)]^c$.

The Pareto distribution is widely used for big losses in property and liability insurance. An oddity of the distribution is the fact that only moments exist up to c. Moreover, we find that the smaller the shape parameter, the higher the frequency of big losses compared with small losses. Typical values

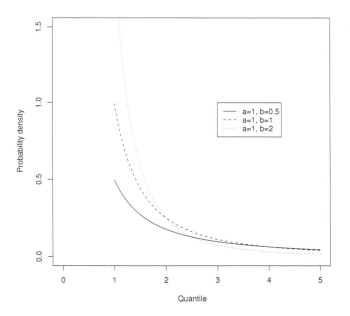

FIGURE 3.13: Pareto distribution densities.

for c are: ≈ 1 for earthquake and storm, ≈ 1.5 for fire in industry, ≈ 1.8 for general liability, ≈ 2 for occupational injuries and ≈ 2.5 for motor liability. Here we see that often not even the variance exists.

Vilfredo Pareto was an economist interested in income distributions. Because no reliable data regarding income below a certain threshold was available in his time, he analysed the income y above the level a (Woll, 1978, 341). The *Pareto law* states that the number of earners N making y beyond a follow the equation $N = \beta \cdot y^{-\alpha}$ or equivalently $\log(N) = \log(\beta) - \alpha \log(y)$. Given this form, on double-logarithmic graph paper the distribution becomes a straight line.

We propose truncating the Pareto distribution at a certain maximum. Therefore, when we write $Pareto(a, c, B)$ we mean that $Pr(X < x \mid X > B) = Pr(X < B)$ or more explicitly

$$F(x) = \begin{cases} 1 & \text{if } x \geq B \\ 1 - [a/x]^c & \text{if } a \leq x < B. \end{cases} \tag{3.34}$$

The expected value of the truncated distribution is for $c \neq 1$ and $c \neq 2$:

$$\mu = \frac{c}{c-1}\left[a - B(\frac{a}{B})^c\right] + B\left[\frac{a}{B}\right]^c. \tag{3.35}$$

Analogously, the standard deviation of the truncated Pareto distribution is given as (also for $c \neq 1$ and $c \neq 2$)

$$\sigma^2 = \frac{c}{c-2} \left[a^2 - B^2 (\frac{a}{B})^c \right] + B^2 \left[\frac{a}{B} \right]^c . \qquad (3.36)$$

For the case $c = 1$ the expectation is simply $a(\log(B/a)+1)$, for $c = 2$ follows $2a - a^2/B$.

A noteworthy feature of the Pareto distribution is that it keeps its shape, i.e., c, even when subject to some excess condition. In short, the following holds:

$$Pr(X < x \mid x > D) = 1 - Pr(X > x \mid x > D)$$
$$= 1 - \frac{Pr(X > \max[x, D])}{Pr(X > D)} \qquad (3.37)$$

and this obviously yields to

$$Pr(X < x \mid x > D) = \begin{cases} 1 - [D/x]^c & \text{if } x > D \\ 0 & \text{else.} \end{cases} \qquad (3.38)$$

Generalised Pareto

Distribution:	$F(x) = 1 - \left[1 - c\frac{x-a}{b} \right]^{1/c}$
Inverse:	$x_p = a + b \{ 1 - (1-p)^c \} / c$
Density:	$f(x) = \frac{1}{b}(1 + c\frac{x-a}{b})^{1/c-1}$
Range:	$a \leq x < \infty,\, a > 0,\, c < 0$
MLE:	$0 = \sum\limits_{i=1}^{N} \left[\frac{1}{\hat{c}^2} \log \left(1 + \hat{c}\frac{x_i-a}{\hat{b}} \right) - (\frac{1}{\hat{c}}+1) \frac{(x_i-a)/\hat{b}}{1+\hat{c}(x_i-a)/\hat{b}} \right]$
	$N = \sum\limits_{i=1}^{N} (\frac{1}{\hat{c}}+1)\, \hat{c}\frac{(x_i-a)/\hat{b}}{1+\hat{c}(x_i-a)/\hat{b}}$
	with the threshold a fixed in advance.
Mean:	$\mu = a + \frac{b\Gamma(c^{-1}-2)}{c^3\Gamma(1+1/c)}$ for $c < 1$
Variance:	$\sigma^2 = \frac{2b^2\Gamma(c^{-1}-1)}{c^2\Gamma(1+1/c)} - \mu^2$ for $c < 0.5$

Some authors prefer to set $c \leftarrow -\varsigma$. For $c = 0$ the GPD has also been defined as $F(x) = 1 - exp\{-(x-a)/b\}$. For $c > 0$ the range of x becomes $0 \leq x < b/c$. The Generalised Pareto becomes the exponential with the shape parameter going to 0. This follows from $\lim_{n\to\infty}(1 + x/n)^n = e^x$ with $x \in \Re$. For $c_{GPD} > 0$, $c_{Pareto} = 1/c_{GPD}$ and $b = ca$ the GPD becomes the Pareto distribution.

The generalised Pareto distribution is the limiting distribution for observations exceeding a high threshold. This is true for a wide class of underlying distributions. It contains many into account some limits of parameters.

In order to assess the threshold, one can make use of the fact that the *mean excess function* $E(X - u \mid X > u)$ is a linear function of the threshold u according to

$$E(X - u \mid X > u) = \frac{b}{1+c} - \frac{-c}{1+c}u. \tag{3.39}$$

As is the case with the Pareto distribution, here too if $X \sim GPD(a, b, c)$ then the excess $X - u \mid X > u$ obeys a $GPD(a, b', c)$. The GPD and the GEV dis-

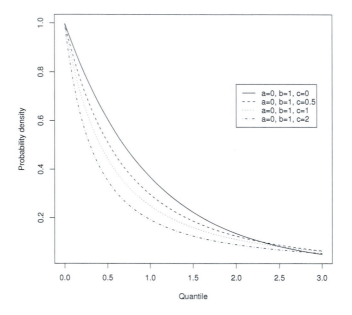

FIGURE 3.14: Generalised Pareto distribution densities.

tribution are related. A Taylor series expansion of e^x is $1 + x + \frac{x^2}{2} + \ldots + \frac{x^n}{n!}$. With just the first two terms as an approximation, each can be transformed into the other. In Figure 3.15 we summarise the presented distributions and put them into context. The random variable X is used to represent each distribution. There are three types of relationship: transformations and special cases – both indicated with a solid line – and limiting distributions, indicated by a dotted line. We can discern two centres of gravity, i.e., the normal distribution and the exponential. The "exponential" ones are used for the severity of individual losses, while Gamma, lognormal and inverse Gaussian model the aggregate losses.

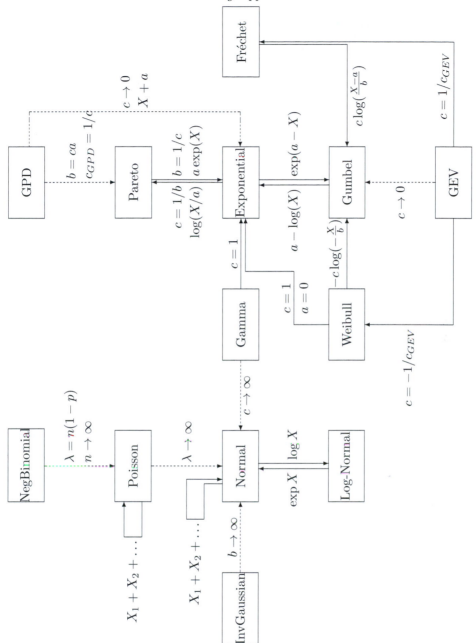

FIGURE 3.15: Synopsis of distributions.

3.4.3.3   Creating New Distributions

There are many ways to create new distributions. The easiest is to take a known probability distribution and apply a sensible function $g(x)$ to the dependent variable x. Thus the new distribution $F_2(x) = F_1(g(x))$. Functions used are *inter alia* (see Panjer (2006) and Klugman et al. (1998) for more details):

- multiplication with a constant,

- raising to a power and

- exponentiation.

Example 3.5. We just take a Pareto distribution $F_1(x) = 1 - [a/x]^c$ for $x > 0$ and choose $g(x) = x^{1/\tau}$ with $\tau > 0$. From that follows $F_2(x) = 1 - [a/x^{1/\tau}]^c$. This cumulative density function is known as Burr distribution and widely used to model income distributions.   △

In Figure 3.15 on page 117 the reader may discover some analogous relationships.

Yet another method consists of *mixing distributions*. A random variable Y is a so-called two-point mixture of random variables X_1 and X_2 if its cumulated distribution function is given by

$$F_Y(y) = w_1 F_{X_1} + w_2 F_{X_2}, \qquad (3.40)$$

where $w_1 > 0$ and $w_2 > 0$ and $w_1 + w_2 = 1$. The generalisation to n-points is too obvious to be mentioned. The rationale of this method is that there are two sub-samples, i.e., X_1 and X_2, in the data with different distributional characteristics.

Using this technique we immediately see that the number of parameters to be fitted to available data increases rapidly. Not only we have to account for the number of parameters of the two distributions but the weights w_i are also to be considered, thus for a two-point mixture one more parameter has to be estimated. This may lead to a better fit but also to a severe lack of robustness.

Example 3.6. Klugman et al. (1998, 106) give an example of mixing two Pareto distributions with the special feature that the shape parameter c_2 of the second distribution is conditioned to be $c_1 + 2$. We know from previous discussions that a small shape parameter implies more severe losses or equivalently a thicker and longer tail. This mixture thus combines small losses with large losses. Formally, using our favourite representation of the Pareto distribution we infer:

$$F(x) = w[1 - (a_1/x)^{c_1}] + (1 - w)[1 - (a_2/x)^{c_1+2}]$$
$$= 1 - w(a_1/x)^{c_1} - (1 - w)(a_2/x)^{c_1+2}$$

This resulting distribution has only 4 parameters.   △

Besides transformation and mixing there is a third possibility, viz. *splicing*. For a definition of the density of a so-called k-component spliced distribution see Panjer (2006, 92). Graphically it can be easily understood that different pieces are put together to form a valid cumulative distribution function. Thus one could model a specific body with another specific tail, or combine a parametric distribution with an empirical. The latter situation is quite common as there may be abundant data below a certain threshold while above it the data must be conjectured. See Figure 3.16 where a lognormal body is spliced with an exponential tail.

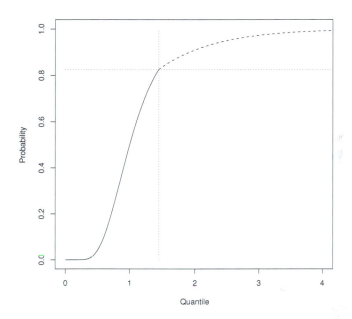

FIGURE 3.16: A spliced cdf, lognormal body with exponential tail.

A single transformation of the normal distribution (or other symmetrical distribution) due to Tukey (1977) leads to a family of distributions with variable skewness and leptokurtosis. Let Z be a random variable with standard normal distribution. Then the proposed standard distribution is:

$$Y = T_{g,h}(Z) = \frac{1}{g}(e^{gZ} - 1)e^{hZ^2/2} \tag{3.41}$$

with $g \neq 0$, $h \in \Re$. More general with an additional shift and scale parameter:

$$X = T_{g,h,A,B}(Z) = A + B \cdot Y \tag{3.42}$$

with $A \in \Re$ and $B > 0$. These functions are called *g-and-h* distributions. The parameter g determines the skewness while h controls the kurtosis; X is leptokurtic – more acute peak around the mean than the normal distribution and fatter tails – for $h < 0$ and platykurtic for $h > 0$. The magnitude of h determines the amount of kurtosis. Values for g are typically in the range (-1,1); for $g = 0$, the distribution is symmetric. There are two special cases: $h = 0$ is the lognormal and $h = g = 0$ is the normal distribution. The probability function cannot be given explicitly; the density is with the respective quantile values y_p and z_p

$$t_{g,h}(y_p) = \frac{e^{-(h+1)z_p^2/2}}{\sqrt{2\pi}\left[e^{gz_p} + \frac{h}{g}z_p(e^{gz_p} - 1)\right]}. \tag{3.43}$$

The g-and-h distribution is a heavy-tailed distribution which was applied to similar problems as the extreme value distributions, i.e., high wind speeds (Field, 2004) but also to financial data (Fischer, 2006) and loss severities of operational risk. Dutta and Perry (2006) analysed data from the 2004 Loss Data Collection Exercise with this distribution. They found (Dutta and Perry, 2006, 4) that "... the g-and-h distribution results in a meaningful operational risk measure in that it fits the data and results in consistently reasonable capital estimates. Many researchers have conjectured that one may not be able to find a single distribution that will fit both the body and the tail of the data to model operational loss severity; however, our results regarding the g-and-h distribution imply that at least one single distribution can indeed model operational loss severity without trimming or truncating the data in an arbitrary or subjective manner." Jobst (2007) finds this distribution useful for the task.

Example 3.7. The following snippets encode the quantile, probability and density function of the g-and-h distribution. We use it to plot some characteristic samples.

```
# the quantile function
qTgh<-function(p,g,h){zp<-qnorm(p);
        if (g==0) zp else
        (exp(g*zp)-1)*exp(0.5*h*zp^2)/g}
# the density function
dTgh<-function(z,g,h){u<-exp(-0.5*(h+1)*z^2);
            dum<-exp(g*z);
        if (g==0)denom<-sqrt(2*pi) else
        denom<-sqrt(2*pi)*(dum+z*(dum-1)*h/g);
        u/denom}
# the probability function
pTgh<-function(p,g,h){eps<-1e-6;
            r<-uniroot(function(t){qTgh(t,g,h) - p},
                interval=c(eps,1-eps));
            r$root}
#$ plot the probability function
q<-0:400*.0025
g<-0.;h<-2.
```

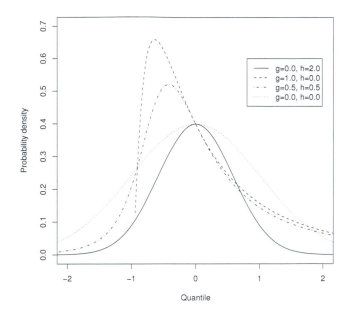

FIGURE 3.17: Some examples of the g-and-h distribution density.

```
# plot the density
y<-qTgh(q,g,h); z<-qnorm(q)
plot(y,dTgh(z,g,h),xlab="Quantile",ylab="Probability density",
   ylim=c(0,.7),xlim=c(-2,2),type="l",lty=1)
```

$\triangle$

Although the following has nothing to do with the title of the section, we want to mention yet another distribution family with four parameters, viz. *stable distributions*. These have been proposed for many types of physical and economic systems too. In the field of operational risk it is mainly the heavy-tail property and the skewness which make these distribution appealing. The lack of closed formulae for density and distribution function is overcome by the notorious use of numerical algorithms. The stable distribution is a generalisation of the normal distribution.

The term stable stems from the fact that the shape of the distribution is retained besides scale and shift under addition. One possible definition (Nolan, 2010, 7) is as follows: if $X, X_1, X_2, \ldots, X_n$ are independent, identically distributed stable random variables, then for every $n > 1$:

$$X_1 + X_2 + \ldots + X_n \overset{d}{=} c_n X + d_n$$

for some constants $c_n > 0$ and d_n. The symbol $\stackrel{d}{=}$ means here equality in distribution. The class of all distributions that satisfy this definition is char-

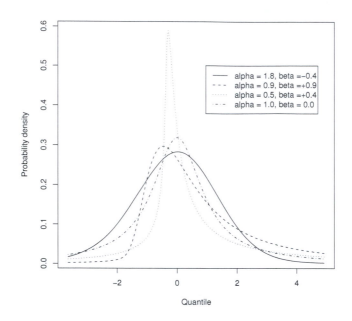

FIGURE 3.18: Some examples of the stable probability density.

acterised by four parameters, which are designated by $(\alpha, \beta, \gamma, \delta)$. Moreover, c_n must be of the form $n^{1/\alpha}$. The parameter α is called the index of stability; it must be in the range $0 < \alpha < 2$. The smaller α the heavier the tail. The parameter β is called the skewness and must be in the range $-1 < \beta < 1$; $\beta = 0$ the distribution is symmetric. The parameter $\gamma > 0$ is a scale parameter. The parameter δ is a location parameter. This is the parametrisation according to Nolan (2010, 8). The Figure 3.18 shows some different densities. They were created with the function `dstable(x, alpha, beta, gamma, delta, pm=0)` from the R-package `fBasics`. Operational risks as modelled by stable distributions are studied in Chernobai and Rachev (2004). Additional references can be found in Chernobai et al. (2007, 155).

3.4.3.4 Tails of the Distributions

The tail behaviour of large losses is critical because the more mass they contain, the higher the risk. Distributions differ in respect to the heaviness of the tails. The distributions proposed above cover the whole range of tail behaviour, as can be seen in Table 3.3. This classification stems in part from Wang (1998, 29). In the compounding of frequency and severity distributions,

TABLE 3.3: Tail behaviour of some distributions

Heavy-tailed	Moderate-tailed	Light-tailed
Pareto	Inverse Gauss	Gamma
Lognormal		Exponential
Weibull $(0 < c < 1)$		Gumbel
Fréchet		Poisson
		Negative binomial

the tail of the resulting distribution is in essence determined by the heavier of the two distributions. As both Poisson and the negative binomial distributions are light-tailed one has to choose a heavy-tailed severity distribution in order to get a heavy-tailed aggregate distribution.

The tail may be also characterised by its slope. Mack (2002, 89) defines this as

$$\gamma(x) = -\frac{d\log(f(x))}{d\log(x)}. \tag{3.44}$$

For typical individual losses in insurance for large X this "slope" should become constant in the interval $3 < \gamma(x) < 1$. Let us compare the slopes of the Pareto and the Weibull distributions:

$$\gamma_{\text{Pareto}}(x) = c + 1, \tag{3.45}$$

$$\gamma_{\text{Weibull}}(x) = 1 - c + c\left(\frac{x}{b}\right)^c. \tag{3.46}$$

If the slope of the distribution density decreases too rapidly with increasing loss, this means that the tail is too short. Therefore only a Weibull with shape parameter less than 1 is a good model for big losses. The slope for the lognormal is $1 + (\log(x) - a)/b^2$, i.e., increasing with x and thus – like the inverse Gaussian and the Gamma – not an appropriate model for individual losses.

Another way to define heavy tails is to compare the slope in the tail with the slope of the exponential distribution. Hence the term *subexponential distribution*.

The tail of a distribution with distribution function $F(x)$ is described by $\overline{F}(x) = 1 - F(x)$. The property is given, if:

$$\lim_{x \to \infty} \frac{\overline{F}(x)}{e^{-ax}} = \infty$$

for all $a \geq 0$. With subexponential distribution the sum of independent, identically distributed random variables above a certain threshold (i.e in the tail) is mainly determined by the biggest summand.

3.4.3.5   Moments of the Compound Distribution

Definition 3.6. The *moments about the origin*, and especially the expectation with $k = 1$ are defined as

$$\mu_k' = E(X^k) = \begin{cases} \sum_x x^k p_X(x) & \text{for the discrete case,} \\ \int_{-\infty}^{\infty} x^k f_X(x)dx & \text{continuous case,} \end{cases} \quad (3.47)$$

with f_X the probability density function and p_X the probability mass function. Furthermore, we define the symbol μ for the mean

$$\mu = \mu_1' = E(X). \quad (3.48)$$

Analogously the *moments about the origin* and especially the variance with $k = 2$, are defined as

$$\mu_k = E([X - \mu]^k) = \begin{cases} \sum_x (x - \mu)^k p_X(x) & \text{for the discrete case,} \\ \int_{-\infty}^{\infty} (x - \mu)^k f_X(x)dx & \text{continuous case.} \end{cases} \quad (3.49)$$

Specifically we define the variance to be

$$Var(X) = \sigma_X^2 = \mu_2 = E([X - \mu]^2). \quad (3.50)$$

□

The conditional expectation $E[X|Y = y]$ is defined as

$$E[X|Y = y] = \begin{cases} \sum_x x p_{X|Y}(x|y) & \text{for the discrete case,} \\ \int_{-\infty}^{\infty} x f_{X|Y}(x|y)dx & \text{continuous case.} \end{cases} \quad (3.51)$$

From here the following formula leads us to the expectation:

$$E[X] = \begin{cases} \sum_y p_Y(y)E[X|Y = y] & \text{for the discrete case,} \\ \int_{-\infty}^{\infty} f_Y(y)E[X|Y = y])dy & \text{continuous case.} \end{cases} \quad (3.52)$$

We can consider the conditional expectation to be a function $h(y)$, that is $h(y) = E[X|Y = y]$. Now we can make this function dependent upon the random variable Y by replacing y by Y. Thus, define $h(Y) = E[X|Y]$. Note that $h(Y)$ is a random variable whereas $h(y)$ is a real function. Taking the expectation of $h(Y)$ yields the so-called total or iterated expectation

$$E[X] = E_Y[E[X|Y]]. \quad (3.53)$$

Proofs of the above relationships can be found in Klugman et al. (1998, 391). Let us return to the compound losses characterised by the random sum of individual losses as

$$U = \sum_{j=1}^{M} X_j. \qquad (3.54)$$

The expected value of the compound losses is with Eq(3.53):

$$E\left[U\right] = E_N\left[E\left[U\,|N\right]\right] = E_N\left[N \cdot E\left[X\right]\right] = E\left[N\right] \cdot E\left[X\right] = \bar{N}\bar{X}. \qquad (3.55)$$

The variance is analogously given by its definition and the use of the iterated expectation in the form

$$\begin{aligned} Var(U) &= E_M\left[Var[U\,|\,M]\right] + Var_M\left[E\left[U\,|M\right]\right] \\ &= E_M\left[M \cdot Var[X]\right] + Var_M\left[M \cdot E\left[X\right]\right] \qquad (3.56) \\ &= E\left[M\right] \cdot Var\left[X\right] + Var\left[M\right] \cdot (E\left[X\right])^2. \end{aligned}$$

In a later section (3.6.5 starting on page 167) we shall also determine the covariance of two compound loss distributions.

Somewhat related are the moments of a *mixture of distributions* introduced on page 95. A new distribution and density can be constructed as a linear combination according to

$$g(x) = \sum_{i=1}^{N} q_i f_i(x) \qquad (3.57)$$

with $1 \leq q_i \leq 0$ and $\sum_{i=1}^{N} q_i = 1$. It has to be stressed that this has nothing to do with a distribution of linear combinations of losses, i.e $f(\sum q_i X_i)$. We will use the mixture for modelling the aggregate losses by lines of business, as the mapping table contains the above-mentioned weights (or probabilities) q_i. The mean of $g(x)$ is the weighted average of the means $E(f_i(x)) = \mu_i$, i.e.,

$$E[g(x)] = \sum_{i=1}^{N} \mu_i = \bar{\mu}. \qquad (3.58)$$

For the variance we have with $\sigma_i^2 = Var(f_i(x))$ for $i = 1, \ldots, N$

$$Var[g(x)] = \sum_{i=1}^{N} q_i \left[\sigma_i^2 + (\mu_i - \bar{\mu})^2\right]. \qquad (3.59)$$

For a derivation see Hogg and Klugman (1984, 50).

3.4.4   Data Fitting

The techniques used to find the distribution family and the appropriate
parameters is not actually part of the model. Indeed, the model assumes this
information to be available as input. Nonetheless, we believe that a short
overview is appropriate, especially as the distributions to be analysed are not
common in standard statistics text books. In-depth coverage can be found
inter alia in Klugman et al. (1998), Hogg and Klugman (1984) and Law and
Kelton (1991).

3.4.4.1   Parameter Estimation

Data fitting often means finding the most suitable, or "best," set of parameters for a particular distribution function. "Best" needs quantifying and,
depending on the definition of "good," the parameters may be different. The
object function of an optimisation problem can imply some model assumptions.

The standard objective functions or estimating functions are:

- moment matching MM, with L-moments as sub-category,

- percentile (quantile) matching,

- Maximum Likelihood Estimator (MLE),

- Minimum chi-square Estimator(MCE).

All matching methods lead to a system of equations to be solved. The
number of parameters of the distribution restricts the number of moments to
be used. In moment matching most often both mean and variance are matched.
The *moment matching equations* simply estimate the moments using sample
(empirical) moments

$$\hat{\mu}'_k = E(X^k) = \frac{1}{n} \sum_{i=1}^{n} x_i^k = m'_k \tag{3.60}$$

for $k = 1, \ldots, r$ where r is the number of parameters. From there the parameters θ_i are found as

$$\hat{\theta}_i = f(m'_1, m'_2, \ldots, m'_k) \quad \text{for} \quad i = 1, \ldots, r. \tag{3.61}$$

When appropriate, moment methods have the advantage of simplicity. The
disadvantage is that they do not possess the optimality properties of maximum
likelihood and least squares estimators.

L-moments use probability weighted moments, especially for flood data
using the very same distributions that we propose for compound losses. "The
main advantage of L-moments over conventional moments is that L-moments,
being linear functions of the data, suffer less from the effects of sampling variability: L-moments are more robust than conventional moments to outliers

in the data and enable more secure inferences to be made from small samples about an underlying probability distribution. L-moments sometimes yield more efficient parameter estimates than the maximum-likelihood estimates" (Hosking, 1990).

Percentile matching tries to estimate the r parameters of a distribution to be fitted $F_X(x|\theta)$ from a system of of r equations. The equations take the following form:

$$p_j = F_X(x_j|\theta) \quad \text{for} \quad i = 1, \ldots, r. \tag{3.62}$$

If we use more equations than there are parameters we once again face a problem of "fit" (or of finding a generalised inverse).

The *maximum-likelihood estimator* starts by modelling the joint distribution function of the say n independent random variables X_i as a function of the estimates' vector $\hat{\theta}$, i.e.,

$$L(\theta) = \prod_{i=1}^{n} f(X_i|\theta) \quad \Leftrightarrow \quad l(\theta) = \sum_{i=1}^{n} \log\left[f(X_i|\theta)\right]. \tag{3.63}$$

This function represents the probability of observing the given data as a function of the parameter θ. We can maximise this probability by choosing the parameter accordingly, i.e., taking the derivative (or gradient) with respect to θ and equating it with zero. We arrive at the parameters by solving these equations.

Sometimes the expression for the ML-estimators of distributional parameters are quite simple; sometimes you need a solver to calculate the parameters; and generally you need a numerical procedure to maximise the above unconstrained equation in order to approximate the parameters.

Example 3.8. Let us find the ML-estimator for a sample of X_i of occurrences to fit a Poisson distribution and compare it to the MM-estimator. The latter is obviously $\hat{\lambda}_{MM} = \sum_{i=1}^{N} X_i/N = \bar{X}$. With the definition of the Poisson mass density at page 104 and its logarithm:

$$l(\lambda) = \sum_{i=1}^{n} (-\lambda + X_i \log \lambda - \log X_i!). \tag{3.64}$$

Setting the derivative at zero yields

$$\frac{\partial l(\lambda)}{\partial \lambda} = 0 = \sum_{i=1}^{N} [-1 + X_i/\lambda], \tag{3.65}$$

$$\lambda_{ML} = \frac{1}{N} \sum_{i=1}^{N} X_i = \bar{X}. \tag{3.66}$$

Both estimators are equal. This is a special case. △

Maximum likelihood provides a consistent approach to the problem of estimating parameters. This means that MLE can be used for a large variety of estimation situations, especially noteworthy is the fact that it works for censored data.

MLE has some interesting properties. Firstly, the estimates are (very) consistent, i.e., they converge in probability with sample size to the true value of the parameters. Secondly, although they are not unbiased, they become minimum variance unbiased estimators as the sample size increases. Thirdly, they are asymptotically normally distributed. Therefore, confidence bounds of hypothesis tests can be performed. Within the class of asymptotically normally distributed estimators, MLEs are the most efficient. Finally, these estimators have an invariance property under transformation, i.e., if $\phi = h(\theta)$ for a function $h()$ then $MLE(\phi) = \hat{\phi} = h(\hat{\theta})$. Obviously this cannot be true for unbiased estimators.

On the other hand, the likelihood equations need to be specifically derived for a given distribution. This is non-trivial, often requiring numerical methods. MLE can be heavily biased for small samples and the optimality is not very pronounced for small samples.

The *method of least squares* originates from astronomic and geodetic measurements where the "true" location parameters must be determined by too large of a number of measurements containing errors. A precursor to least squares was the minimisation of the least absolute deviation proposed by Roger Boskovich and Pierre Simon Laplace. Formally:

$$\min_{a_1,\dots,a_m} \sum_{i=1}^{n} \mid y_i - F(x_i, a_1, \dots, a_m) \mid. \tag{3.67}$$

It is believed that both A.M. Legendre and C.F. Gauss discovered (or invented?) the method independently. Suppose we have n data points of a empirical cumulated distribution function $\{y_i, x_i\}$ with $i = 1, \dots, n$ for which we want to fit a distribution $F(x, a)$ with m parameters $a_1, \dots, a_m$. The expression to be minimised is

$$\min_{a_1,\dots,a_m} \sum_{i=1}^{n} [y_i - F(x_i, a_1, \dots, a_m)]^2. \tag{3.68}$$

If one supposes that each data point y_i has a normally distributed measurement error, then it can be shown that the least square estimator is a maximum likelihood estimator (Press et al., 1992, 652). In addition, it can be shown that of all linear combinations of measurements estimating an unknown, the least square estimator has the least variance. As an interesting note C.F. Gauss developed an equivalent system to the more commonly taught principles of d'Alembert, Lagrange and Hamilton in mechanics, based on the least square foundation. It is known as *principle of least constraint* (Szabó, 2001, 61). [3]

The main difference between these two object functions is in their treatment of those outliers which are assigned too much weight in the least squares minimisation. To reconcile the two object function, Huber (1981) introduced the so-called M-estimators. In essence, special weighting functions Ψ are sought that first increase with deviation and then decrease. The minimisation becomes

$$\min_{a_1,\ldots,a_m} \sum_{i=1}^{n} \Psi \left[y_i - F(x_i, a_1, \ldots, a_m) \right]^2. \tag{3.69}$$

See also Rice (1995, 366) and Press et al. (1992, 695), who consider those estimates the most relevant class for model fitting.

The least square method has an intimate relationship with regression. To solve the problem above we need numerical procedures. The standard for this task is the so-called Levenberg-Marquardt method (Press et al., 1992, 678).

There are also some tests to assess the probability that the given data sample is drawn from a given distribution. The most used *Goodness-of-Fit*-tests are the Chi-square-test and the Kolmogorov-Smirnov test.

Now let us take a closer look at a real world situation and show how we divide up the data for a given event and eventually determine the necessary distributions.

Example 3.9. Let us fit a distribution for the large losses of a given event, say fraud. In the first step we take the raw data and "eye ball" at what threshold the large losses start. The artificial but random data is as follows in Table 3.4 on page 130. A scatter-plot (Figure 3.19 on page 131) better reveals the structure than the table. We now decide to set the threshold at 2. (Again, this is the result of eye-balling, i.e., a visual inspection and guess. It is important to remember that our model only simulates the compound losses on an individual basis in order to make them available for insurance. Therefore the threshold has an additional restriction. With those data-points beyond the threshold, we shall run the fitting exercise by trying the distributions as proposed above.

The code looks like this:

```
library (ismev,MASS,stepfun,evd)
event<-read.table("z:/bass/event.dat",header=TRUE)
x<-event$Loss
y<-sort(x[x>2])
# MLE-Fitting
# Gumbel
gumb<-gum.fit(y)
# General Extreme Value
```

---

[3] In German it is called "Prinzip des kleinsten Zwanges," formally a variational principle for $\min = \int \left[\mathbf{a} - d\mathbf{F}/dm \right]^2 dm$ with acceleration $\mathbf{a}$, force $\mathbf{F}$ and mass m.

TABLE 3.4: Losses from an event category

Number	Year	Loss	Number	Year	Loss	Number	Year	Loss
1	1992	10.705	19	1996	0.004	37	2000	0.865
2	1992	0.014	20	1996	0.001	38	2000	3.094
3	1992	0.128	21	1996	0.003	39	2001	0.078
4	1992	0.002	22	1996	0.075	40	2001	0.025
5	1993	0.095	23	1996	0.005	41	2001	0.039
6	1993	0.082	24	1997	0.033	42	2002	0.004
7	1993	2.064	25	1997	0.690	43	2002	0.024
8	1993	1.409	26	1997	0.137	44	2002	0.530
9	1993	2.399	27	1997	9.380	45	2002	0.077
10	1993	0.568	28	1997	0.097	46	2002	0.424
11	1994	0.649	29	1998	0.076	47	2003	4.056
12	1994	0.066	30	1998	0.074	48	2003	0.002
13	1994	0.003	31	1999	0.045	49	2003	0.001
14	1994	0.002	32	1999	1.026	50	2003	0.700
15	1995	1.757	33	1999	1.249	51	2003	0.026
16	1995	0.011	34	1999	0.025	52	2003	3.000
17	1995	0.001	35	1999	6.293			
18	1996	0.307	36	1999	1.714			

```
extr<-gev.fit(y)
# Weibull
weib<-fitdistr(y,"weibull")
#Frechet
frech<-fitdistr(y,dfrechet,start=list(loc=0,scale=1,shape=1))
# Plotting
z<-1:500; d=(max(y)*1.5)/max(z); z<- (z-1)*d
plot(ecdf(y),do.points=FALSE,verticals=TRUE,
     xlim=c(0,1.5*max(y)),main="",ylab="Probability",
     xlab="Loss Amount",lty=1)
lines(z,pgumbel(z,gumb$mle[1],gumb$mle[2]),lty=2)
lines(z,pweibull(z,weib$estimate[1],weib$estimate[2]),lty=3)
lines(z,pgev(z,extr$mle[1],extr$mle[2],extr$mle[3]),lty=4)
legend(8.,0.6,legend=c("ECDF","Gumbel","Weibull","GEV"),
                 lty=c(1,2,3,4))
```

It produces a plot as depicted in Figure 3.20. We see that there are big differences, especially at the tail. The Generalised Extreme Value distribution (which is identical with the Fréchet in this instance) is very heavy tailed. △

Only data concerning losses of a certain severity is collected. This threshold may vary from bank to bank, as there is no industry standard. In the aftermath of the Basel initiative banks rushed to install their loss data base and may have opted for too low a threshold compared with the amounts they would

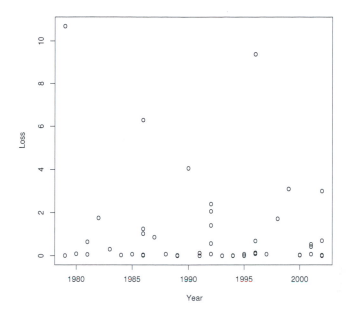

FIGURE 3.19: Data as scatter-plot.

deem irrelevant. The records lead to a distribution of losses which is *truncated from below*. Now we are interested in the true distribution. We assume that the amounts x_i of X are recorded under the assumption $x \geq d$ thus yielding a new variable Z:

$$Z = \begin{cases} X & \text{for} \quad x > d, \\ \text{not defined} & \text{for} \quad x \leq d. \end{cases} \qquad (3.70)$$

The empirical distribution of Z is given. Now we must determine the distribution of the variable X from the ground up. The following holds (Hogg and Klugman, 1984, 129):

$$F_Z(x) = \begin{cases} \frac{F_X(x) - F_X(d)}{1 - F_X(d)} & \text{for} \quad x > d, \\ 0 & \text{for} \quad x \leq d. \end{cases} \qquad (3.71)$$

Now we can proceed as described above, e.g., fitting according to Eq.(3.68, 3.67 or 3.63). In the following we give a brief numerical example.

Example 3.10. We generate some random variates according to a specific distribution with a given set of parameters. Then we truncate these variates, fit it to the same distribution and compare the outcome. Our primary goal is not to demonstrate perfect fit, but rather to highlight the procedure itself.

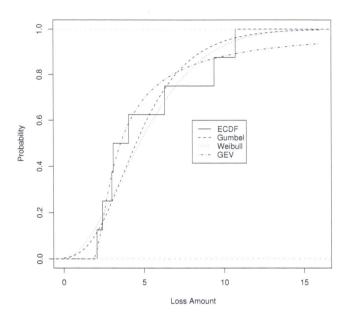

FIGURE 3.20: Large loss severity fit.

The function `hogg2` calculates the sum of the squared deviations. `optim` is a general purpose optimisation program.

```
hogg2<-function (c,x,y,d){   # least square to be minimised
   Fxd=pweibull(d,shape=c[1],scale=c[2]);
   imFxd<-1-Fxd;
   cf<-length(x)/(length(x)+1);
   sum( ((pweibull(x,shape=c[1],
            scale=c[2])-Fxd)/imFxd - y*cf)^2  )
}

m<-100                              # of variates
c0<-c(2,1)                          # parameters on input
z<-rweibull(m,shape=c0[1],scale=c0[2])  # random variates

d<-0.2;   c<-c0                     # the parameters
x<-sort(z[z>d])                     # truncated values | ties)
y<-1:length(x) ; y<-y/length(x)     # sample distribution (no
s<-optim(par=c(c),hogg2,x=x,y=y,d=d,method = "Nelder-Mead")
s$par[1]; s$par[2]
```

The resulting parameters are shape $= 1.8850340$ and scale $= 0.9998162$ to be compared with 2 and 1 on input. The higher the threshold, the smaller the number of data points and the poorer the fit. △

If we take the derivative of the distribution of Eq.(3.71), i.e., $f_Y(x) = f_X(x)/(1 - F_X(d))$ for $x > d$, we can also apply, numerically, the maximum likelihood estimator for the parameters (see Hogg and Klugman, 1984, 139). Now let us move on to the frequencies.

Example 3.11. Having determined the severity distribution, we may go on to assess the frequency distribution. In the case of the Poisson distribution the parameter estimation is very simple: there have been 10 occurrences within 13 years. Therefore the intensity parameter is $10/13 = 0.769$. To calculate the ML-estimator for the negative binomial is cumbersome. We determine the moments, i.e., mean and variance. The mean is equal to the intensity. The mean is 0.833, the variance 0.697. For this constellation the negative-binomial cannot be calculated, because this distribution combines like a mixing the Poisson with a gamma. As in the Poisson the mean and the variance are equal, but on account of the mixing, the variance will increase, so that we cannot accommodate the situation where the variance is smaller than the mean. △

Example 3.12. Again, we want to estimate the frequency of an event class. Let us use the table below, catastrophic losses in aviation.

TABLE 3.5: Losses from aviation accidents

Year	Occurrences	Total Loss	Year	Occurrences	Total Loss
1980	2	347	1991	19	1227
1981	2	160	1992	16	1262
1982	8	1253	1993	28	1614
1983	11	1370	1994	31	2930
1984	3	183	1995	22	1281
1985	13	3148	1996	20	2254
1986	5	233	1997	32	1411
1987	10	1465	1998	29	1817
1988	9	1805	1999	29	1667
1989	18	2096	2000	21	1117
1990	12	1967	2001	11	4521

A first shot at the frequency distribution would yield either Poisson with intensity $\lambda = 15.95$ or a negative binomial with parameters $n = 3.292$ and $p = 0.171$, the latter from the moment matching. One would think that the number of aviation disasters depends on some variable that accounts for the temporal change of conditions. In this case it could be "Miles flown" or "Number of Departures." Investigation and research reveal that more than 70% of accidents happen during take-off or landing, making the second indicator more appropriate. Nonetheless, it is only a *correlative indicator*. Thus, it is clear how we might transfer this idea of normalization over to banking.

In order to estimate the frequency for the next year indexed by $N+1$ we must first rescale the indices so that the index estimate for the next year equals one. If one assumes the index value $\hat{I}_{N+1}$ is 1.0 to the base year 1980, then the indices to be used from the table I_j are transformed to $K_j = I_j/\hat{I}_{N+1}$.

In the Poisson case we have a new estimate: $\hat{\lambda} = \hat{I}_{N+1}\sum_{j=1}^{N}n_j/\sum_{j=1}^{N}I_j$ with n_j the number of occurrences in year j and I_j the index of the same year. Numerically $\hat{\lambda}_{N+1} = 1 \times 351/24.69 = 14.2$.

Year	Index	Year	Index	Year	Index	Year	Index
1980	1	1986	1.031	1992	1.147	1998	1.310
1981	0.958	1987	1.035	1993	1.327	1999	1.245
1982	1.172	1988	1.071	1994	1.319	2000	1.131
1983	0.981	1989	1.038	1995	1.186	2001	1.036
1984	0.986	1990	1.164	1996	1.295		
1985	0.943	1991	1.039	1997	1.273		

The use of indices introduces an additional parameter to accommodate the fact that parameters are time varying and dependent on underlying factors like business volume, etc. Figure 3.21 shows the result. △

Example 3.13. Now we turn to the attritional losses that are modelled as annual aggregates from Example 3.9 on page 129. As a first step we have to prepare the data by aggregating the data-points below the threshold by year of occurrence.

Year	Aggregate Loss	Year	Aggregate Loss
1992	0.144	1998	0.150
1993	2.154	1999	2.345
1994	0.720	2000	0.865
1995	0.012	2001	0.142
1996	0.395	2002	1.059
1997	0.957	2003	0.729

As suggested above, we fit the three distribution candidates to this data. Again, we use the R-language for convenience.

```
#  load the library
library(MASS)
#  read data points into x
x<-c(0.144,2.154,0.720,0.012,0.395,0.957,0.150,2.345,
      0.865,0.142,1.059,0.729)

#  fit Gamma distribution
```

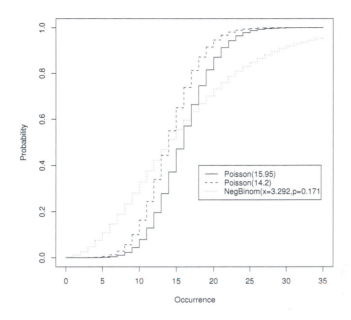

FIGURE 3.21: The fitted frequency distributions.

```
y<-fitdistr(x,"gamma",start=list(shape=1,scale=1))
#   fit lognormal distribution
z<-fitdistr(x,"lognormal",start=list(meanlog=10,sdlog=1))

#   determine the maximum likelihood estimates for
#   the Inverse Gaussian
loc<-sum(x)/length(x)
scale<-length(x)/(sum(1/x)-1/loc)

#   Define the Inverse Gaussian cdf
invGaussCDF<-function(x,lambda,a) {
   d1<-  sqrt(lambda*x/a^2.)-sqrt(lambda/x);
   d2<-  -sqrt(lambda*x/a^2.)-sqrt(lambda/x)
   d3<- exp(2*lambda/a);
   ret<-pnorm(d1)+d3*pnorm(d2);
   ret}
# plot the distributions
t<-1:500; d=(max(x)*1.5)/max(t); t<- (t-1)*d
plot(ecdf(x),do.points=FALSE,verticals=TRUE,xlim=
   c(0,max(x)*1.5),lty=1,xlab="Loss Amount",ylab=
   "Probability",main="")
lines (t,pgamma(t, shape=mean(y$estimate[1])
```

```
        ,scale=mean(y$estimate[2])),type="l",lty=2)
lines(t,plnorm(t, meanlog=z$estimate[1],
        sdlog=z$estimate[2]),type="l",lty=3)
lines(t,invGaussCDF(t,scale,loc),type="l",lty=4)
legend(1,0.4,legend=c("empirical","Gamma","Log-normal",
        "Inverse Gaussian"),lty=c(1,2,3,4))
```

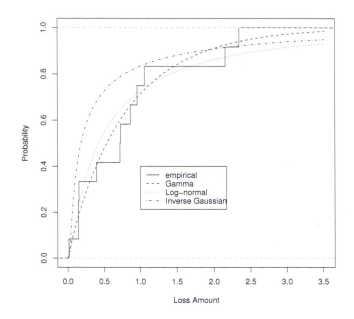

FIGURE 3.22: Insurance parameters.

We have chosen the maximum likelihood estimator. Now we must choose according to the maximum value. Visually, we receive the impression that the gamma-distribution would be best. But what about the figures? We follow up this question in Example 3.16 on page 141.   △

Bayesian statistics can be used to estimate parameters, merging samples and mixing distributions, allowing us to smoothly incorporate new information and data.

A *prior distribution G* is said to be a conjugate prior for a family of distributions H if the prior and the posterior distribution are of the same family G. In the following Table 3.6 you find some conjugates that are of interest in our context. Conjugate priors are very useful when updating the sample or adding information because they do not change the distributional family. Only the parameter values of the prior – called hyper-parameters – change.

TABLE 3.6: Some conjugate priors

Family	Conjugate Prior
Negative Binomial (n,θ)	$\theta \sim Beta(\alpha, \lambda)$
Poisson(θ)	$\theta \sim Gamma(\delta, \gamma)$
Exponential(λ)	$\lambda \sim Gamma(\delta, \gamma)$
Weibull(α, θ)	$\theta \sim Gamma(\delta, \gamma)$
Normal(μ, σ^2)	$\mu \sim Normal(\nu, \zeta^2)$
Normal(μ, σ^2)	$\sigma^{-2} \sim Gamma(\delta, \gamma)$

Bayes' theorem (see Footnote 2 on page 96) is a formula for learning. Suppose we have collected observations $x_1, x_2, \ldots, x_n$. Then

$$p(\theta \mid x) = \frac{p(x \mid \theta)\pi(\theta)}{\int p(x \mid \theta)\pi(\theta)d\theta}. \tag{3.72}$$

Now we collect an additional realisation x_{n+1}. Then

$$p(\theta \mid x, x_{n+1}) = \frac{p(x_{n+1} \mid \theta)p(\theta \mid x)}{\int p(x_{n+1} \mid \theta)\ p(x \mid \theta)d\theta}. \tag{3.73}$$

So our prior in the current up-dated sample is the posterior from our previous sample — hence the usefulness of conjugate priors.

For our count processes for the number of occurrences of losses the beta distribution is often used as a prior. This is motivated by the fact that this distribution is very versatile and can cover a broad range of shapes. Secondly, the support encompasses the interval from 0 to 1.

Example 3.14. The Poisson distribution of the number of losses k is given as

$$p(k, \theta) = \frac{e^{-\theta}\theta^k}{k!}. \tag{3.74}$$

Now let us assume a prior distribution of the parameter θ belonging to the Gamma family, i.e.,
$\theta \sim Gamma(\alpha, \beta)$ and thus

$$p(\theta) = \frac{e^{-\theta/\beta}\theta^{\alpha-1}}{\Gamma(\alpha)\beta^\alpha}. \tag{3.75}$$

The posterior distribution of the parameter is

$$p(\theta \mid k) = \frac{p(\theta)p(k, \theta)}{\int p(\theta)p(k, \theta)d\theta} \propto \theta^{k+\alpha-1}e^{-\vartheta(1+1/\beta)}. \tag{3.76}$$

From this we infer that the parameter given the model distribution has changed to the Gamma distribution's parameters according to

$$\theta \,|\, k \sim Gamma(\alpha + 1, \frac{\beta}{\beta + 1}).$$

(3.77)

$\triangle$

Example 3.15. Now let us calculate the predictive distribution $p(y)$ by integrating

$$p(y\,|\,k) = \int_0^\infty p(y\,|\,\theta)p(\theta\,|\,x)d\theta$$

$$= \frac{1}{\Gamma(\alpha)\beta^\alpha}\frac{1}{y!}\int_0^\infty e^{-\theta}\theta^y e^{-\theta/\beta}\theta^{\alpha-1}d\theta$$

$$= \frac{1}{\Gamma(\alpha)\beta^\alpha}\frac{1}{y!}\int_0^\infty \exp\{-\theta(1+1/\beta)\}\theta^{y+\alpha-1}d\theta$$

$$= \frac{1}{\Gamma(\alpha)\beta^\alpha}\frac{1}{y!}\int_0^\infty \exp\{-\xi\}\xi^{y+\alpha-1}(1+1/\beta)^{y+\alpha-1}\frac{d\xi}{1+1/\beta}$$

$$= \frac{1}{\Gamma(\alpha)\beta^\alpha}\frac{1}{y!}(1+1/\beta)^{-(y+\alpha)}\int_0^\infty \exp\{-\xi\}\xi^{y+\alpha-1}d\xi$$

$$= \frac{1}{\Gamma(\alpha)\beta^\alpha}\frac{1}{y!}(1+1/\beta)^{-(y+\alpha)}\Gamma(y+\alpha)$$

$$= \binom{\alpha+y-1}{y}\left(\frac{\beta}{\beta+1}\right)^y\left(\frac{1}{\beta+1}\right)^\alpha.$$

(3.78)

Therefore $y\,|\,k \sim NB(\alpha, 1/(\beta+1))$. For the above derivation we used the Eulerian and Gaussian definitions (for non-negative or zero integers) of the Gamma function

$$\Gamma(\alpha) = \int_0^\infty t^{\alpha-1}\exp\{-t\}\,dt = \lim_{v\to\infty}\frac{v!v^{\alpha-1}}{\alpha(\alpha+1)(\alpha+2)\ldots(\alpha+v-1)}.$$

(3.79)

The fact that the predictive distribution stems from the negative binomial family is critical. The negative binomial distribution is also the result of mixing a Poisson distribution with the weights of a Gamma distribution for the parameter uncertainty.

$\triangle$

In general the posterior distribution Eq.(3.72) and especially its denominator cannot be calculated easily or analytically. This means that numerical integration must be used – either classical quadrature or Monte Carlo simulation – in order to calculate the predictive distribution or other characteristics, e.g., the mean, etc. Coles (2001, 171) gives a good first sketch of the numerical treatment.

3.4.4.2 Goodness-of-Fit Tests

There are several tests to estimate the confidence about a data sample belonging to a postulated distribution function. The two most prominent tests are:

- Chi-Square test and

- Kolmogorov-Smirnov test.

More information on these standard tests can be found in Rubinstein (1981) and Law and Kelton (1991). As a reminder we summarise the inferential test procedure in the Algorithm 1. The *critical value* (one sided test) or values

---

1. Formulate hypothesis H_0 and alternative H_A

2. Calculate test statistic T from sample

3. (i) Determine critical region of size α or (ii) determine p-value

4. (i) Check whether T is in critical region and if yes, reject or (ii) check whether p-value is less-equal α and if yes reject.

ALGORITHM 1: Hypothesis testing

(two sided test) determine the region of rejection. The so-called *p-value* or *level of significance* is the smallest value of α for which the null-hypothesis will be rejected.

Chi-square test

The chi-square test, dating back to 1900 and devised by Pearson, is the best known goodness-of-fit test. It is a one-sample test that examines the frequency distribution of observations grouped into classes. The observed counts in each class are compared to the expected counts from the hypothesized distribution. The test statistic with sample size N is defined as

$$Y = \sum_{j=1}^{n} \frac{(N_j - Np_j^0)^2}{Np_j^0}, \tag{3.80}$$

where N_j is the observed number of outcomes in class j and Np_j^0 is the expected number of outcomes if the null-hypothesis were true.

Under the null hypothesis that the sample with unknown distribution $F_X(x)$ stems from the hypothesized distribution $F_0(x)$, it has a distribution with $n-1$ degrees of freedom. For any significance level α, reject the null-hypothesis if it is greater than the critical value for which $Pr(Y > \hat{\chi}^2_{(1-\alpha)}) = \alpha$.

When the sample being tested comes from a continuous distribution, one factor affecting the outcome is the choice of the number of classes. This becomes particularly important when the expected count in one or more classes falls below 1 or the average expected counts per class falls below five.

Chi-square tests apply to continuous variables, discrete variables or a combination of the two. On the other hand, for large sample sizes, if the hypothesized distribution is discrete, only the chi-square test may be used. Moreover, the chi-square test can easily be applied when the parameters of a distribution are estimated. However, particularly for continuous variables, information is lost by grouping the data. When the hypothesised distribution is continuous, the Kolmogorov-Smirnov test is more likely to reject the null hypothesis when it should; therefore, it is more powerful than the chi-square test.

Kolmogorov-Smirnov test

Suppose $F_0(x)$ and $F_X(x)$ are two distribution functions. In the one-sample situation, $F_X(x)$ is the empirical distribution function and $F_0(x)$ is a hypothesized cdf.

The two-sided hypothesis is:

$$H_0: \quad F_X(x) = F_0(x) \text{ for all } x \text{ versus}$$
$$H_A: \quad F_X(x) \neq F_0(x) \text{ for at least one } x.$$

The Kolmogorov-Smirnov test tests this hypothesis. The KS-test statistic is given by:

$$D = \sup_{x \in R^1} |F_X(x) - F_0(x)|. \tag{3.81}$$

If the test statistic is greater than some critical value, the null-hypothesis is rejected. The appropriate distribution is tabulated.

When the parameters are estimated from the sample, rather than specified in advance, the tests described above are no longer adequate. Different tables of critical values are needed. In fact, for the KS test, the tables vary for different distributions, parameters estimated, methods of estimation and sample sizes.

The main differences between the tests are that (1) the KS-Test is only applicable to continuous distribution functions whereas the Chi-square is more flexible. (2) KS uses all information from the sample, but the Chi-square does not, due to the classes into which the empirical data-point must be sorted.

Yet another test, known as the Anderson-Darling-Test uses the Cramer-von Mises statistic. This is beyond our scope. See Law and Kelton (1991) and Rubinstein (1981) for further information.

Example 3.16. We continue the Example 3.13 from page 134 and test the goodness of the fit. We also establish the log-likelihood for the fits. We determine the so-called *Kolmogorov-Smirnov one-sample statistic* D_N which is the supremum of the absolute value of the difference between the given frequency and the fitted distribution (Rubinstein, 1981, 28). This can be done easily with the script below.

```
# calculate log-likelihood
s1<-0;s2<-0;s3<-0;
for (s in x) {
        s1<-s1+log(dgamma(s,y$estimate[1],y$estimate[2]));
        s2<-s2+log(dlnorm(s,z$estimate[1],z$estimate[2]));
        s3<-s3+log(invGauss(s,loc,scale))}
s1;s2;s3


# perform Kolmogorov-Smirnov Test
ks.test(x,"pgamma",shape=y$estimate[1],scale=y$estimate[2])
ks.test(x,"plnorm",z$estimate[1],z$estimate[2])
ks.test(x,"invGaussCDF",lambda=loc,a=scale)
```

We summarise the results in the table below. As is evident, our prediction that the gamma-distribution fits best was accurate.

	Log-Likelihood	D_N
Gamma	-9.67	0.183
lognormal	-10.97	0.228
inverse Gaussian	-207.2	0.750

In this example we get a consistent ranking from the two methods. This need not always be the case. △

Now we start from the premise that a set of reliable and validated distributions is known. We want to find the distribution from this set that fits the data best. For this purpose we can apply those two measures to compare the fitted distributions.

3.4.4.3 Graphical Analysis

In addition to the rigorous testing of a data sample with a distribution one can apply graphical methods in order to acquire a feeling for the fitting process. There are two widely used graphical representations, namely the *quantile plot* (or quantile-quantile plot or Q-Q plot) and the *probability plot* (or probability-probability plot or P-P plot).

In the fitting context we are often faced with a sample and thus with an empirical distribution function and distribution estimates representing the population. In a Q-Q plot the quantiles of one distribution are plotted against the quantiles of the other. Suppose a sample of ordered observations $x_{(j)}$ for $j = 1, \ldots, N$ with an empirical cumulated distribution function $\hat{G}(x)$ and a candidate distribution function estimate $\hat{F}(x)$. Then the Q-Q plot consists of the data points.

The empirical distribution was introduced on page 91, thus $G^{-1}(x_{(j)}) = j/N$. Some prefer to take either $j/(N+1)$ or $(j-0.5)/N$ in order not to have a value of 1 for a finite value of x.

If $\hat{F}(x)$ were identical with the true underlying distribution $F(x)$ and if the sample size is at least moderately large then $\hat{G}(x)$ and $\hat{F}(x)$ will be close together and the Q-Q plot will show a straight line with slope 1 and intercept 0. Therefore the graphical analysis of a fit consists in judging whether data points form a straight line.

Analogously, the P-P plot is defined as:

$$\left(\hat{F}(x_{(j)}), \hat{G}(x_{(j)}) \right) \quad for \quad j = 1, \ldots, N. \tag{3.82}$$

The Q-Q plot amplifies the differences of the tails of the distributions whereas the P-P plot evidences differences in the middle (Law and Kelton, 1991, 375).

Example 3.17. We pick up the Example 3.13 at page 134 to construct the Q-Q plot (see Figure 3.23). As can be seen the small sample size is a problem irrespective of the method used for estimating the distribution.

```
# read data points
x<-c(0.144,2.154,0.720,0.012,0.395,0.957,0.150,2.345,0.865,
0.142,1.059,0.729)
x2<-sort(x)
q<-ppoints(x2)
# fit gamma and lognormal distribution
y<-fitdistr(x,"gamma",start=list(shape=1,scale=1))
z<-fitdistr(x,"lognormal",start=list(meanlog=10,sdlog=1))
# plot data
plot(qgamma(q,shape=y$estimate[1],scale=y$estimate[2]),y=x2,
    pch=1,xlab="Empirical Quantile",ylab="Estimated Quantile",
    xlim=c(0.,max(x)),ylim=c(0,max(x)))
lines(qlnorm(q, meanlog=z$estimate[1],sdlog=z$estimate[2]),y=x2,
    type="p",pch=15)
abline(0,1)
legend(1.5,.5,legend=c("Gamma","Log normal"),pch=c(1,15))
```

△

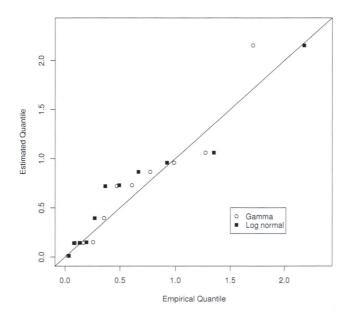

FIGURE 3.23: Quantile-quantile plot.

3.4.4.4 Extreme Value Theory

Extreme value theory (Embrechts et al., 1997; Coles, 2001; Kotz and Nadarajah, 2000) is used primarily to quantify the stochastic behaviour of a process at unusually large or small values. Such analyses often require estimation of the probability of events that are more extreme than any previously observed.

Many fields have been using extreme value theory for a long time including meteorology, oceanography and increasingly finance, because operational risk shares some features with those fields.

Several large or excessively frequent losses occurring in the same year may be a threat to the survival of a financial institution. Such a scenario is a possible outcome of an extreme event comparable to a large natural disaster.

One important theorem due to Fisher and Tippett (1928) dating back quite a long time states that the maxima of blocks of observations, under very general conditions, are approximately distributed as the generalized extreme value distribution. This distribution has three forms: (1) *Gumbel* (light tail), (2) *Fréchet* (heavy tail), and (3) *Weibull* (bounded tail).

In terms of the tail of a distribution, the corresponding theorem from Balkema, de Haan and Pickands from the 1970s states that the observations exceeding a high threshold, under very general conditions, approximate the generalized Pareto distribution. The exponential (light tail) and Pareto (heavy tail) are special cases of this distribution.

Both of these propositions are *limit theorems* comparable to the Central Limit Theorem.

We are not so interested in, say, the annual maxima, but rather in the "excess over threshold" part of EVT. Thus, we focus more on the Generalised Pareto distribution. From insurance modelling we know that the (two parameter) Pareto distribution is most often used for representing natural catastrophic events. Because of this theoretical affinity to the situation of rare but great losses we have included these distributions in the inventory of sensible functions to test against the available data. ETV can be used to improve the estimate of the quantile at levels as high as 99.9% confidence. One must bear in mind, however, that although these distributions are called "extreme value" this does not preclude their use without an extreme value model.

3.4.5   Random Number Generators

In a Monte Carlo setting we are must generate random variables that conform to some chosen distribution, either discrete or continuous. In most generators a *uniform random generator* is used. Nowadays, the most widely used are so-called *linear congruential generators* based on the following algorithm (Rubinstein, 1981, 21):

$$X_{i+1} = (aX_i + c)(mod \quad m). \tag{3.83}$$

The meaning of the modulo *mod* is as follows: $X_{i+1} = (aX_i + c) - km$ with $k = (aX_i + c)/m$ the largest positive integer in $(aX_i + c)/m$. The recursive character of the formula obviously indicates that a fundamental request of random numbers, namely their independence, is not given. With the given parameters a, c and m the successive number is completely determined by its predecessor. (The term "congruent" outside geometry was introduced by Gauss to represent two numbers a and b whose remainder when divided by c is equal.)

The uniform pseudo-random variable U_i is calculated from

$$U_i = \frac{X_i}{m}. \tag{3.84}$$

Pseudo-random number generators have some problems. Firstly, it can be shown that every sequence gets into a cycle. Secondly, the dependence may be an insurmountable problem for some applications. Fast and reliable new generators called "Mersenne Twister" have been introduced (Matsumoto and Nishimura, 1998). For our purposes it seems plausible that the choice of generators is not so important as we are focused on statistics (means) rather than random aspects of the number in question. (The name has something to do with the use of Mersenne numbers, i.e., primes of the form $2^n - 1$ of which at the moment 41 are known.)

Non-uniform random numbers can be produced in several ways. The simplest way is the so-called "inverse method" where $X = F_X^{-1}(U)$ with U a uniform random number in the range $[0, 1]$ (Rubinstein, 1981, 39). This method is efficient only when the inverse can be expressed analytically. Thus, it is applicable for the following severity distributions of our set (see Section 3.4.3.2 on page 107):

- Weibull,

- Fréchet,

- Gumbel,

- Generalised Extreme Value,

- Exponential,

- Pareto and

- Generalised Pareto.

So, generally speaking, the following Algorithm 2 can be used to generate random variates for the above continuous distributions. Again, $U(0, 1)$ denotes a uniform distribution on the interval $[0, 1]$.

---

1. Draw u from $U(0, 1)$,

2. Return $F^{-1}(u)$

ALGORITHM 2: Inverse method for random variate generation

For discrete distributions the recipe looks like Algorithm 3. It is also pos-

---

1. Draw u from $U(0, 1)$,

2. Find k such that $p_k \leq u < p_{k+1}$

3. Return k

ALGORITHM 3: Inverse method for discrete random variate generation

sible to approximate the probability function and then use these algorithms. This is often done for the normal distribution. Also commonly employed is the acceptance-rejection method of John von Neumann, outlined in Law and

Kelton (1991, 478). Specific algorithms can be found in Rubinstein (1981), Devroye (1986), Law and Kelton (1991) and Evans et al. (1993). The following example is such an algorithm.

Example 3.18. The generation of inverse Gaussian distributed random variates is numerically extremely difficult. There is a standard algorithm (see Algorithm 4) due to Michael et al. (1976) for generating random variates $r \sim IG(a, b)$. $\triangle$

---

1. Set $e = a/2b$, $f = ae$, $c = 4ab$, $d = a^2$,

2. Draw v from $\Phi(0, 1)$,

3. Set $y = v^2$,

4. Compute $x = a + ey - f\sqrt{cy + dy^2}$,

5. Draw u from $U(0, 1)$,

6. If $u < a/(a + x)$ then return x, else return d/x

ALGORITHM 4: Inverse Gaussian random variates

For the sake of completeness we also mention the Algorithm 5 for a two-point mixture according to Eq.(3.40) on page 118.

---

1. Draw u from $U(0, 1)$,

2. if $u < w_1$ return $F_1^{-1}(u)$ else $F_2^{-1}(u)$

ALGORITHM 5: Two-point mixture

Because random variables are used so extensively with Monte Carlo and its applications in physics, publicly available libraries abound (see for example `http://www.netlib.org/random`).

3.5 Run-Through Example: Quantitative Impact Study Data

In March 2003 the Basel Committee on Banking Supervision (BCBS) published its third impact study concerning operational risk with respect to 89 participating banks reporting the losses incurred during the year 2001 (BCBS, 2003a). The participants were asked to report especially on individual events and losses above a threshold of Euro 10,000.

The data shows some clustering around lines of business and events. Whether this clustering reflects reality as such, or the way in which data is collected, is difficult to determine. The study yields information on an array of summary statistics. Three tables are of particular interest and are therefore reported here (see Table 3.8), viz. total loss amounts cross-classified by lines of business and events, number of loss events by the same classification and the stratification of gross losses by amount (Table 3.7). From Table 3.7 we see

TABLE 3.7: Stratification of losses

Gross Loss Amount	Number of Events	Percentage	Value of Losses	Percentage	Average Loss
0 - 10	313	0.7%	1934	0.02%	6
10 - 50	36745	77.7%	720489	9.2%	20
50 - 100	4719	10.0%	324783	4.2%	69
100 - 500	4217	8.9%	847645	10.9%	201
500 - 1000	563	1.2%	387818	5.0%	689
1000 - 10000	619	1.3%	1748752	22.4%	2825
10000 +	93	0.2%	3764104	48.3%	40474
	47269	100.0%	7795525	100.0%	

that there is a strong discrimination between high frequency/low severity (the many small account for very low total amounts) and the few high severity are far beyond averages calculated from Table 3.8 on page 150.

If we plot the so-called *Lorenz diagram* (Figure 3.24 on page 148) the fact of a high concentration of small severities and of the very high concentration with extreme losses is evidenced by the marked departure from the diagonal. At the lower end we have a very flat polygon, i.e., below average loss per occurrence and at the upper end a very steep curve indicating extremely high loss per occurrence. This strong concentration also indicates a high dispersion and thus we have to reckon with substantial risk capital. For further evidence we calculate the *Gini index* G which is defined as the portion of the area between the two curves and the total area, thus[4]

$$G = \sum_{j=1}^{k} (x_j - x_{j-1})(y_j + y_{j-1}) \quad \text{with} \quad x_0 = y_0 = 0. \qquad (3.85)$$

Here x stands for the cumulated percentage of occurrences and y for the cumulated part of losses of Table 3.7. If G is close to zero then we have

---

[4]Note that it is a kind of PRE-measure as discussed in Section 2.10.2.

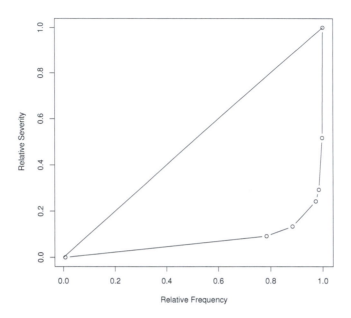

FIGURE 3.24: Lorenz curve of number of events by value of losses.

homogeneous data; at the other end, if G is in the limit equal to one, then we have largely varying magnitudes. Numerically we find $G = 0.852$, which again indicates a massive departure from the average loss. The Lorenz curve is yet another example for the heuristic called *Pareto principle* where more than 80% of loss cases do not make up more than 20% of the loss amount. This is not a coincidence, as Pareto developed his distribution to model income distributions of an economy. We carry this forward in an example.

Example 3.19. We can write the Gini-index in a continuous form by taking the limit to yield

$$G = 2 \int_0^1 y \, dx. \tag{3.86}$$

Now in the Lorenz curve we show *relative* quantities, i.e., obviously the frequency and here the relative severity y. This means for a Pareto-distributed severity that we need

$$y = \frac{\int_0^p x_p(t) \, dt}{\int_0^1 x_p(t) \, dt} = 1 - (1-p)^{1-1/c}, \tag{3.87}$$

where the inverse x_p is according to page 112 $x_p = a(1-p)^{-1/c}$. Without too much work we can rewrite the integral of Eq.(3.86) and then solve for G:

$$G = \frac{1}{2c - 1}.$$

In our very example we have a Gini-index of 0.852 and therefore $c = \frac{1}{2}(G^{-1} + 1) = 1.087$. We know and it conforms to the Lorenz-diagram, that this is a very dangerous parametrisation. In Figure 3.25 the two curves are shown for comparison. The axis are inverted. $\triangle$

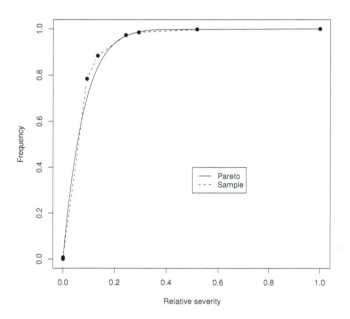

FIGURE 3.25: Lorenz curve and Pareto fit.

Thus, our belief that it make sense to differentiate between annual aggregate loss generators and compound distributions for substantial losses is reinforced.

From the study we model an average bank by imposing the following restrictions:

- we aggregate the data by Events;

- the average losses correspond to the totals from the study divided by 89;

- the frequencies are also divided by 89;

TABLE 3.8: Summary of the QIS data for 89 banks

Gross Loss Amounts	Internal	External	Empl.	Clients	Physical	Disr.	Exec.	Na	Total
Corporate Finance	49.4	5	2.5	157.9	8	0.5	49.6	0.6	273.5
Trading & Sales	59.5	40.4	64.8	193.4	87.9	17.6	698.4	1.1	1163.1
Retail Banking	331.9	787.1	340	254.1	87.5	26.5	424.5	37.4	2289
Commercial Banking	21.2	324.9	20.4	156.4	1072.9	18.2	619.4	23.2	2256.6
Payment & Settlement	23	21	11.6	10.5	15	78.6	93.5	0.3	253.5
Agency Services	0.2	3.9	7.6	5	100	40.1	174.1	0.8	331.7
Asset Management	6.4	4.6	10.2	77	2.3	2.3	113.2	0.05	216.05
Retail Brokerage	61.5	1.2	50.7	158.6	513.2	28	97.1	3.4	913.7
Not Assigned	10.5	23.4	18.7	11.5	6.7	0.7	22.7	3.8	98
Total	563.6	1211.5	526.5	1024.4	1893.5	212.5	2292.5	70.65	7795.15

Number of Losses	Internal	External	Empl.	Clients	Physical	Disr.	Exec.	Na	Total
Corporate Finance	17	20	73	73	16	8	214	2	423
Trading & Sales	47	95	101	108	33	137	4603	8	5132
Retail Banking	1268	17107	2063	2125	520	163	5289	347	28882
Commercial Banking	84	1799	82	308	50	47	1012	32	3414
Payment & Settlement	23	322	54	25	9	82	1334	3	1852
Agency Services	3	15	19	27	8	32	1381	5	1490
Asset Management	28	44	39	131	6	16	837	8	1109
Retail Brokerage	59	20	794	539	7	50	1773	26	3268
Not Assigned	35	617	803	54	13	6	135	36	1699
Total	1564	20039	4028	3390	662	541	16578	467	47269

- we choose thresholds for separating the aggregate from the compound losses such that the bulk falls into the aggregate loss generators according to Table 3.9.

TABLE 3.9: Thresholds in millions

Event	Threshold	Event	Threshold
Internal	0.50	Physical	2.00
External	0.10	Disruption	0.50
Employment	0.35	Execution	0.25
Clients	0.50	Extreme	50.00

Further detail is not necessary at this point. We already have appropriate and realistic input for the further development of the example we present below in Table 3.10 the data model. We have chosen the most obvious distri-

TABLE 3.10: Data model

	Attritional losses	Severity	Frequency
Internal	Gamma(1.5,2.50)	Pareto(2.39,0.5,250)	Poisson(1.64)
External	Gamma(1.5,4.52)	Pareto(5.0,0.1,50)	Poisson(22.17)
Employment	Gamma(1.5,2.8)	Pareto(6.22,0.35,100)	Poisson(1.14)
Clients	Gamma(1.5,2.7)	Pareto(5.41,0.5,500)	Poisson(7.0)
Physical	Gamma(1.5,1.23)	Pareto(1.08,2.,500)	Poisson(1.35)
Disruption	Gamma(1.5,1.05)	Pareto(7.0,0.5,100)	NegBin(0.74,5.)
Execution	Gamma(1.5,5.3)	Pareto(2.69,0.25,75)	Poisson(11.37)
Extreme	N.A.	Pareto(2.31,50,2000)	Poisson(0.1124)

butions; there is no reason to conjecture anything more complex. (We have included one negative binomial distribution in order to later demonstrate that the correlation mechanism we propose is not limited to Poisson distributed random variables.)

The severity for the extreme events spans a range from 50 million to 2 billion. Its frequency is very low: approximately once every ten years.

The expected losses as defined by the distributions and their parameters from Table 3.10 are depicted in Figure 3.26. The total expected loss equals 58.6 millions EUR, corresponding to EUR 7795.15 from Table 3.8 on page 150 divided by 89.

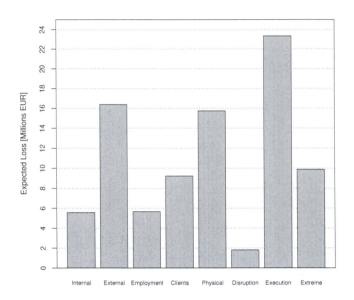

FIGURE 3.26: Expected loss by events.

3.6   Correlation of Losses

3.6.1   Motivation

We want to describe how to generate correlated random vectors with discrete marginal distributions for integrating dependency into our Monte Carlo framework. To this end we need to introduce some additional statistical terms. The Basel 2 accord foresees the use of correlation provided its estimates and usage is sound. Verbally BCBS (2004a, 145)

Risk measures for different operational risk estimates must be added for purposes of calculating the regulatory minimum capital requirement. However, the bank may be permitted to use internally determined correlations in operational risk losses across individual operational risk estimates, provided it can demonstrate to the satisfaction of the national supervisor that its systems for determining correlations are sound, implemented with integrity and take into account the uncertainty surrounding any such correlation estimates (particularly in

TABLE 3.11: Standard deviation and coefficient of variation

	Standard Deviation σ	CV
Internal	3.418	0.614
External	5.568	0.340
Employment	3.446	0.609
Clients	3.727	0.405
Physical	43.391	2.756
Disruption	1.509	0.832
Execution	6.710	0.288
Extreme	720.586	72.964

periods of stress). The bank must validate its correlation assumptions using appropriate quantitative and qualitative techniques.

Which kind of entities are connected by the correlation is an open question. It can be assumed that the relation is either between Events or Lines of Business (or both).

The summing to be applied, if correlations cannot be implemented, embodies the strongest dependence model there is, namely full dependence. This translates into the highest capital requirement possible. Therefore, a defensible correlation modelling is of utmost importance. The modellers' preferred position is to assume independence but here this choice is not available. Therefore, correlated losses represent the second best solution.

3.6.2 Measures of Association

Correlation, like probability, is a difficult term. It was introduced in 1888 by Francis Galton while investigating the height of parents and children. The term "regression" was employed to describe to estimate the height of parents based on their children's height.

Definition 3.7. For random variables X and Y, the *Pearson correlation coefficient*

$$Corr(X,Y) = \frac{Cov(X,Y)}{\sqrt{Var(X)Var(Y)}}. \tag{3.88}$$

The *covariance* of the random variables X and Y is the expectation (see Eq.(3.6) on page 124):

$$\begin{aligned} Cov(X,Y) &= E([X - \mu_X][Y - \mu_Y]) \\ &= E(XY) - \mu_X\mu_Y. \end{aligned} \tag{3.89}$$

For n random variables X_j, $j = 1, \ldots, n$ the *correlation matrix* $\mathbf{R} = [r_{ij}] \in \Re^{n \times n}$ is defined by its elements $r_{ij} = Corr(X_i, X_j) = Corr(X_j, X_i) = r_{ji}$. □

The Pearson correlation is a measure of how well a *linear* equation describes the relation between two random variables X and Y measured jointly. Note that $-1 \leq Corr(X, Y) \leq 1$, but that $Corr(X, Y) = 1$, if and only if there is a linear relationship between $Y = aX + b$ for some constants $a > 0$ and b. These coefficients can be calculated by minimising the square distance $E([Y - aX - b]^2)$, respectively $E([X - cY - d]^2)$ as:

$$a = \frac{Cov(X, Y)}{Var(X)}, \quad b = E(Y) - aE(X),$$

$$c = \frac{Cov(X, Y)}{Var(Y)}, \quad d = E(X) - cE(Y).$$

This brings us to another expression of the correlation, viz.

$$Corr(X, Y) = \sqrt{ac} \tag{3.90}$$

to be the square root of the product of the two slopes of the regression lines.

Example 3.20. We apply the correlation to a data set $\{x, y\}$ as shown in Figure 3.27 and perform a linear regression. In the given situation of a parabola-like function the correlation turns out to be nearly zero, here 0.15. We show both regression lines from x on y and vice-versa. Actually, one could try to transform y in such a way as to make it linear.   △

Without a linear relationship the permissible range of $Corr(X, Y)$ is further restricted. The correlation matrix $\mathbf{R}$ is symmetric and positive semi-definite, its diagonal term $Corr(X_j, X_j) = 1$ because of the identity $Cov(X_j, X_j) = Var(X_j)$. Now, *positive semi-definite* means that $\mathbf{x}^T \mathbf{R} \mathbf{x} \geq 0$ for all $\mathbf{x} \in \Re^n$. Correlation should not be mistaken for neither *causality* nor *dependency*. The latter two terms can only be assessed, if at all, logically. It is difficult, if not impossible, to prove that a "necessary antecedent" (the cause) produced a certain effect. However, it may be possible to exclude causes leading to a given effect.

In order to understand additional features it is now high time to introduce some terms related to multivariate distribution.

Definition 3.8. Let X and Y be random variables. The *joint probability density* $f(x, y)$ for the continuous case or the *joint probability mass function* $p(x, y)$ in the discrete case are defined as

$$Pr(a \leq X \leq b, c \leq Y \leq d) = \int_a^b \int_c^d f(x, y) dx dy,$$

$$p(x, y) = Pr(X = x, Y = y).$$

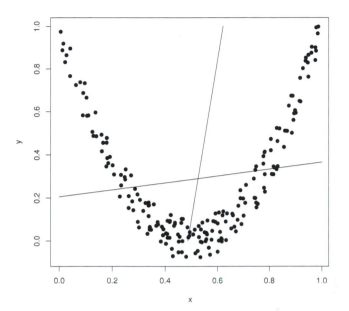

FIGURE 3.27: Linear regression on non-linear function.

We call $F(a, b)$ as below the *joint distribution function*:

$$F(a, b) = \begin{cases} \sum\limits_{x=-\infty}^{a} \sum\limits_{x=-\infty}^{b} p(x, y) & \text{for the discrete case} \\ \int\limits_{-\infty}^{a} \int\limits_{-\infty}^{b} f(x, y) dx dy & \text{for the continuous case.} \end{cases} \tag{3.91}$$

The *marginal probability mass function* respectively the *marginal probability density function* of a random variable is given by keeping all other random variables of the multivariate distribution fixed, i.e.,

$$f_X(x) = \int\limits_{-\infty}^{\infty} f(x, y) dy \quad \text{for} \quad -\infty < x < \infty,$$

$$p_X(x) = \sum_y p(x, y).$$

□

From the definition we can find $f(x, y)$ from the joint distribution function $F(x, y)$ as

$$f(x, y) = \frac{\partial^2 F(x, y)}{\partial x \partial y}, \tag{3.92}$$

if the partial derivatives exist. The term *independence* means that the joint probability $p(x, y)$ or $f(x, y)$ of the random variables X and Y obeys

$$p(x, y) = p(x)p(y) \quad \text{respectively} \quad f(x, y) = f_X(x)f_Y(y) \tag{3.93}$$

or with $F_X(x) = F(x, \infty)$ and $F_Y(y) = F(\infty, y)$:

$$F(x, y) = F_X(x)F_Y(y). \tag{3.94}$$

Coming back to correlation, it can be shown that if two random variables are independent then their correlation is zero. The reverse is not true, other than in the special case where the variables follow a bivariate normal distribution.

For modelling dependent (or at least correlated) random variables it is not sufficient to know or estimate the correlation. We need to discover a joint distribution, be it the "true" or a modelled one. Most joint distributions are not observable and thus not known. This kind of modelling is at the heart of the following sections.

In the following the term *copula* for "dependence function for random variables" may be helpful. For a good overview on copulae, please refer to Frees and Valdez (1998).

Definition 3.9. A *copula* function C is a multivariate probability distribution with standard uniform $[0, 1]$ distributed margins. $\quad\square$

Assume two random variables X and Y with distribution functions $F(X)$ and $G(Y)$. Obviously, these functions (random variables) are both uniformly distributed $F(X), G(Y) \sim U(0, 1)$. Now a copula, say $H(x, y) = C(F[x], G[y])$, has a density (if the derivatives exist) of

$$\begin{aligned} h(x, y) &= \frac{\partial^2 C}{\partial F \partial G} \frac{\partial F}{\partial x} \frac{\partial G}{\partial y} \\ &= \frac{\partial^2 C}{\partial F \partial G} f(x)g(y) \\ &=: c(x, y)f(x)g(y). \end{aligned} \tag{3.95}$$

In the latter expression, copula may best be understood as coupling, especially if the case of independence is considered where the density of the copula is $c(x, y) = 1$.

We have mentioned that the correlation need not lie in the interval $[-1, 1]$ but its attainable range including the origin is further restricted to $-1 \leq \rho_L \leq Corr(X, Y) \leq \rho_U \leq 1$. The extremal correlation $\rho = \rho_L$ is attained if and only if X and Y are *counter-monotonic*; $\rho = \rho_U$ is attained if and only if X and Y are *co-monotonic*.

Definition 3.10. Two random variables X and Y are *co-monotonic* if there exists a random variable T such that $X = g(T)$ and $Y = h(T)$ with probability one where the functions $g(.)$ and $h(.)$ are non-decreasing. $\quad\square$

A theorem by Fréchet (Wang, 1998, 5) states that for any bivariate probability distribution $F(x,y)$ with marginals $F_X(x)$ and $F_Y(y)$ the following holds

$$\max\left[F_X(x) + F_Y(y) - 1, 0\right] \leq F(x,y) \leq \min\left[F_X(x), F_Y(y)\right]. \qquad (3.96)$$

Analogously, because the copula is also a probability function and extends to higher dimensions, there are corresponding copulae for the best and the worst cases, namely

$$C_U(u_1, u_2, \ldots, u_N) = \min[u_1, u_2, \ldots, u_n], \qquad (3.97)$$

$$C_L(u_1, u_2, \ldots, u_N) = \max\left[\sum_{i=1}^{N} u_i - (N-1), 0\right]. \qquad (3.98)$$

We see immediately that these two copulae do not need any parameters. To assume the worst dependency is still better than to assume complete dependence and then add the risk figures of the single events or cells. Secondly, such copulae could be used in a standardised scenario in order to make the diversification effect of the chosen correlation structure evident.

What is the density of the co-monotonic copula? With the help of "higher" mathematics we can express the partial derivative according to Eq.(3.95) as

$$c_U(u_1, u_2) = \frac{\partial^2 C}{\partial u_1 \partial u_2} = \frac{\partial J_{\{u_1 < u_2\}}}{\partial u_2} = \delta(u_1 - u_2) \qquad (3.99)$$

where $J_{\{u_1 < u_2\}}$ is the indicator variable taking either the values 1 or 0 in dependence of the inequality's truthvalue and $\delta(a - b)$ is the so-called *Dirac function* defined as

$$\delta(\tau) = \begin{cases} 1 & \text{if} \quad \tau = 0, \\ 0 & \text{otherwise}, \end{cases}$$

with $\int_{-\infty}^{\infty} \delta(\tau)d\tau = 1$. You may think of this function as the standard normal probability density $\phi(0, s^2)$ where we take the limit of the variance $s^2 \to 0$.

This means that the whole probability mass is concentrated in the point where the independent variables are equal. Otherwise, the probability is zero. This leads to a very simple recipe for generating co-monotonically dependent random variables according to Algorithm 6. Now let us determine these upper correlations ρ_U given that we can model the joint distribution function with this copula, namely

$$f(x,y) = f_X(x)f_Y(y)\delta(F_X(x) - F_Y(y)). \qquad (3.100)$$

1. Draw u from $U(0,1)$

2. Calculate $X_i \leftarrow F_{X_i}^{-1}(u)$ for $i = 1, 2, \ldots, n$

3. Go to Step 1.

ALGORITHM 6: Generation of co-monotonic random variates

The covariance can be determined with the joint density as Eq.(3.100) to yield

$$Cov_U(XY) = \int_{-\infty}^{\infty}\int_{-\infty}^{\infty} xyf(x,y)dxdy - \mu_X\mu_Y \qquad (3.101)$$

$$= \int_0^1 F_X^{-1}(u)F_Y^{-1}(u)du - \mu_X\mu_Y.$$

Run-Through Example 1. We continue our example from the QIS data model. Let us determine the upper and lower Fréchet bounds. Instead of simulating we will use the method given in Section 3.6.4 on the following page 159. The result is shown for the sake of economy in a matrix-like array where the upper triangular values correspond to the upper bounds and vice versa.

$$\begin{pmatrix}
1.00 & 0.96 & 0.93 & 0.96 & 0.94 & 0.97 & 0.96 & 0.68 \\
-0.94 & 1.00 & 0.94 & 0.99 & 0.95 & 0.95 & 0.99 & 0.63 \\
-0.80 & -0.91 & 1.00 & 0.94 & 0.94 & 0.94 & 0.94 & 0.73 \\
-0.92 & -0.98 & -0.89 & 1.00 & 0.95 & 0.96 & 0.99 & 0.65 \\
-0.84 & -0.92 & -0.80 & -0.90 & 1.00 & 0.95 & 0.95 & 0.68 \\
-0.84 & -0.92 & -0.78 & -0.90 & -0.81 & 1.00 & 0.96 & 0.72 \\
-0.93 & -0.99 & -0.90 & -0.98 & -0.91 & -0.91 & 1.00 & 0.64 \\
-0.43 & -0.57 & -0.36 & -0.54 & -0.39 & -0.38 & -0.55 & 1.00
\end{pmatrix}$$

What we see quite clearly is that the bounds depend on the frequency random numbers being of the same order of magnitude. For the first seven values the bounds are close to 1. But for the last column and row we have situations where the extreme losses have a frequency of 0.1134, whereas the others are of order of magnitude ten times larger.

3.6.3   Model Choices

In the actuarial so-called "collective risk" model used by the AM-approach, it is implied that the loss amounts of each severity distribution are independent from each other. Therefore, the dependency can only be introduced by the frequencies in order not to be inconsistent. For the correlation with a loss from another event would make it a latent variable as discussed above and thus imply a correlation also between two losses from the same severity distribution. A more rigorous argumentation can be found in Frachot et al. (2004).

The central idea of modelling the joint distribution is based on the transformation of the input distributions into normal correlated variates. Practically speaking, the greatest effort must be expended on the transformation of the input correlation to the Gaussian correlation of the transformed variables, especially when the correlation is not "feasible" and requires approximation.

In the frequency-severity approach used here the frequencies are distributed either according to a Poisson or a negative binomial distribution.

If only Poisson distributions are present, then one may construct dependent variates by exploiting the fact that this special distribution has an additivity property. This so-called stochastic representation constructs a bivariate Poisson as $(Y_1, Y_2) = (Z_1 + Z_{12}, Z_2 + Z_{12})$ where Z_1, Z_2 and Z_{12} are independent Poisson variables (Powojowski et al., 2002). The dependence is introduced by the common variable Z_{12}. However, this approach is impossible with higher dimensions and has the additional disadvantage that it allows only for positive correlation.

Realistically, the normal and the t-copula are the only two functions to be used. Other functions have obscure parameters that we think cannot be estimated to the satisfaction of the regulators. As the estimation of the correlation may be the most challenging task of all, there is really no reason to use them. William Ockham would certainly protest.

There exist general procedures for generating dependent random variates with arbitrary marginal distributions and a feasible correlation matrix. Feasible means that the transformed correlation is a positive definite. Such a procedure has been presented by Cario and Nelson (1997). Let us examine this closely while restricting our attention to the above mentioned Poisson and negative binomial marginal distributions. They called their method NORTA for "normal to anything" transformation.

3.6.4 The Normal Transformation

The NORTA transformation $\mathcal{T}$ takes *standard multivariate normal random vector* $\mathbf{Z} = (Z_1, Z_2, \ldots, Z_n)^T$ and its known correlation matrix $\mathbf{R} = Corr(\mathbf{Z})$, and transforms it in such a way to achieve the desired marginal distributions for the components of the input vector $\mathbf{X} = (X_1, X_2, \ldots, X_n)^T$ with the associated correlation matrix $\mathbf{Q} = Corr(\mathbf{X})$. The correlation matrix $\mathbf{Q}$ can be realized by iteratively adjusting the correlation matrix $\mathbf{R}$.

The NORTA transformation can be viewed as a two-step process, first transforming a multivariate normal vector $\mathbf{Z}$ into a multivariate uniform vector $\mathbf{U}$, then transforming the multivariate uniform vector into the desired input vector $\mathbf{X}$. The joint distribution of $\mathbf{U}$ is a copula.

Formally

$$\mathbf{X} = \left(F_{X_1}^{-1}(\Phi(Z_1)), F_{X_2}^{-1}(\Phi(Z_2)), \dots, F_{X_{1n}}^{-1}(\Phi(Z_n))\right)^T, \qquad (3.102)$$

where Φ is the (univariate) standard normal cdf.

In our discrete cases we have the probability mass function for each variable indexed by i, $i = 1, 2, \dots, n$, defined by the mass points $(x_1^{(i)}, \dots, x_m^{(i)})$ and their probability masses $(p_1^{(i)}, \dots, p_m^{(i)})$ such that $F_{X_i}(x) = \sum_{k:x \leq x_k} p_k^{(i)}$. The inverse is $F_{X_i}^{-1}(\xi) = \inf\{x | F_{X_i}(x) \geq \xi\}$. In our context the probability masses are given by

$$p_k^{(i)} = \frac{\exp\{-\lambda_i\}\lambda_i^k}{k!}$$

for Poisson distributed variables and

$$p_k^{(i)} = \binom{n_i + k - 1}{k} q_i^k (1 - q_i)^k$$

for negative binomial distributed variables.

The discrete transformation is given by

$$z_k^{(i)} = \Phi^{-1}\left(\sum_{n=1}^{k} p_n^{(i)}\right). \qquad (3.103)$$

Now the neuralgic question is: which $\mathbf{R}$ corresponds to the given $\mathbf{Q}$? Were $\mathbf{R}$ known, it would be easy to draw correlated normally distributed variates and to transform them back. Figure 3.28 shows the transformation procedure schematically.

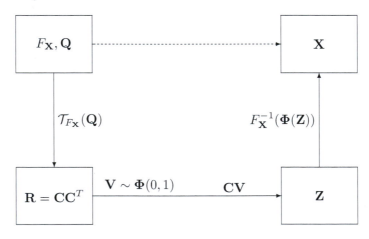

FIGURE 3.28: The transformation.

3.6.4.1 The Correlation Transformed

Again, we follow here Cario and Nelson (1997). $\mathbf{X}$ is a function of $\mathbf{Z}$ of the form $\mathbf{X} = (X_1(Z_1), X_2(Z_2), \ldots, X_n(Z_n))^T$ with $X_i(Z_i) = F_{X_i}^{-1}(\Phi(Z_i))$. The (Z_i, Z_j) *shall* have a bivariate normal density ϕ_2 with correlation $r_{ij} = Corr(Z_i, Z_j)$. The correlation

$$q_{ij} = Corr(X_i, X_j) = \frac{E\left[X_i X_j\right] - E\left[X_i\right]E\left[X_j\right]}{(Var\left[X_i\right]Var\left[X_j\right])^{1/2}} \qquad (3.104)$$

has the only unknown term $E\left[X_i X_j\right]$ as mean and variance are given.

Now, from the above it follows for the continuous case that

$$E\left[X_i X_j\right] = \int\limits_{-\infty}^{\infty} \int\limits_{-\infty}^{\infty} X_i(u) X_j(v) \phi_2(u, v, r_{ij}) du dv. \qquad (3.105)$$

The correlation q_{ij} is a function of r_{ij} only.

Between $z_k^{(i)}$ and $z_{k+1}^{(i)}$ the $x_{k+1}^{(i)}$ are constant. Thus the integral of Eq.(3.105) simplifies to

$$E\left[X_i X_j\right] = \sum_k \sum_l x_k^{(i)} x_l^{(j)} \int\limits_{z_{k-1}^{(i)}}^{z_k^{(i)}} \int\limits_{z_{l-1}^{(j)}}^{z_l^{(j)}} \phi_2(u, v, r_{ij}) du dv. \qquad (3.106)$$

For later numerical calculation let us truncate the distribution at ten standard deviations from above and ten from below the standard normal distribution, i.e., $z_0^{(i)} = -10$ and $z_{m^{(i)}}^{(i)} = 10$. The number of points to be considered is $m^{(i)} = nint[F_{X_j}^{-1}(\Phi(10))]$ where $nint(.)$ stands for the nearest integer. The term to be evaluated thus reads

$$E\left[X_i X_j\right] \approx \sum_{k=1}^{m^{(i)}} x_k^{(i)} \sum_{j=1}^{m^{(j)}} x_l^{(j)} \left(\int\limits_{z_{k-1}^{(i)}}^{z_k^{(i)}} \int\limits_{z_{l-1}^{(j)}}^{z_l^{(j)}} \phi_2(u, v, r_{ij}) du dv \right). \qquad (3.107)$$

This integral sum has to be evaluated in order to calculate $\mathbf{Q}$ from the original given correlation $\mathbf{R}$. The integral function in Eq.(3.107) can be approximated as described in Hull (1997, 260) or Genz (1992).

In order to find the correct r_{ij} that corresponds to the input correlation $q_{ij} = f(r_{ij})$ a numerical iteration procedure must be used, e.g bisection, regula falsi, etc. as the function cannot be inverted analytically.

Example 3.21. Numerically we have calculated the function $q(r)$ for the Poisson distributed frequencies of External Fraud $\lambda = 22.17$ and Extreme Losses $\lambda = 0.1124$. We have chosen the highest and the lowest values. The result is depicted in Figure 3.29. The Fréchet bounds can be read off at the lower and upper ends to be $\rho_U = 0.63$ and $\rho_L = -0.57$. The graph is a slightly convex curve. △

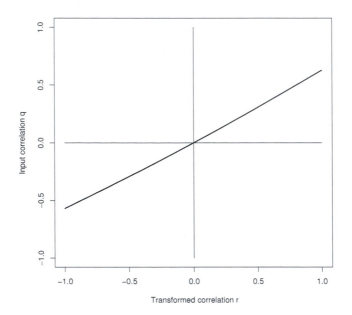

FIGURE 3.29: The transformed correlation as a function of the input correlation.

We present Algorithm 7 to be used in order to draw dependent random variates. We suppose that the appropriate correlation **R** is known, positive semi-definite and feasible.

The algorithm has just one additional step – step 4 – above and beyond the standard procedure (Scheuer and Stoller, 1962; Rubinstein, 1981, 65; Law and Kelton, 1991, 505) for generating correlated multivariate standard normal variates.

3.6.4.2   Feasibility

It may be that the transformed correlation **R** is not positive definite although the input correlation **Q** is. This is a manifest weakness of this approach. In this case we have to look for an approximated **R*** that is positive definite and comes as close as possible to **R**. This problem can be defined as a programming problem with an objective function and constraints (Ghosh and Henderson, 2002, 2000).

1. Cholesky decomposition of $\mathbf{R} = \mathbf{C}\mathbf{C}^T$

2. Generate $\mathbf{V} = (V_1, V_2, \ldots, V_n)^T$ random iid standard normal variates

3. Set $\mathbf{Z} \leftarrow \mathbf{C}\mathbf{V}$

4. Calculate $X_i \leftarrow F_{X_i}^{-1}(\Phi(Z_i))$ for $i = 1, 2, \ldots, n$

5. Go to Step 2.

ALGORITHM 7: Generation of correlated discrete random variates

From Henderson (2003) we know that for most cases a very simple but still good approximation exists, viz. to set all negative eigenvalues to a small positive value ϵ. To produce $\mathbf{R}^*$ we apply the following algorithm:

1. Eigenvalue decomposition of $\mathbf{R} = \mathbf{U}\mathbf{L}\mathbf{U}^T$ where $\mathbf{L} = diag(\lambda_i)$ with λ_i the eigenvalues of $\mathbf{R}$.

2. set $\lambda_i^* = \max(\lambda_i, \varepsilon)$, $\varepsilon = 0.00001$, $\mathbf{L}^* = diag(\lambda_i^*)$

3. $\mathbf{H} := \mathbf{U}\mathbf{L}^*\mathbf{U}^T =: [h_{ij}]$

4. re-normalise $\mathbf{R}^* = \left[h_{ij}/\sqrt{h_{ii}h_{jj}}\right]$.

ALGORITHM 8: Correction of correlation

The re-normalisation is done to force the diagonal terms to 1, a necessary condition for a correlation matrix. Now this new structure is valid. Rebonato and Jäckel (1999) give some more sophisticated approaches for constructing a valid matrix. Ghosh and Henderson (2002) mention the so-called semi-positive programming approach to construct the matrix with an optimisation strategy.

If we want to generate correlated variates from a sample of several different observations $\{\mathbf{x}_1, \mathbf{x}_2, \ldots, \mathbf{x}_n\}$ then the matrix $\mathbf{R}$ can be calculated in a simple way. Suppose the empirical cumulated density functions $\hat{G}_i$ are determined then $\mathbf{R} = Corr[\Phi^{-1}(\hat{G}_i(\mathbf{x}_i)), \Phi^{-1}(\hat{G}_j(\mathbf{x}_j))]$.

Di Clemente and Romano (2004, 199) describe an algorithm to estimate a Student's t-copula from observations. Then Algorithm 9 is used to generate dependent variates.

3.6.4.3   Extension beyond Normality

A quite natural extension to the algorithm above is to use t-distributed variates in lieu of normally distributed variates. A student's t-distributed random number T with ν degrees of freedom can be described as follows (Rubinstein, 1981, 94):

$$T = \frac{Z}{\sqrt{S/\nu}} \tag{3.108}$$

where $Z \sim \Phi(0,1)$, $S \sim \chi^2_\nu$ and Z and S are independent.

Interestingly, the correlation between two random variables Z_1 and Z_2 is identical to the correlation between the variables $T_1 = \frac{Z_1}{\sqrt{S/\nu}}$ and $T_2 = \frac{Z_2}{\sqrt{S/\nu}}$ (Bluhm et al., 2003, 106), i.e.,

$$Corr(Z_1, Z_2) = Corr(T_1, T_2). \tag{3.109}$$

Therefore, the Algorithm 7 on page 163 can be extended to the generation of t-distributed random variates by adding just one line:

---

1. Cholesky decomposition of $\mathbf{R} = \mathbf{C}\mathbf{C}^T$

2. Generate $\mathbf{V} = (V_1, V_2, \ldots, V_n)^T$ random iid standard normal variates

3. Draw S with $S \sim \chi^2_\nu$

4. Set $\mathbf{T} \leftarrow \mathbf{C}\mathbf{V}\sqrt{\frac{\nu}{S}}$

5. Calculate $X_i \leftarrow F_{X_i}^{-1}(t_\nu(T_i))$ for $i = 1, 2, \ldots, n$

6. Go to Step 2.

ALGORITHM 9: Random variates with t-distribution

Note that the degree of freedom ν must be greater than 4. For $\nu \geq 30$ the t-distribution is well-approximated with the normal distribution.

This extension can potentially lead to a higher correlation of the tail events as the t-distribution, in dependence from the degrees of freedom, has a thicker tail and thus more mass with the extremal events. It is known that for a normal copula the correlation in the tail drops to zero.

For this extensions one can simply take the transformed correlation matrix $\mathbf{R}$ as before.

Example 3.22. We want to explore the difference between the normal and the Student-t distributed marginals. We have claimed that the t-distribution should lead to higher correlation in the tail. We take the random variates of the frequencies of two events, i.e., "External Fraud" and "Execution, Delivery & Process Management" for which a correlation of 0.5 was requested. We plot the conditional correlation $Corr(X_1, X_2 \mid X_1 > F_{X_1}^{-1}(\alpha), X_2 > F_{X_2}^{-1}(\alpha))$. Obviously for $\alpha = 0$ it must equal 0.5. Numerically, the bigger the quantile, the

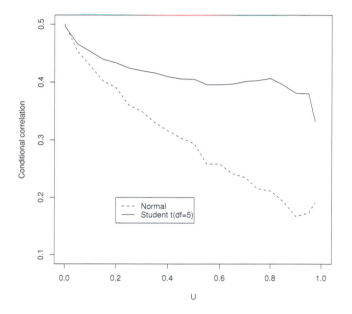

FIGURE 3.30: Conditional correlation (tail dependence) of two frequencies.

fewer points in the sample and thus the greater the uncertainty demonstrates. $\triangle$

Now we have presented a method to generate correlated loss frequencies. The compounding will also lead to a correlation between the total losses by events and lines of business. These can be calculated as described farther below.

Run-Through Example 2. We have shown above an method to impose a correlation structure on the frequencies of the number of losses. Because empirical correlations will be nearly impossible to obtain and because this method lends itself to sensitivity analysis and dependency scenarios this seems a defensible choice. From the quantitative impact study no input can be inferred. Therefore we just choose for the base case the following correlations with quite small numbers. Because of symmetry of the matrices we give only the lower triangle in order to increase readability.

1							
0.3	1						
0.1	0.1	1					
0.1	0.2	0.1	1				
0.1	0.1	0.1	0.1	1			
0.1	0.1	0.1	0.1	0.2	1		
0.2	0.1	0.1	0.1	0.1	0.1	1	
0.0	0.0	0.0	0.1	0.1	0.1	0.0	1

Below we show the output summary of 100,000 generated numbers of loss occurrence according to the distributions of Table 3.10 and the correlation matrix from Table 4.9 on page 314. The implementation of this algorithm is quite demanding because an implicit integral equation must be solved numerically.

```
number of iterations 100000
normal copula used
input data----------------------------------------------------
number of processes 8
1   Poisson: lambda      1.640   mean=  1.640  variance=  1.640
2   Poisson: lambda     22.170   mean= 22.170  variance= 22.170
3   Poisson: lambda      1.140   mean=  1.140  variance=  1.140
4   Poisson: lambda      7.000   mean=  7.000  variance=  7.000
5   Poisson: lambda      1.350   mean=  1.350  variance=  1.350
6   Neg bin: p= 0.740 x= 5.000   mean=  1.757  variance=  2.374
7   Poisson: lambda     11.370   mean= 11.370  variance= 11.370
8   Poisson: lambda      0.1124  mean= 0.1124  variance= 0.1124
correlation matrix on input: is positive definite
    1.000
    0.600  1.000
    0.200  0.200  1.000
    0.200  0.400  0.200  1.000
    0.200  0.200  0.200  0.200  1.000
    0.200  0.200  0.200  0.200  0.400  1.000
    0.400  0.200  0.200  0.200  0.200  0.200  1.000
    0.100  0.100  0.100  0.200  0.200  0.200  0.100  1.000
correlation matrix transformed
    1.000
    0.629  1.000
    0.224  0.216  1.000
    0.212  0.405  0.217  1.000
    0.222  0.214  0.227  0.215  1.000
    0.221  0.212  0.226  0.214  0.438  1.000
    0.421  0.202  0.217  0.203  0.214  0.213  1.000
    0.169  0.165  0.172  0.327  0.329  0.324  0.166  1.000
output section--------------------------------------------------
1   mean      1.641   variance      1.630
2   mean     22.181   variance     22.017
3   mean      1.138   variance      1.140
```

```
4    mean       7.018   variance      7.058
5    mean       1.351   variance      1.357
6    mean       1.758   variance      2.390
7    mean      11.378   variance     11.310
8    mean       0.114   variance      0.113
  sample correlation
     1.000
     0.597  1.000
     0.203  0.202  1.000
     0.200  0.399  0.201  1.000
     0.201  0.200  0.200  0.201  1.000
     0.203  0.203  0.200  0.201  0.401  1.000
     0.399  0.199  0.206  0.201  0.201  0.200  1.000
     0.100  0.099  0.105  0.207  0.200  0.199  0.101  1.000
  relative deviation
     0.000
     0.005  0.000
    -0.014 -0.010  0.000
     0.002  0.002 -0.005  0.000
    -0.007  0.000  0.000 -0.006  0.000
    -0.013 -0.013  0.001 -0.007 -0.002  0.000
     0.003  0.007 -0.030 -0.003 -0.006 -0.002  0.000
     0.000  0.009 -0.045 -0.035  0.000  0.003 -0.005  0.000
  summed quadratic deviation    0.000311073614
```

The differences between the imposed correlations and the sample correlations are very small. Nonetheless, with a simulation approach the result will never be perfect. Moreover, the transformed correlation matrix departs substantially from the imposed input correlation.

Example 3.23. For completeness we show in Figure 3.31 some 250 random losses for which the analogous correlation methodology is employed. The implementation is slightly more complex as now a true two-dimensional integral must be solved. We do not make use of this feature in the run-through example. This possibility may be of special interest for the insurance industry. △

3.6.5   Correlation of the Total Losses

The compound losses are characterised by a random sum from a random number of occurrences and random loss amounts as

$$U = \sum_{j=1}^{M} X_j \quad \text{and} \quad V = \sum_{j=1}^{N} Y_j. \tag{3.110}$$

It is assumed that both the $X_j = X$ and the $Y_j = Y$ are independent and identically distributed. The covariance is by definition Eq.(3.89) of page 153 and the over-bar indicating expectations:

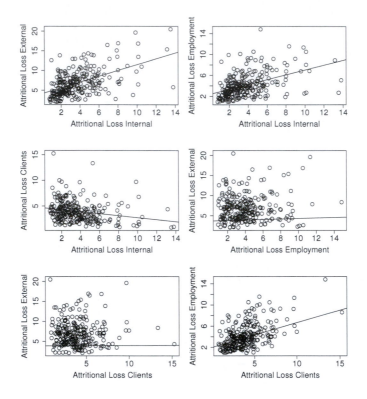

FIGURE 3.31: Correlated random attritional losses.

$$Cov(U, V) = E\,[UV] - \bar{U}\bar{V} \tag{3.111}$$

and the variance as special case (see Eq.(3.56)).

$$Var(U) = E\,[M] \cdot Var\,[X] + Var\,[M] \cdot (E\,[X])^2 \tag{3.112}$$

With

$$
\begin{aligned}
E(UV) &= E_X\,[E_Y\,[E\,[MXNV\,|X,Y]]] \\
&= E\,[X] \cdot E\,[Y] \cdot E\,[MN] \\
&= \bar{X}\bar{Y}E\,[MN]
\end{aligned} \tag{3.113}
$$

the covariance is

$$Cov(U, V) = E\,[X]\,E\,[Y]\,Cov(M, N). \tag{3.114}$$

The correlation can be written as

$$Corr(U,V) = E\left[X\right] E\left[Y\right] Corr(M,N) \left[\frac{Var(M)Var(N)}{Var(U)Var(V)}\right]^{0.5} \qquad (3.115)$$

or more explicitly

$$Corr(U,V) = Corr(M,N)\times$$

$$\left[\frac{Var(M)Var(N)\bar{X}^2\bar{Y}^2}{\left[\bar{M}\cdot Var(X) + Var(M)\cdot\bar{X}^2\right]\left[\bar{N}\cdot Var(Y) + Var(N)\cdot\bar{Y}^2\right]}\right]^{0.5}.$$

Suppose the count processes are both Poisson distributed. Then $Var(M)$ equals $\bar{M}$. Thus, the correlation simplifies to

$$Corr(U,V) = Corr(M,N)\frac{\bar{X}\bar{Y}}{\sqrt{E\left[X^2\right] E\left[Y^2\right]}}. \qquad (3.116)$$

The formulae Eq.(3.115) and Eq.(3.116) can be used to calibrate the correlations of the occurrences $Corr(M,N)$ if the data of the annual losses is considered more reliable.

We can rewrite Eq.(3.116) by replacing $E\left[X^2\right]$ by $Var(X) + \bar{X}^2$ and in a second step use

$$\frac{\bar{X}}{\sqrt{E\left[X^2\right]}} = \frac{\bar{X}}{\sqrt{Var(X) + E\left[X\right]^2}} = \frac{1}{\sqrt{1 + Var(X)/\bar{X}^2}} = \frac{1}{\sqrt{1 + CV(X)^2}} \qquad (3.117)$$

to give

$$Corr(U,V) = Corr(M,N)/\sqrt{[1 + CV(X)^2]\,[1 + CV(Y)^2]}. \qquad (3.118)$$

We have made use of the definition of the so-called coefficient of variation $CV := \sqrt{Var(X)}/E(X)$. Now we see that the correlation between the total losses is always smaller than the correlation between the according frequencies, that is

$$Corr(M,N) \geq Corr(U,V), \qquad (3.119)$$

because $1/\sqrt{1 + CV(X)^2} \leq 1$ holds. This is true also for a negative binomially distributed frequency.

3.6.6 Estimation of the Correlations

The estimation of the correlations proves to be difficult. One could argue that it is a second moment like the variance, with the variance being estimated from the same sample. There are some difficulties nonetheless. Firstly, we do not estimate the variances explicitly as they result from the fitting. Secondly, from actuarial experience and analogy to insurance we have a set of tried and true distributions. Thus, with correlations we should know something about a joint distribution. But the data may only be good enough for the marginals. Third, not knowing the distribution makes it impossible to define outliers in the data. But the correlation estimate is very much influenced by outliers or better by some apparently not fitting data point. Last of all, if the normality assumption is not tenable then a correlation definition based on ranks instead of scores may work better.

The sample estimate of the (empirical) correlation r is calculated from the n data points $\{x_i\}$ and $\{y_i\}$ in accordance with Eq.(3.104) on page 161:

$$r = \frac{\frac{1}{n}\sum_{i=1}^{n} x_i y_i - \bar{x}\bar{y}}{\sqrt{s_x^2 s_y^2}}, \tag{3.120}$$

where we use $\bar{x} = \frac{1}{n}\sum_{i=1}^{n} x_i$ and $s_x = \frac{1}{n}\sum_{i=1}^{n}(x_i - \bar{x})^2$ and analogously for $\bar{y}$ and s_y.

Now r is a biased estimator for the correlation. Under the assumption that the two or more variables involved are bivariate or multivariate normally distributed then the bias $E(r) - \rho$ can be reduced. According to Zimmermann et al. (2003) Ronald A. Fisher proposed (Fisher, 1915)

$$\hat{\rho}_{Fisher} = r\left[1 + \frac{1-r^2}{2n}\right]. \tag{3.121}$$

There is a newer correction due to Olkin and Pratt (1958) that reads as follows

$$\hat{\rho}_{OP} = r\left[1 + \frac{1-r^2}{2(n-3)}\right]. \tag{3.122}$$

The normality assumption is present in these corrections. From the formulae we see that the bias is very much a function of the sample size n. In operational risk we generally have to reckon with a very limited number of points.

Embrechts et al. (1999) have compiled a nice little primer on typical misunderstandings and flaws related to correlation. The reader is exhorted not to naively trust the correlation when there is no model. Instead, they suggest trying to model the dependency. But to model means to choose and to choose implies preferring an alternative. And there we go again, back to Ockham.

An alternative approach is to have experts estimate the correlation, or more generally, the dependence, of several variables. Using the divide-and-conquer strategy common in risk management, it is again preferable to separate judgments about individual variables from judgments about relationships among variables. This is a feasible modelling approach, using copulae. Clemen et al. (2000) have done some descriptive work on the assessment of dependence measures for modelling expert knowledge, employing six different methods. Whether expert estimates constitute a sound estimate as required by the regulators remains to be tested.

3.6.7 Testing for Correlation

There are statistical tests, both parametric and non-parametric, that can be applied to hypothesized correlations between pairs of observations. The first assume a bivariate normal distribution of the two data sets. The latter could be called more appropriately a *distribution-free* method. For guidance on testing see Algorithm 1 on page 139.

Tests also depend on the *measurement scale* of the variables. Recall the following classification:

- *nominal:* the variable is measured in terms of whether it belongs to a certain category. Gender is a typical example where "female" can be encoded as 1 and "male" as 0.

- *ordinal:* the variables measured can be ranked in terms of which has less and which has more, but still they do not allow us to determine "how much more." Say "upper middle class" is ranked higher than "middle class."

- *interval:* such measurement can not only be ranked but allow us to compare sizes and differences between them. The temperature scale in degrees Celsius (or Fahrenheit) belongs to this class.

- *ratio:* unlike interval scale properties, also the ratio between single values has an interpretation. Think of the absolute temperature scale in degrees Kelvin (or Rankine). A temperature of $300K$ is twice as much as $150K$.

Correlations can exist between differently scaled variables and methods applicable only to certain classes of variables.

3.6.7.1 Parametric Tests

Parametric tests assume a bivariate normal distribution among the paired data sets x_i, y_i for $i = 1, \ldots, n$. This means that (1) both X and Y must have a (marginal) normal distribution and (2) for each X (respective Y) the according Y (respectively X) must be normally distributed (array distribution) and (3) their variances must be homogeneous (so-called homoscedasticity). It is almost impossible to test this beforehand, especially as the correlation should be known in order to infer the correlation. One must often be content to check the marginal distribution on normality. Computer experiments have shown that the following tests have a certain robustness with respect to departure from the assumptions.

The procedures below are only applicable to variables that are at least ordinally measurable.

Recall the sample correlation r from Eq.(3.120). To test whether the null-hypothesis $H_0 : \rho = 0$ must be rejected in favour of the alternative, say

$H_1 : \rho > 0$ we have to construct the statistic

$$T = \frac{r\sqrt{n-2}}{\sqrt{1-r^2}} \tag{3.123}$$

and compare it to the Student's t-distribution with $n-2$ degrees of freedom.

If the null-hypothesis is $H_0 : \rho = c$ with a constant c, then one should use *Fisher's Z-transform*

$$Z(r) = \frac{1}{2} \log(\frac{1+r}{1-r}) = \tanh^{-1}(r) \tag{3.124}$$

which has for large n a normal distribution with mean $0.5 \log((1+\rho)/(1-\rho))$ and variance $1/(n-3)$ (Hogg and Tanis, 1993, 544). In contrast to r, Z is ratio scaled. If we compare an $r_1 = 0.4$ with $r_2 = 0.8$, then the correlation has not doubled, but increased from $Z_1 = 0.42$ to $Z_2 = 1.1$, i.e., almost tripled. Similarly, an increase in correlation of 0.05 is much more significant with higher values of r than with lower values.

3.6.7.2   Rank-Based Tests

Rank tests sacrifice some information as the size of the values is not specifically considered. This loss in information must be gauged against the gain in generality due to the complete absence of any distributional assumption. In the case of the loss frequencies where the (marginal) distributions are assumed to be either Poisson or negative binomially distributed, these are the only applicable tests.

Spearman's rank correlation

Again, we have n pairs of observations $\{x_i, y_i\}$. Let u_i be the rank of the ith value of the x-sample and v_i accordingly for y_i. Now, Spearman's rank correlation coefficient r_S is the same as the sample correlation applied to the ranks, i.e.,

$$r_S = \frac{\frac{1}{n} \sum_{i=1}^{n} u_i v_i - \bar{u}\bar{v}}{\sqrt{s_u^2 s_v^2}}, \tag{3.125}$$

where we use $\bar{u} = \frac{1}{n} \sum_{i=1}^{n} u_i$ and $s_u = \frac{1}{n} \sum_{i=1}^{n} (u_i - \bar{u})^2$ and analogously for $\bar{v}$ and s_v.

Intuitively, one might assume Pearson's correlation could answer the question: "How well can the relationship in the data be represented by a linear function?" whereas Spearman's correlation would answer the question: "How well can the relationship in the data be represented by a *monotonic but otherwise arbitrary function?*" Interestingly, when the underlying relationship is linear, then the two measures coincide.

The null-hypothesis "no population correlation between ranks" $\rho_S = 0$ can be tested against the alternatives $H1 : \rho_S > 0$, $H1 : \rho_S < 0$ and $H1 : \rho_S \neq 0$

by subjecting the statistic r_S to the critical values $r_{S,\alpha}$. In the same order we have the rejection regions $r_S \geq r_{S,\alpha}$, $r_S \leq -r_{S,\alpha}$ and $\mid r_S \mid \geq r_{S,\alpha/2}$. The critical values are tabulated, see Kokoska and Nevison (1989, 86). The following table contains just an excerpt. From this table we see very clearly

TABLE 3.12: Critical values $r_{S,\alpha}$

	α			
n	0.05	0.01	0.005	0.001
5	0.9000			
10	0.5636	0.7455	0.7939	0.8788
15	0.4429	0.6036	0.6536	0.7536
20	0.3805	0.5203	0.5699	0.6617
30	0.3063	0.4251	0.4670	0.5488
40	0.264	0.3681	0.4051	0.4788

that in order to reject the null-hypothesis with a small sample we need a very high sample correlation coefficient.

For large n there is the following normal approximation: $Z = r_S \sqrt{n-1}$ has a standard normal distribution.

Kendall's tau

Kendall's tau is equivalent to Spearman's rho with regard to the underlying assumptions. It is comparable in terms of its statistical power. But while ρ_S is interpreted as a Pearson correlation applied to ranks, Kendall's tau is a probability, that is, it is the difference between the probability that in the measured data the two variables are in the same order and the probability that they are not. So, concordant (discordant) means that two x-values x_i and x_j are (not) in the same order than the corresponding y_i and y_j. Kendall's tau is defined as (Press et al., 1992, 637), (simplified for no ties in the data):

$$\tau = \frac{concordant - discordant}{concordant + discordant} \tag{3.126}$$

or more specific (Wang, 1998, 7):

$$\hat{\tau} = \frac{2}{n(n-1)} \sum_{i<j} \text{sign}[(x_i - x_j)(y_i - y_j)]. \tag{3.127}$$

For large n the calculation of this measure can be quite demanding, as $0.5n(n-1)$ combinations must be assessed.

Also for the distribution of τ there exists a normal approximation with mean zero and $\sigma^2 = (4n+10)/(9n(n-1))$ (Press et al., 1992, 637).

Example 3.24. We have two fields of application in mind for the correlation

TABLE 3.13: Critical values τ_α

n	0.05	0.01	0.005	0.001
5	0.6000	0.8000		
10	0.4222	0.5556	0.6000	0.7333
15	0.3143	0.4476	0.4857	0.5619
20	0.2632	0.3684	0.4105	0.4842
30	0.2138	0.2966	0.3287	0.3885
40	0.1821	0.2538	0.2821	0.3359

(header: α)

test: firstly, we have to demonstrate to the regulator that the correlations of losses of different event types or lines of business used are sound and secondly, to test the validity of indices or indicators of risk.

We use data from Table 3.5 on page 133, but only the last 15 points. We assume that the number of accidents is independent from the loss amount. For the test we use the following R-script:

```
## Number of serious losses in aviation (including
## Twin Towers) from 1987 to 2001
x <- c(10, 9, 18, 12, 19, 16, 28, 31,
                    22, 20, 32, 28, 29, 21, 11)
## Average loss per accident
y <- c(146.5, 200.6, 116.4, 163.9, 64.6, 78.9, 57.6, 94.5,
                    58.2, 112.7, 44.1, 62.7, 57.5, 53.2, 411.0)

## H0: no correlation, H1: r != 0 (two sided test)
## (for cor.test the confidence level of 0.95 is default)
## Pearson
cor.test(x, y, method="pearson", alternative="two.sided")
## Spearman
cor.test(x, y, method="spearman",alternative="two.sided")
## Kendall
cor.test(x, y, method="kendall", alternative="two.sided")
```

For the Pearson's product-moment correlation we get the test statistic $t = -3.1066$, a p-value $p^* = 0.00834$ and the correlation $r = -0.6527404$. The critical value is from a table $t_{\alpha/2,13} = 2.1788$. We reject the null-hypothesis as $|t| = 3.1066 > 2.1788$, or $p^* < \alpha = 0.05$.

The Spearman's rank correlation test yields a statistic $S = 1009$, a critical value $p^* = 0.0005169$ and a correlation $\rho_S = -0.8025026$. Here also, the null-hypothesis can be rejected.

The third test for Kendall's rank correlation τ gives a test statistic value of $z = -3.2818$, $p^* = 0.001031$ and the $\tau = -0.6315862$. A glance at the significance level suffices to convince us that it should be rejected.

Obviously, the result is not quite as expected. In our context, the actuarial standard approach rests on the assumption that frequency and severity are independent.

$\triangle$

3.6.7.3 Dependencies between Events and Business Lines

TABLE 3.14: QIS number of losses

	Internal	External	Employment	Clients	Physical	Disruption	Execution	Total	max i
Corporate	17	20	73	73	16	8	214	421	73
Trading	47	95	101	108	33	137	4603	5124	137
Retail	1268	17107	2063	2125	520	163	5289	28535	17107
Commer.	84	1799	82	308	50	47	1012	3382	1799
Settlement	23	322	54	25	9	82	1334	1849	322
Agency	3	15	19	27	8	32	1381	1485	32
Asset	28	44	39	131	6	16	837	1101	131
Brokerage	59	20	794	539	7	50	1773	3242	794
Total	1529	19422	3225	3336	649	535	16443	45139	20395
max j	1268	17107	2063	2125	520	163	5289	28535	

In Section 2.10.2, especially starting on page 72, we have touched upon association in the context of causality. We have pointed out that an example involving the Goodman-Kruskal lambda would follow. We are actually very much interested in the question of whether there is a relevant relation between the number of losses occurring in a specific line of business and a given event type associated with that loss. The analysis starts with the so-called *contingency table* from the QIS study (BCBS, 2003a, 6), reproduced and simplified below as Table 3.14.

In a first step we apply the Goodman-Kruskal lambda (Goodman and Kruskal, 1954), an asymmetric PRE-measure, to the so-called contingency table of Table 3.14. The indices $i = 1, \ldots, I$ and $j = 1, \ldots, J$ stand for the rows and the columns. In addition we set f_{ij} for the number in the cell $\{i, j\}$ and $f_{i\cdot} = \sum_{j=1}^{J} f_{ij}$. Further n is the grand total of all cells. The prediction error can be set to the deviation $n - \max_i f_{i\cdot}$, while the conditional error, after considering the information of the second dimension, is $\sum_{j=1}^{J} (f_{\cdot j} - \max_i f_{ij})$. Simplifying, we become

$$\lambda_{xy} = \frac{\sum_{j=1}^{J} \max_i f_{ij} - \max_i f_{i\cdot}}{n - \max_i f_{i\cdot}}. \tag{3.128}$$

Applying this formula to our table we get

$$\lambda_{xy} = \frac{20395 - 19422}{45139 - 19422} = 0.0378. \tag{3.129}$$

According to this measure the association is very weak, if not vanishing. Let us try the other way, i.e., inferring from event on lines of business. We calculate

$$\lambda_{yx} = \frac{28535 - 17107}{45139 - 17107} = 0.4077, \tag{3.130}$$

which tells us a completely different story. We have to consider that λ is best suited to nominally scaled values and low dimensional tables. However, another classical procedure is to test for independence (Rice, 1995, 489; Hogg and Tanis, 1993, 563; Miller and Miller, 1999, 439). This is done with the *chi-square test*. For simplicity's sake, we define the relative frequency of values as $\hat{\pi}_{ij} = f_{ij}/n$, which in turn can be thought of as probability estimates. The null hypothesis is that for all cells $\{i, j\}$ the probabilities are $\pi_{ij} = \pi_{i.}\pi_{.j}$.

The asymptotic test is based on the statistic

$$S = \sum_{i=1}^{I} \sum_{j=1}^{J} \frac{(f_{ij} - f_{i.}f_{.j}/n)^2}{f_{i.}f_{.j}/n}. \tag{3.131}$$

The degrees of freedom here are $df = (I-1)(J-1) = 7 \times 6 = 42$. The rejection region of the null hypothesis is $S > \chi^2_{1-\alpha,df}$. With $1 - \alpha = 95\%$ we read off a table or with the R-function qchisq(p=.95,df=42) we retrieve $\chi^2_{.95,42} = 58.12404$. We calculate $S = 19557$. Thus we have a massive departure from independence. This means that certain types of events are much more frequent for certain lines of business than others. For example, there are a huge number of external frauds in retail banking but very few in trading. This is hardly surprising.

3.6.7.4   Sound Correlations?

As already mentioned, banks are permitted to use internally determined correlations in operational risk losses, provided they can demonstrate that their systems for determining correlations are sound, implemented with integrity and take into account the uncertainty surrounding any such estimates. Now the question remains whether such correlation assumptions can be validated by quantitative techniques. We have shown in the preceding sections how to use correlations for simulation and in some standard tests for validating – or, more precisely, not refuting – the hypothesised associations.

Reason and experience suggest that if there are correlations between the number of loss occurrences, then they must be rather low. Now, if a relationship between the variables in question is "objectively" rare in the population, there is no way to identify such a relation in a small sample. Even if our small sample were in fact perfectly representative of the population, the effect would not be statistically significant. On the other hand, if a relation is "objectively" very strong or frequent in the population, then it can be found to be highly significant even in a study based on a very small sample.

In operational risk management we consistently deal with small samples, corresponding to a short history and presumably small correlations or weak associations. Thus we find ourselves in a situation where it is statistically impossible – even for the non-parametric measures – to refute the hypothesis of zero correlation. Whether this fact can be reconciled with the soundness demand of the regulator remains open to debate.

In order to satisfy the regulator, we have to assume some correlation which can be defended on the grounds of the stress regime. In times of particular stress the uncertainty surrounding these estimates must be considered. Caution demands that we posit some positive correlation, in order to increase the economic capital figure.

3.7 Risk Measures and Allocation

3.7.1 Risk Measures

Definition 3.11. A *risk measure* is a function that maps random variables describing risk to real numbers. ☐

This definition is somewhat generic and becomes useful only after having defined what property the risk measure should have. There are many risk measures in use depending on the type of investment and on tradition. Financial institutions often use quantiles (also "percentile") as known from Value at Risk and from EC for credit are widely used. This is due to the fact that RiskMetrics, at the time a product of J.P. Morgan, is the *de facto* standard or at least the most important benchmark. Later we will discuss why bankers like this measure.

Risk measures have been around for quite a long time. For an overview see Balzer (1994). With X either a value, a loss or a return risk is measured as:

- Standard deviation σ_X, "Volatility" as risk measure for stocks,

- Utility functions

$$U(X) \approx U(\mu) + U'(\mu)(X - \mu) + U''(\mu)\frac{(X - \mu)^2}{2} + \dots \qquad (3.132)$$

$$E(U(X)) = U(\mu) + U''(\mu)\frac{\sigma^2}{2} + \dots \qquad (3.133)$$

The expansion around mean μ does not seem very plausible for risk management as we are focusing on the tail. But implicit in the so-called

Capital Asset Pricing Model is the assumption of a quadratic utility function if the returns are not normally distributed (Haugen, 1993, 201). The dependent variable is the value of the portfolio.

- Relative Lower Partial Moments (RLPM)

$$RLPM_n = E(|X - q|^n \,|X > q)$$ (3.134)

This is a general definition embodying special cases like expected shortfall for $n = 1$ and the so-called "semi-variance" with $n = 2$.

- Maximum shortfall MS (Worst Conditional Expectation WCE)

$$MS(X, q) = \max\{|X - q| \,|X > q\}$$ (3.135)

on finite support X.

$$WCE(X, \alpha) = \max\{X \mid A, Pr(A) > \alpha\}$$ (3.136)

with A the set of "scenarios." Maximum shortfall embodies information about the extremes (beyond threshold $q = F^{-1}(\alpha)$ or quantile) but no additional information on the distribution $F(x)$.

- Value at risk

$$x_\alpha = VaR(\alpha, x) = \inf\{x\,|F(x) > \alpha\}$$ (3.137)

In a narrow (original) sense for normally distributed investment returns:

$$x_\alpha = \gamma_\alpha \sqrt{\mathbf{e}^T \mathbf{S} \mathbf{e}}$$ (3.138)

with $\mathbf{e}$ an exposure vector, $\mathbf{S}$ covariance matrix, α confidence and γ_α the critical value of the standard normal distribution at confidence level α, i.e., $\phi^{-1}(\mu = 0, \sigma^2 = 1, x = 1 - \alpha)$.

- "Economic Capital" EC derived from VaR

$$EC(\alpha, X) = x_\alpha - E(X)$$ (3.139)

In credit risk management the shift by the amount $E(X)$ is motivated by the belief that the probability of default is always incorporated into the spread in addition to a risk-free interest rate. This has no foundation in the realm of operational risk as one cannot expect the bank to set up general provisions equalling this amount. Therefore, we understand EC in the following as a synonym for

$$EC(\alpha, X) = x_\alpha.$$ (3.140)

- Expected shortfall (tail conditional expectation TCE, tail value at risk, TailVaR, expected tail loss ETL)

$$ES(\alpha, X) = E\left[X \mid X \geq x_\alpha\right]$$

$$= x_\alpha + \int_{x_\alpha}^{\infty} [x - x_\alpha]\, dF_X(x). \tag{3.141}$$

A generalisation for X not having a continuous distribution is (Acerbi and Tasche, 2001):

$$GES = \frac{1}{1-\alpha}\left[E(X - x_\alpha \mid X > x_\alpha) + x_\alpha\left(1 - \alpha - \Pr(X \geq x_\alpha)\right)\right]. \tag{3.142}$$

- Expected Policyholder Deficit (EPD)

$$EPD(x_\alpha, X) = \int_{x_\alpha}^{\infty} (x - x_\alpha)dF_X(x) \tag{3.143}$$

Obviously

$$ES(\alpha, x) = \frac{1}{1-\alpha}EPD(x_\alpha, X) + x_\alpha. \tag{3.144}$$

Interestingly, it can be shown that the following relation holds:

$$VaR(\alpha, X) \leq WCE(\alpha, X) \leq ES(\alpha, X). \tag{3.145}$$

If X has a continuous distribution then

$$WCE(\alpha, X) = ES(\alpha, X). \tag{3.146}$$

We will concentrate on Value-at-Risk, or better yet, on a quantile-based risk measure and on expected shortfall for allocation purposes. Therefore, the inequality $x_\alpha < ES(\alpha, X)$ will be of interest in later discussions.

3.7.2 Coherent Risk Measures

Some years ago Artzner et al. (1998) presented a set of properties for risk measures that met with great interest in the risk management community.

Definition 3.12. A risk measure η is called *coherent* if it satisfies the following four axioms (Artzner et al., 1998; Meyers, 2000, 2):

1. *Sub-additivity* : for all random losses X and Y

$$\eta(X + Y) \leq \eta(X) + \eta(Y), \tag{3.147}$$

2. *Monotonicity* : for each pair X and Y with $X \leq Y$

$$\eta(x) \leq \eta(Y), \qquad (3.148)$$

3. *Positive Homogeneity* ; for all $\lambda \geq 0$ and random loss X

$$\eta(\lambda X) = \lambda \eta(X), \qquad (3.149)$$

4. *Translation Invariance*

$$\eta(X + \alpha) = \eta(X) + \alpha. \qquad (3.150)$$

$\square$

Implicitly, meeting these conditions is regarded as a necessary minimum for a good risk measure. The first axiom means that diversification of several risks is reflected. The second point requires that those losses that are "almost surely" lower than a benchmark risk get a lower value for risk. The third property is also very natural: think of changing currency from USD to EUR by multiplying the losses by the exchange rate. The risk should not be influenced by this operation. The last item states that if there is a "riskless" component α like a fixed payment then it should not influence the risky part. It implies that $\eta(X - \eta(X)) = 0$.

Now we shall investigate whether the most used risk measures conform with these items and if not, what can be done. We will start with value-at-risk, by which we understand a quantile of given confidence level.

Example 3.25. Let us assume two independent loans A and B, each with a default probability p, $0.006 \leq p < 0.01$. We define Portfolio C as containing both loans A and B. The VaR of A and respectively of B is $VaR(99\%, A) = VaR(99\%, B) = 0$, because for each the probability of default is less than the confidence level chosen. But if we look at the portfolio, the VaR of C is greater than 0 because $\Pr(X_C = 0) = (1 - p)^2 < 99\%$. This means that the risk as measured by VaR of the ensemble of the constituents is not necessarily less than or equal to the sum of the risk of the ensemble's parts. VaR is thus in contravention to the subadditivity property. This example is taken from (Bluhm et al., 2003, 168). $\triangle$

Example 3.26. Is maximum shortfall coherent? We look at a table of scenarios of outcomes of two random variables x_1 and x_2 in addition to the constructs $x_1 + x_2$ respectively $2x_1$ and $x_2 + 1$. We check the four properties according to the definition above.

1. sub-additive: $15_{column4} \leq 8_{column2} + 9_{column3} = 17$

2. monotone: obvious

3. homogeneous: $2 \times 8_{column2} = 16_{column5}$

TABLE 3.15: Some scenarios

Scenario	x_1	x_2	$x_3 = x_1 + x_2$	$x_4 = 2x_1$	$x_5 = x_2 + 1$
1	3	2	5	6	3
2	0	3	3	0	4
3	1	3	4	2	4
4	8	7	15	16	8
5	3	0	3	6	1
6	5	1	6	10	2
7	0	9	9	0	10
8	7	3	10	14	4
9	6	2	8	12	3
10	3	4	7	6	5
MS	8	9	15	16	10

4. translation invariant: $9_{column3} + 1 = 10_{column6}$

Thus, we can confirm that maximum shortfall is a coherent measure. △

Because expected shortfall is a special expectation, it is quite simple to verify its properties, but only for X having a continuous distribution (Acerbi and Tasche, 2001).

We can summarize the coherence properties for some of the aforementioned popular risk measures in Table 3.16. In a simulation setting with equally likely

TABLE 3.16: Synopsis of coherence

Property	MS	ES	VaR	StdDev	EPD
sub-additive	Y	Y	N	Y	Y
monotone	Y	Y	Y	N	Y
homogeneous	Y	Y	Y	N	Y
trans. invariant	Y	Y	Y	Y	N

draws, the expected shortfall is a (almost) coherent measure whereas VaR is not at all coherent.

3.7.3 Comparing VaR with Expected Shortfall

The main two contenders on the scene are value-at-risk and expected shortfall. The latter definitely contains the more elegant mathematical properties (see Table 3.16 on page 181). Nonetheless, other arguments must be reviewed, including some in favour of VaR, before deciding which method is superior:

- VaR is popular with market risk and credit risk due to historic and regulatory developments,

- Besides VaR and ES no other concept is used in the industry,

- VaR is one point of the distribution whereas ES embodies through the integration domain more information,

- VaR defines "good" and "bad" whereas ES quantifies "bad": "how bad is bad?,"

- ES is coherent whereas VaR needs an additional coherent allocation mechanism,

- VaR is computationally easier than ES and converges better,

- VaR is always less than ES at the same confidence $1 - \alpha$.

From a numeric point of view, the difference between the two for the same distribution depends on the tail. With close to normal distributions they are almost identical, whereas with heavy-tailed distributions, e.g., a Pareto distribution, the discrepancy is marked. See Yamai and Yoshiba (2002) for a deeper analysis.

VaR is not sub-additive, i.e., the important property that the risk of an aggregated portfolio is smaller than the sum of the risks of its components may be violated. The conclusion is that VaR is an inappropriate risk measure for allocating capital charges among organisational units of a bank. In addition, VaR is not consistent with diversification and can lead to sub-optimal solutions for optimisation or hedging and insurance buying.

Loosely speaking, VaR is the minimal loss in the $\alpha\%$ "bad" cases, but it does not say anything about the expected loss in the $\alpha\%$ "bad" cases. Now this second statement is of interest to depositors, deposit insurance institutions, creditors and potentially, tax-payers and the lender of last resort.

If banking supervision should try to maintain stability of the financial system and represent the interest of the stakeholders and the economy as a whole, then it must choose expected shortfall as the measure and not value-at-risk. VaR is "blind" toward risks that create large losses with a very small probability (below the critical probability level $1 - \alpha$).

Not so surprisingly, VaR is the best risk measure at the firm-level from the viewpoint of shareholders and management. Shareholders' costs associated with bankruptcies, i.e., legal costs, loss of goodwill, etc. are almost independent of the size of the loss that triggered the default. The same is true for the costs borne by the management, which are also primarily related to the default event itself.

Thus, there is an inherent conflict of interest between shareholders and management on the one hand and creditors, depositors and tax payers, on the other, when it comes to the choice of proper risk measure.

Now, the Basel initiative has opted for a VaR-style (i.e., quantile) of risk measure. This may be due to the de facto standard and its wide spread use in the industry. If not optimal, it does at least provide continuity, making the Basel 2 proposal less disruptive than it might otherwise have been.

3.7.4 Allocation

Not only losses but also appropriate risk capital must be assigned to some *cost objective* like processes, products and departments (here Lines of Business). The purposes and effects of such an endeavour are numerous, inter alia (Horngren and Foster, 1987, 412):

- aiding management decisions and encouraging motivation within the organisation,

- measuring efficiency,

- justifying costs charged to customers.

In cost accounting one differentiates between *direct cost* and *indirect cost*. The first is identified specifically with a single cost objective, i.e., re-financing interest for a credit. The latter is characterised by the fact that it cannot be assigned to a specific object, i.e., an insurance premium to cover the premises or benefits to the supervisory board and especially the cost of keeping capital to cover risk. But for the reasons cited above, an assignment is more than sensible. Because of the lack a logical stringency, the allocation scheme can be chosen according to different philosophies, e.g.,

- cause and effect,

- benefits received,

- fairness and,

- ability to bear.

Orthogonal to the allocation mechanism stands its purpose. Most often it is simply assumed that the allocation should serve the economic decision-making and resource allocation processes. However, allocation can be used to steer the business toward certain goals or to encourage specific behaviours while discouraging others. In many corporations it is the latter that is recognised as the primary purpose of allocation.

The simplest approach is to try to define the causation and to attribute the losses accordingly. The proportions of the losses per LoB must be calculated based on some quantification and measure. For example, "benefits received" tries to attribute cost to those objectives that profit from the cost incurred. This can be the case with "cross-selling" where it is often assumed that certain activities benefit others in rather an indirect way. The ability to bear relates to the utility philosophy where those who generate most profits should also have some marginally decreasing utility from it. This is in blatant contrast with the practice of how bonuses are doled out.

Mathematicians are often wooed into treating allocation axiomatically. This can differ considerably from the real world situation.

3.7.5   Some Properties

One has to make a distinction between "coherent measure" and "coherent allocation." There have been some proposals as to what constitutes a good allocation scheme (Delbaen and Denault, 1999).

Definition 3.13. An allocation is called *coherent* if it has the following four properties:

1. *Full allocation*:

$$\sum_{i=1}^{n} K_i = K \tag{3.151}$$

2. *No undercut*:

$$K_a + K_b + \ldots + K_z \leq \eta(X_a + X_b + \ldots + X_z) \tag{3.152}$$

   for any subset $\{a, b, c, \ldots, z\}$ of $\{1, 2, 3, \ldots, n\}$

3. *Symmetry*: Within any decomposition, a substitution of a risk X_a with an otherwise identical risk X_b will have no effect on the allocation.

4. *Riskless allocation*: The allocation to a risk that has no uncertainty is zero.

With K_i we designate the capital allocated to an objective j and η is the risk measure.   □

No undercut means explicitly that any decomposition of the total risk will not increase the capital beyond the stand-alone risk. Especially $K_a \leq \eta(X_a)$, i.e., the contribution should not exceed the stand-alone risk.

This definition is reasonable and intuitive but it does not suffice to characterise a single allocation method.

Example 3.27. In the following we want to illustrate that the allocation has a formal correspondence with a situation in game theory, viz. the so-called coalition game (Denault, 1999). A coalition game consists of N coalitions of n players and a cost function c that associates a real number $c(S)$ to each subset S of N coalitions. The goal of each player is to minimize the cost by accepting or declining to take part in coalitions. Given the sub-additivity of c, the players of a game have an incentive to form the largest coalition N. They need only to find a way to allocate the cost $c(N)$ of the full coalition N among themselves.

Given some differentiability conditions on the cost function c, the right way of allocating risk capital is through the Aumann-Shapley price. It is given by the gradient (first derivative) of c with respect to the "fractional player" (portfolio size). Formally (Aumann and Shapley, 1974):

$$K_i = -E(X_i \mid X < q). \qquad (3.153)$$

A closer look at the formula above reveals immediately that it is identical – besides the negative sign for costs – with the definition of Expected Shortfall (Eq.(3.141) on page 179). Moreover, the so-called Aumann-Shapley value leads to a coherent allocation in the above mentioned sense. The table below shows the correspondence between the game perspective and risk management.

Game	(Fractional) Player	Cost function
Risk allocation	(Divisible) Portfolio	Risk measure

$\triangle$

3.7.6 Allocating Methods

Regulators have defined a quantile (call it Value-at-Risk style) as the measure for the risk capital. It cannot be allocated according to an inner logic as it has mathematically speaking unattractive properties with regard to coherence. Therefore, an auxiliary method is needed to perform the task. We look into two methods. Actually, we recommend the use of the expected shortfall method as allocation mechanics.

3.7.6.1 Covariance Principle

A quantile measure has unattractive properties especially with regard to coherence and thus allocation. The lack of sub-additivity forces the use of an allocation scheme stemming from another risk measure.

In credit risk and in insurance the "covariance principle" is used to allocate VaR or a similar quantile measure. The idea is to take the gradient of the variance and define the VaR as variance times a normalization factor called "Capital Multiplier" CM_α (Bluhm et al., 2003, 172)

$$CM_\alpha = \frac{x_\alpha}{\sigma}. \qquad (3.154)$$

The gradient turns out to be the covariance. We start with the variance of the linear combination:

$$\sigma^2 = \mathbf{e}^T \mathbf{S} \mathbf{e} = \sum_{i=1}^{N} \sum_{j=1}^{N} e_i e_j \sigma_{ij} \qquad (3.155)$$

and take the derivative with respect to a component e_i:

$$2\sigma\frac{\partial\sigma}{\partial e_i} = 2\sum_{j=1}^{N} e_j\sigma_{ij} = 2Cov\,[e,e_i] = 2\beta\,[e,e_i]\,\sigma. \tag{3.156}$$

We have made use of the setting $e = \sum_{i=1}^{N} e_i$ and the rule $Cov(e_i,e_k) + Cov(e_j,e_k) = Cov(e_i + e_j, e_k)$. Thus, we define the partial derivative as β_i in analogy to the CAPM

$$\beta_i = \beta\,[e,e_i] = \frac{\partial\sigma}{\partial e_i} = \frac{1}{\sigma}Cov\,[e,e_i]. \tag{3.157}$$

The allocation is thus

$$K_i = \beta_i \times CM_\alpha \tag{3.158}$$

with

$$\sum_{i=1}^{n} K_i = K = x_\alpha \tag{3.159}$$

as $\sum_{i=1}^{N} \beta_i = \sigma$. This approach is easy and computationally efficient but problematic for one-sided and skewed distributions (pure risk).

3.7.6.2   Expected Shortfall

We could try yet another risk measure, viz. expected shortfall. In a similar way we define the capital multiplier CM_{ES} to be the ratio between EC and ES

$$CM_{ES} = \frac{x_\alpha}{ES(\alpha,X)}. \tag{3.160}$$

ES is always greater than the quantile x_α as the second is the lower integration bound of the expectation. Now, we could postulate that the capital multiplier should be approximately one, i.e.,

$$ES(\alpha^*, X) \overset{!}{=} x_\alpha. \tag{3.161}$$

The following holds in general:

$$\alpha^* < \alpha. \tag{3.162}$$

Now the factors become $CM_{ES} = x_\alpha/ES(\alpha^*, X)$. This procedure has some advantages: the numerical stability of the allocation measure increases and the sampling error decreases as the number of losses greater than α^* increases. The allocation takes the following form

$$K_i = ES_i \times CM_{ES} \quad \text{for} \quad i = 1,\dots,N. \tag{3.163}$$

The *contributory expected shortfall* ES_i of a sub-entity i is defined analogously to ES with Eq.(3.141) on page 179:

$$ES_i(p, X_i) = E\,[X_i\,|X \geq x_p]. \tag{3.164}$$

We see immediately that the sum of the contributory expected shortfall over all sub-entities i equals the expected shortfall. The following holds

$$ES = E\left[X \mid X \ge x_p\right] = \sum_{i=1}^{N} E\left[X_i \mid X \ge x_p\right]$$

$$= E\left[\sum_{i=1,}^{N} X_i \mid X \ge x_p\right] = E\left[X \mid X \ge x_p\right]. \tag{3.165}$$

What about the allocation objectives? Actually, the regulatory categories of Events and Lines of Business are obvious candidates as the losses to be collected must be mapped onto these classes.

Definition 3.14. By *net loss* we understand the (gross) loss after insurance indemnification. If there is none, then net loss equals gross loss. □

Because we consider insurance as risk mitigant there will be a distribution before application of insurance and another after. We call these two different losses "gross" and "net" (or net of insurance). This means that for an annual total loss S there is X^g and X^n. Therefore, there also exist a gross economic capital $x_{\alpha*}^g$ and a net economic capital $x_{\alpha*}^n$. Obviously, the net values are relevant for the capital determination. But in order to determine the contribution of insurance in the same terms as the other allocation objectives, we should break down the gross losses to net losses plus insurance. Because $X^g = X^n + (X^g - X^n) = X^n + \sum_{k=1}^{K} C_k$, i.e., the difference of gross and net is by definition the insurance payment and with the linearity of the expected shortfall, it follows that

$$ES^g(X^g) = \sum_{i=1}^{N} ES^g(X_i^g) = \sum_{i=1}^{N} ES^g(X_i^n) + \sum_{k=1}^{K} ES^g(C_k) \tag{3.166}$$

where C_k are the claims paid by the insurance policy k and $ES^g(X_i) = E\left[X_i \mid X^g \ge x_{\alpha*}^g\right]$. On the other hand, the contributions after insurance would be determined as

$$ES^n(X^n) = \sum_{i=1}^{N} ES^n(X_i^n). \tag{3.167}$$

These are the relevant values. Note that

$$ES^g(X_i^n) \approx ES^n(X_i^n), \tag{3.168}$$

i.e., the values differ slightly as the condition $X > x_{\alpha*}$ leads to a slightly different set of outcomes.

We could have selected the losses Y_j by lines of business $j = 1, \ldots, M$ and found the expected shortfall to be

$$ES^n(X^n) = \sum_{j=1}^{M} ES^n(Y_j^n), \tag{3.169}$$

because $\sum_{i=1}^{N} X_i = \sum_{j=1}^{M} Y_j$ must hold.

Run-Through Example 3. We show in Table 3.17 the allocation for the QIS example. We can see both gross and net allocation for the events and net for the lines of business. The allocations for the events do not differ by very much, thus confirming Eq.(3.168).

TABLE 3.17: Allocation

Event	gross/net	net/net	Line of Business	net
Internal	5.56	5.47	Corporate	33.6
External	13.3	13.3	Trading	99.6
Employment	5.89	5.88	Retail	163
Clients	9.52	9.52	Commercial	188
Physical	230	230	Payment	37
Disruption	2.31	2.29	Services	30.3
Execution	24.9	24.9	Asset	33
Extreme	367	374	Brokerage	81.3
FI	21.4			
Terrorism	1.07			
Liability	5.02			
D&O	7.04			
Property	1.45			

Now we have two major allocation schemes, viz. gross and net. The difference between them is of interest because it shows the effect that the insurance programme has on the allocation objectives i. The difference in allocated capital $K_i^g - K_i^n$ can be further analysed by the tautological transformation $(K_i^g - W) + (W - K_i^n)$. Now we can choose W to be $K_i^g \times K^n/K^g$. The latter coefficient, call it χ, is the overall scaling effect due to insurance. If all objectives were to profit at the same rate, then it would be by this amount, say S_i. But the other part of the variance can be attributed to the "mix" M_i. From

$$S_i = K_i^g - \chi K_i^g \quad , \quad M_i = \chi K_i^g - K_i^n$$

we can record for each objective, be it an event or a line of business, both *scale* and *mix effects*. Obviously, $\sum_i S_i = K^g - K^n$ and $\sum_i M_i = 0$.

However, this number may be important when it comes to "optimising" insurance, as we need to measure the impact of the policy's characteristics on insurance.

3.7.7 Simulating the Risk Measures

In the simulation setting we determine from the simulated sample $D_n^{(0)}$ the EC as quantile x_α from ordering the data sample according to its magnitudes

x and choosing the outcome x whose ascending order number k is the first to fulfil the condition $k/n > \alpha$, i.e.,

$$EC(\alpha, x) = \min(x_{k:n} \,|\, k = 1, \ldots, n; k/n > \alpha) \qquad (3.170)$$

with the order statistic $x_{1:n} < x_{2:n} < \cdots < x_{k:n} < \cdots < x_{n:n}$ and the confidence level α, we find the point estimate for the risk capital.

For the allocation we first find α^* and then calculate the contributions as expected shortfall (ES) for the objective (either event or lines of business) according to:

$$K_i = E\left[X_i \,|\, X \geq x_{\alpha^*}\right] \qquad (3.171)$$

$$= \frac{1}{n} \sum_{j=1}^{n} J_{\{x_{\alpha^*} < X_j\}} \cdot x_j$$

where $J_{\{a<b\}}$ is an indicator variable with values:

$$J_{\{a<b\}} = \begin{cases} 1 & \text{if} \quad a < b \\ 0 & \text{if} \quad a \geq b. \end{cases}$$

From an implementation point of view, expected shortfalls require some book-keeping. It is obvious that the ES can only be calculated when all draws have been done. But of course, this is also true for the EC.

3.8   Insurance Modelling and Mitigation

3.8.1   Insurance

Caveat emptor, buyer beware, is a general rule in common law. It means that it is the buyer's responsibility to investigate the quality of merchandise. However, today consumer regulations require certain information to be disclosed. But while financial contracts are governed by this rule, insurance contracts are not. Insurance is ruled by *uberrima fide*, utmost good faith, implying that the insured must declare all relevant facts to the insurer, even if the insurer does not ask. This difference in legal perspective means that insurance companies generally investigate potential misrepresentations: whether the event has happened and whether it is covered by the contract, before paying any claim. Thus both whether and when payment is effected may be open. In financial contracts the event triggering the payment is objectively observable.

Insurance can potentially take a variety of designs, wordings and mechanics. There may be a captive insurance company that in turn buys reinsurance cover through specific reinsurance products. Nonetheless, we look at two basic types covering different mechanics.

Insurance may be applied to either the single occurrence of claim and loss per risk or the aggregate of losses. Risk in the parlance of insurers means the insured object and not the potential loss or damage.

In a first step we describe the major quantitative characteristics of insurance contracts. In a second step we go into greater detail concerning the pricing of policies.

3.8.1.1   Insurance Characteristics

The key features are *deductibles* (or retentions) and *limits* for each claim brought forward and for the annual aggregate of those claims. With a *straight deductible* the insured must pay a certain amount of the loss before the insurer is required to make a payment. In some commercial insurance contracts an *annual aggregate deductible* (AAD) may be used by which all deductions of covered losses are added together until they reach a certain level. If total covered losses are below the aggregate deductible the insurer pays nothing. Once the deductible is exhausted all covered losses incurred thereafter are paid in full by the insurer.

The second important feature is *limits*. There may be stipulations regarding the maximum amount the insurer is required to pay per claim, *each and every* or there may be a limit for the aggregate annual claim. The loss amounts beyond these limits are the insured's responsibility.

Example 3.28. Professional Liability wording:

> *Limit of Liability Each Claim* [straight deductible]: The liability of the Company for damages and/or claim expenses as a result of each claim made against the insured during the policy period or the extended reporting period shall not exceed in total the amount stated in the Declaration.

> *Aggregate Limit of Liability:* The liability of the Company for damages and/or claim expenses as a result of all claims made against the insured during the policy period and/or the extended reporting period shall not exceed in total the amount stated in the Declaration.

> *Deductible Each and Every Claim*: The deductible shall apply separately to each claim and shall be borne by the insured and remain uninsured. The deductible shall apply to all damages and claim expenses. The Company shall not have any obligation to make any payments under this policy for damages or claim expenses until the deductible has been paid.

> *Annual Aggregate Deductible*: Subject to the foregoing, the amount

stated in the Declaration as the Annual Aggregate Deductible shall be the total amount paid by the insured under the Deductible Each and Every Claim for all claims made against the insured during the policy period or the extended reporting period.

$\triangle$

Definition 3.15. The *annual aggregate deductible* AAD is the total amount an insured is responsible to retain for the sum of all losses up to a specified deductible during an annual policy period. Once the annual aggregate deductible has been reached by the accumulation of payment by the insured, the insurer responds to the remaining claims up to the policy limits. $\square$

The AAD must not be confused with a construct also called "Aggregating Specific Deductible" where the insured assumes in addition to the straight deductible an aggregate deductible before the insurance starts to pay any claim amount.

In Figure 3.32 we see the claims pattern of an insurance policy with deductible (2), limit (5) and aggregate limit (9). Both the deductible and the portion above the predetermined and agreed-upon coverage are to be borne by the insured. Suppose there are N insured losses in a specific underwriting

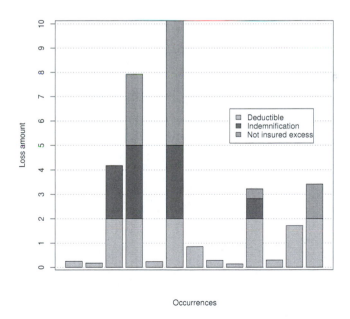

FIGURE 3.32: Losses and claims in time.

year. The total gross loss is given by definition:

$$S = \sum_{j=1}^{N} X_j. \tag{3.172}$$

Suppose insurance covers "each and every," i.e., the number of claims is irrelevant (every). There is a straight deductible of R, an annual aggregate deductible of D, a limit L and an annual limit A. The coverage is defined as $C = L - R$. What is the net loss of the insured?

The total paid by the insurer is simply:

$$Z = \min\left[A, \sum_{j=1}^{N} J_{\{X>R\}} \min\left[X_j, L\right] - \min\left[\sum_{j=1}^{N} J_{\{X>R\}} R, D\right]\right] \tag{3.173}$$

where $J_{\{X>D\}}$ is an indicator variable with values:

$$J_{\{X>D\}} = \begin{cases} 1 & \text{if } X > D \\ 0 & \text{else.} \end{cases}$$

Therefore, the net loss for the institution will be the difference between the total loss and the sum paid by the insurer. This is

$$S_{net} = S - Z. \tag{3.174}$$

From this it is easy to see why insurance has a very non-linear influence on net loss.

The insurance is characterized by $\mathbf{H} = \{R, L, A, D\}$, R the straight deductible, L the straight limit, A the aggregate annual limit and D the annual aggregate deductible. The cover C is $L - R$.

The concept of annual aggregate deductible (and limit) is common in Property and Liability insurance, thus also in Directors & Officers and also in Workers' Compensation. Most often the aggregates are multiples of the limits and deductibles. Say $A = rL$ and $D = sR$ with r and s integers. Then r and s convey a good idea of the protection provided by the aggregates against more frequent losses than expected.

Let us consider some special cases. If there is no annual aggregate deductible the formula Eq.(3.173) is:

$$Z = \min\left[A, \sum_{j=1}^{N} J_{\{X>R\}} \min[X_j - R, L - R]\right]$$

$$= \min\left[A, \sum_{j=1}^{N} \min\left[\max[X_j - R, 0], L - R\right]\right].$$

We note that for a ineffective aggregate deductible or limit we have to set

$D = \infty$, respectively $A = \infty$; for an ineffective deductible we set $R = 0$. For no aggregate features at all the total indemnity becomes

$$Z = \sum_{j=1}^{N} \min\left[\max[X_j - R, 0], L - R\right]. \qquad (3.175)$$

Deductibles are designed to bound the payments of the insured; limits are features to bound the payment of the insurance company. The higher the limit, the costlier the policy. The higher the deductible, the cheaper the policy.

TABLE 3.18: Example: Insurance definition

	Straight Deductible	Limit	Annual Aggregate Limit	Annual Aggregate Deductible
Building/Extended Cover	10	750		
Business Interruption	.05	.2	.45	
Electronics	0.001	0.2		
Valuables	0.01	25		
Fidelity/Forgery	3	50	200	10
General Liability	0.01	50		
Directors and Officers	0	100		
Errors and Omissions	0.01	50	250	

With this incorporation of insurance losses are far better approximated than by proxies like premium or limit.

Example 3.29. We define the insurance policy "Business interruption" with the following characteristics: Straight Deductible 50,000.-, limit 200,000.- and Aggregate Annual Limit 450,000.-

As we can see in Table 3.18 our model makes the impact of insurance transparent. Thus the difference of gross and net loss per Event and Insurance Policy can both be contrasted with the premium and the regulatory requirement. The comparison with the insurance premium should yield in a normal market environment an expected mitigation in the order of magnitude of the premium. A divergence of more than 40% suggests an error in parameter input. △

The regulatory requirement of the draft explicitly stated (BCBS, 2003b, 130):

The bank discloses the reduction of the operational risk capital charge due to insurance.

This has been changed to the following (BCBS, 2005, 151)

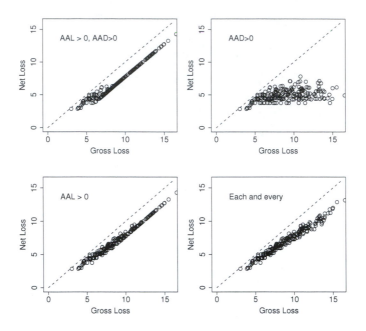

FIGURE 3.33: The fitted compound distribution.

> The bank discloses a description of its use of insurance for the purpose of mitigating operational risk.

The consulted parties apparently found the explicit calculation too oner-ous or difficult. In the Figure 3.34 you can see that the insurance is mostly non-linear. In the lower region "net" equals "gross" because the deductible attributes these losses to the insured. In the intermediate region between 200 and 400,000.- (gross) the net loss depends on the combination of the fre-quency and the severity of losses where the loss mitigation is proportionate to the frequency. The insurance is path-dependent. Beyond the overall limit every marginal increase of loss has to be borne by the insured.

This example shows clearly that an approach similar to credit risk where Loss Given Defaults (LGD) are simply reduced by a recovery factor are by no means appropriate to the relevant insurance modelling.

Insurance cannot be accounted for by just reducing the losses pro-rata by a "recovery factor."

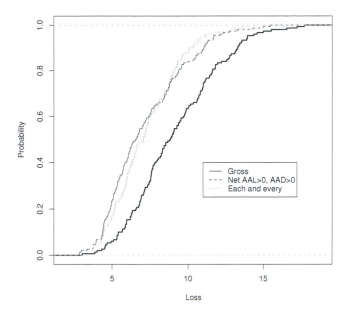

FIGURE 3.34: Effect of insurance on loss distribution.

3.8.1.2 Insurance Pricing

An additional step must be taken to further substantiate the mapping. We look to insurance pricing. Crudely speaking, the premium is made up of the so-called "pure premium" plus mark-ups. For the premium π we write:

$$\pi = \text{Expected Value} + \text{Safety Loading} + \text{Cost} + \text{Profit}$$
$$= E[V] + A[V] + \text{Cost} + \text{Profit}.$$

Let us omit, for the time being, the cost and the profit mark-up and simplify the insurance price as pure premium plus loading

$$\pi = E[V] + A[V]. \tag{3.176}$$

Then the premium is often calculated according the following principles as a function of the claims V and the *risk loading a*:

- *net premium principle* or principle of equivalence $A(V) = 0$,

- *expected value principle* $A(V) = aE[V]$,

- *standard deviation principle* $A(V) = a(Var[V])^{1/2}$,

- *variance principle $A(V) = aVar[V]$* or
- *semi-variance principle $A(V) = aE[(\max[0, V - E[V])^2]$.*

Another interesting class of premium calculation consist of the so-called (zero increase) *expected utility principles*. In practice, the standard deviation principle seems to be the most widely used formula although there are some theoretical disadvantages compared to the variance principle. The loading a is a kind of "price of risk."

Now we must ask ourselves how to determine the parts needed to assess the premium. Roughly speaking we need the expected claims payment and its variance, which depend on insurance characteristics. Furthermore, the loading or some other link to the real world must be calculated. We concentrate on the moments for the time being.

Let us take a closer look at the insurance with no aggregate annual deductible in order to keep things simple. Let X_i, for $i = 1, 2, \ldots, N$ denote the stochastic claims covered by the insurance programme: The number of claims N is also stochastic. The insurance is characterized by $\mathbf{H} = \{R, L, A, D\}$, R the straight deductible, L the straight limit, A the aggregate annual limit and $D = \infty$ the annual aggregate deductible. The cover C is $L - R$.

The total loss to cover with no aggregate deductible or limit is given by Eq.(3.175), viz.

$$Y = \sum_{i=1}^{N} \min\left[\max[X_i - R, 0], C\right]. \tag{3.177}$$

The aggregate claim in turn is given by

$$Z = \min[A, Y]. \tag{3.178}$$

It is worthwhile noting that this two-stage calculation process of premium the reflects the fact that the aggregate features act analogously to a so-called *stop-loss reinsurance* contract being applied to the residuals from a so-called *excess-of-loss treaty*.

For the time being suppose that we know the cumulative density function of Z, denoted $G(z)$. Therefore the two moments (about the origin) are as follow: for the expectation

$$E[Z] = \int_0^A y\, dG(y) + A \int_A^\infty dG(y) = \int_0^A y\, dG(y) + A[1 - G(A)]$$

$$= \int_0^A [1 - G(y)]\, dy \tag{3.179}$$

and

$$E[Z^2] = \int_0^A y^2\, dG(y) + A^2 \int_A^\infty dG(y)$$

$$= \int_0^A y^2\, dG(y) + A^2[1 - G(A)] \tag{3.180}$$

for the second moment. These integrals are widely used in actuarial science and are known as *limited expected value* and more generally *k-th limited moment* (Hogg and Klugman, 1984, 56; Daykin et al., 1994, 79; Klugman et al., 1998, 32). The nomenclature varies.

The variance is derived simply as

$$Var\,[Z] = E\,[Z^2] - E\,[Z]^2. \tag{3.181}$$

As mentioned, to find the two moments of Z we need to calculate the distribution G. Let $F_Y(y)$ denote the transformed distribution of Y. G is then a compound distribution of F_Y given by $G = \sum_{n=0}^{\infty} p_n F^{*n}$, where F^{*n} denote the n-fold convolution of $F_Y(y)$ and $p_k = P(K = k)$. If K is a Poisson, binomial or negative binomial distributed stochastic variable (and only one of these) then the Panjer algorithm (Eq.(3.28) on page 106) can be used to calculate the compound distribution after discretising the distribution.

Now let us take a closer look at the distribution $F_Y(y)$ which must be referred to the original distribution $F_X(x)$. The probability function for $y \geq 0$ is

$$F_Y(y) = Pr(Y < y) = Pr(\min\left[C, \max[X - R, 0]\right] < y). \tag{3.182}$$

There are two relevant cases, namely $0 \leq y \leq C$ and $y > C$.

$$F_Y(y) = \begin{cases} Pr(X - R < y), & \text{for} \quad R < x < L, \\ Pr(C < y), & \text{for} \quad x > L. \end{cases} \tag{3.183}$$

This yields

$$F_Y(y) = \begin{cases} F_X(y + R), & \text{for} \quad y \leq C, \\ 1 & \text{for} \quad y > C. \end{cases} \tag{3.184}$$

Finally the price in dependence of the loading can be calculated by $\pi = E\,[Z] + a \cdot Var\,[Z]$. The loading factor a is determined by the insurance company and is thus estimated by the bank. It can range between 8% and 20% just as an indication.

We have now established that we know how to calculate the price of insurance, although it involves numerical calculations. But there are some difficulties. First of all, we do not have one tidy distribution function, but rather a linear combination of distributions. We may map its matrix with the probability masses determining which gross loss from the events will be channelled to the insurance contract. We need to mix the severity distributions of the events j to obtain an overall loss distribution for policy i.

$$F_i(x) = \sum_{j=1}^{K} \frac{q_{ij} n_j}{B} F_j(x) \quad \text{with} \quad B = \sum_{j=1}^{K} q_{ij} n_j. \tag{3.185}$$

The q_{ij} are the elements of the mapping matrix $\mathbf{Q}$ and n_j are the expected number of losses from event j. From this we can infer once again that using Monte Carlo greatly simplifies matters.

It is worthwhile mentioning the fact that insurance companies use *rate on line ROL*, defined by

$$ROL = \frac{E[N]E[Y]}{C} \tag{3.186}$$

to quote prices for excess layers. In the formula N is the number of claims $Y_i = \max(X_i - R, 0)$ that stem from the original losses X_i. The ROL looks like an interest-rate on a bond with notional C. If you rearrange the formula to read $ROL \times C = E[N]E[Y]$ you can interpret ROL as an adjusted average number of claims hitting the cover with a loss of C. Given that only expectations enter this relative price, it is obvious that the *net premium principle* is concealed by it.

TABLE 3.19: Property catastrophe prices

Cover	Rate on line	Coefficient of variation	Loading
0 ... 250%	20%	1.75	10%
250% ... 500%	12%	3.5	15%
500% ... 1000%	6%	6.5	30%
1000% ...	3%	17.5	50%

In Table 3.19 rate on line is the ratio of premium to cover, the *coefficient of variation* is the standard deviation divided by the expected value and the loading is with respect to the CV. This stylized excess-of-loss reinsurance pricing table shows all features described above: the lower the layer of coverage, the higher the price in relation to the cover (rate on line); moreover, the higher the layer, the greater the portion of volatility in the premium.

3.8.2 Dedicated Provisions

A provision is a liability of uncertain timing or amount. Examples of provisions are: warranty obligations; legal obligations to clean up contaminated land or restore facilities and a retailer's policy to refund customers.

Under IAS 37 ("Provisions, Contingent Liabilities and Contingent Assets") in place in the European Union, a provision is recognised when:

- an entity has a present obligation (legal or constructive) as a result of a past event;

- it is probable that an outflow of economic benefits will be required to settle the obligation and

- a reliable estimate can be made of the amount of the obligation.

A constructive obligation arises from the entity's actions, through which it has indicated to others that it will accept certain responsibilities and as a result has created an expectation that it will discharge those responsibilities.

The amount recognised as a provision is the best estimate of the expenditure required to settle the obligation at the balance sheet date. Provisions are reviewed at each balance sheet date and adjusted to the current best estimate. A provision is used only for expenditures for which the provision was originally recognised. But unfortunately, provisions are not recognised for future operating losses.

There may be legislations which allow for general provisions for potential losses from operational risks. In order to integrate them into the model, there is a filter after the total loss has been assessed. Formally the impact is as follows:

$$S_{tot} = S_{net} - \min\left[P, \max[S_{net}, 0]\right] \qquad (3.187)$$

with P the provision.

In credit risk modelling where a distribution approach is also used, it has become customary to differentiate between the *expected loss* EL and the *unexpected loss* UL. (The latter notion is highly unfortunate and completely misleading as it is not the opposite of *expected*.)

From the quantile read off the distribution the expected value ("mean") is deducted with the following argument: in credit business the interest rate incorporates several components, inter alia the risk-free rate, a liquidity premium and the risk premium. The latter is assumed to compensate for a potential default. A simple premium calculation would consist of the expected value of defaults plus a loading. Therefore it is argued that the expected loss must be deducted from the quantile in order to avoid counting the risk twice and overstating the capital requirements.

In the context of operational risk there is no mechanism for collecting a risk premium. Therefore the expected value of losses cannot be deducted without having set up relevant provisions in the balance sheet.

The regulator states the following to the reduction by the Expected Loss (BCBS, 2005, 147):

> Supervisors will require the bank to calculate its regulatory capital requirement as the sum of expected loss (EL) and unexpected loss (UL), unless the bank can demonstrate that it is adequately capturing EL in its internal business practices. That is, to base the minimum regulatory capital requirement on UL alone, the bank must be able to demonstrate to the satisfaction of its national supervisor that it has measured and accounted for its EL exposure.

This is a bit confusing. We understand that the expected loss has to be incorporated, but that it need not be included more than once is obvious.

Actually, how the exposure of expected loss would be assessed is far from obvious.

3.9   Mapping Events to Insurance Policies

3.9.1   Motivation

In the consultative document to the Accord from 2003 we find the following requirement concerning risk mitigation through insurance (BCBS, 2003b, 130)

> The insurance coverage has been explicitly mapped to the actual operational loss exposure of the institution.

In the final text to the Accord, it has been replaced with the requirement below (BCBS, 2005, 151):

> The risk mitigation calculations must reflect the bank's insurance coverage in a manner that is transparent in its relationship to and consistent with, the actual likelihood and impact of loss used in the bank's overall determination of its operational risk capital.

Now, the later wording is by far less strict and consistent than the first. The requirement about "explicit mapping" can be problematic because insurance policies and coverage do not map directly to the Basel Committee's Loss Event Classification. In some cases, insurance policies extend to multiple event-type categories and in other cases multiple insurance coverage applies to a single event-type. In addition, what is meant by "explicitly" may be open to debate. For a quantitative consideration the mapping is necessary. The requirements read on as:

> The insurance is provided by a third party entity. In the case of insurance through captives and affiliates, the exposure has to be covered by an independent third party entity, for example through re-insurance, that meets the eligibility criteria.

The possibility to use captives where there is a direct link between exposure and insurance may be difficult due to conflicting interpretations.

The determination of insurance coverage is often a detailed process that involves the consideration of numerous circumstantial factors that might give rise to coverage limitations. We also believe that there is more than one reasonable way of mapping. Therefore, *thoroughly documenting* the process will be expedited by a favourable opinion by the regulator.

We see three distinct phases for the mapping. First, we need to produce a starting map with confidence ranges. Then we have to back-test this map with the actual or future situation by comparing its theoretical premium with the paid or quoted premium. When a valid mapping is found and agreed upon, the insurance programme can be "optimised" through scenario analysis.

In a working market there is no "free lunch." In our context this means that the insurance premium charged must be in accordance with the estimated claims to be paid.

3.9.2 Preliminaries

Before we can start the mapping exercise we have to make sure that we have all available relevant data. For this purpose the bank's insurance department must contribute comprehensive *policy data.*

The data collected might be represented by the attributes of Table 3.20. It must present limits and deductibles and also sub-categories where a policy covers multiple perils.

The most important coverages, e.g., blanket bond and D&O, may be layered and transferred to several counter parties (confer Figure 3.35). In this context the only difference from a single policy is the lower counter-parties credit risk.

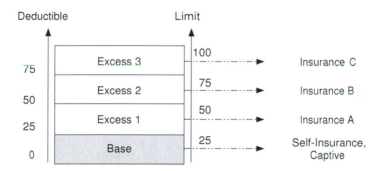

FIGURE 3.35: Insurance layers.

In the analysis one should define a threshold, probably based on a premium, to enforce *materiality.* The premium argument can be defended when the deductibles or limits are relatively small. Then the premium is a proxy for the expected loss. Nonetheless, if the policy covers a high excess layer, the expected loss will be close to zero while the volatility component could be substantial. Then you have to go deeper into the details.

From the attribute "Insurer" we determine whether the risk transferee is a *true third party.* If not, the risk mitigation may be jeopardized. Special care must be give to group-owned captives. These companies may be either licensed insurance companies or re-insurance companies (confer Figure 3.36).

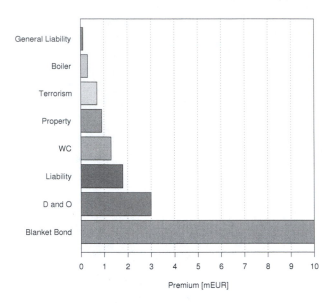

FIGURE 3.36: Policies by premium.

They supply an additional buffer because they are used to retrocede risk to a re-insurance company.

A captive in the form of a re-insurance company needs a fronting insurance. The latter will retrocede almost 100% of the insurance to the bank-owned captive.

There are additional items for analysis. For example there may be several captives with inter-company insurance contracts; and there may be captives doing business with third parties.

From the captive the main re-insurance policy items must be supplied. Here the standard characteristics, i.e., limits and deductibles, may not suffice as the standard re-insurance products comprise more sophisticated features leading to additional data requirements.

3.9.3   Nominal Categories

In our model we use the Event notion on level 1 from the Basel Committee as compound loss generators. These events summarize level 2 events that in turn embrace the level 3 events with the smallest granularity. The latter events can be mapped relatively easily to the insurance wordings and clauses on a yes or no basis. This is shown in Table 3.20. From the outset one is aware that there are more potential policies than events. Moreover, especially with the ten first insurance types (confer Table 3.21) there is a multiple relation between events and insurance. For example, Professional Indemnity may potentially pay losses

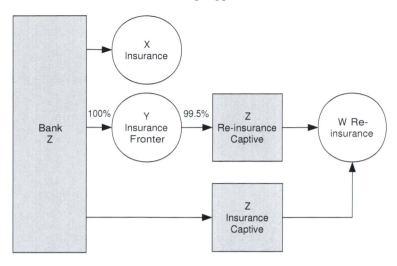

FIGURE 3.37: Risk transfer mechanisms.

from the events Internal Fraud, Clients, etc. and Process Management. On the other hand, Internal Fraud may be covered by a Bankers' Blanket Bond, Electronic Insurance or a policy against Unauthorized Trading.

The situation with Physical Assets is often quite simple. Instead of using only "yes" and "no" experts involved may attempt to substantiate the coverage by using a finer grading, e.g., "never," "rarely," "occasionally," "often," and "always" etc. The historic loss database may give some additional clues. But because the frequency of losses may be quite low it is possible that there are not enough data points for mining.

Now that we have assessed what policies to consider and gained an overview we can tackle specifics.

3.9.4 Quantitative Consideration

Calibrating the mappings with data is a major undertaking. We are trying to find plausible bounds for the weights of the mapping by taking some heuristics and doing some back-testing with the insurance pricing. It seems important to us to reach this kind of granularity in order to make a qualified statement about the efficacy of insurance. We think that the murky statement: "The risk mitigation calculations must reflect the bank's insurance coverage in a manner that is transparent in its relationship to the actual likelihood and impact of loss used in the bank's overall determination of its operational risk capital" is equivalent with an explicit mapping.

TABLE 3.20: Policy data examples

Attributes	Example 1	Example 2
Coverage	Terrorism	General liability
Insurer	Lloyd's London	MAXA
Rating		AA
Policy number	Xyz001	100-00-23
Term	04-01-01 till	04-01-01 till
	04-12-31	04-12-31
Limits	EUR 100m	EUR 5m
	per occurrence	
Aggregate limits	–	EUR 5m (general)
		EUR 2.5m personal injury
		EUR 1m Liquor liability
Deductibles	EUR 5m	EUR 2.5m per occurrence
	per occurrence	EUR 1m Liquor liability
Aggregate deductible	–	EUR 2.5m per occurrence
Premium	EUR 300k	EUR 50k

3.9.4.1   Estimating a Map

Now we have to produce some figures. A first attempt could be as follows: On the level 3 Event categories we have produced entries whether there is coverage or not. If we assume that all events are equally likely, we may just count the entries per insurance. The equal likelihood assumption may become more appealing as more events are defined. Alternatively, weighting by some measure (e.g., average claim size etc.) is another possibility. Then we should assess an overall rate between insured and uninsured.

Example 3.30. From the mapping for "Internal Fraud" we count 27 level 3 Events (see Appendix 5.3 on page 351). We now count the number of items that are normally covered by a specific insurance contract (see Table 3.21 on page 206 for a summary). This count yields the following: Unauthorized Trading 6, Professional Indemnity 2, Property 1, Blanket Bond 18. Assuming that 75% of the losses are insured leads to the quantification in the table below.

Insurance	Count	Percentage	Normalized
Unauthorized treading	6	22.2%	16.7%
Professional indemnity	2	7.4%	5.6%
Property	1	3.7%	2.8%
Blanket bond	18	66.7%	50.0%
Total	27	1	0.75

$\triangle$

In addition to using categorical data to arrive at a starting map, one may also use, inter alia:

- statistics from own historic records,

- statistics from industry data,

- from estimating frequencies for the linguistic assignment from the policy wording.

As always, the historic approach is valid only if the underlying structure has not changed, i.e., the policy wordings and event definitions are stable and secondly there exists a statistically relevant number of claims.

3.9.4.2 Back-Testing

The mapping has a potentially big impact on the insurance payments. Generally speaking, the total payments V_i of a policy i depend on the characteristics $\mathbf{H}_i$ and the loss generators from frequency, severity and the convolved distribution of both. The paid claims V_i by insurance policy are induced by different event losses, which in turn are correlated through the frequencies.

In order to model a fair and economically meaningful insurance policy we have to set it in its proper context. Therefore, back-testing should involve the premium which is the price to be investigated with brokers or directly with insurance companies.

For each policy i we have to calculate a pro-forma premium according to Eq.(3.176) on page 195. Naturally we chose the most often applied standard deviation principle, viz.

$$\pi_i = E[V_i] + a \cdot Var[V_i]^{0.5} \tag{3.188}$$

with V_i the random claims payment of policy i.

We have to compare this theoretical premium with an actual one. This implies "fudging" because the market price may be calculated with different

- distributional assumption (most probably),

- premium principle (less likely) and

- an undisclosed loading.

TABLE 3.21: Standard insurance product mapping

	Internal	External	Employment	Clients	Physical	Disruption	Execution
Special Lines and Major Concerns							
BBB	xx	xx					
Computer Crime		xx					
Commercial General Liab.				xx			
D&O	x				xx		
Electronic Insurance						xx	
Employment Practice Liab.				xx			
Property		(x)			x	xx	
Professional Indemnity	x				xx		xx
Unauthorized Trading	xx						
Non-Financial Property							
Fire Insurance					x		
Extended Coverage					x		
Increased Cost of Working					x		
All Risks (Property)					x		
Construction/Erection All Risk					x		
Glass Insurance					x		
Electronic Insurance					x		
Transit Insurance					x		
Insurance Valuable Items					x		
Travel/Baggage Insurance					x		
Fine Art Insurance					x		
Business Interruption Insurance					x		
Burglary/Theft/Robbery		x			x		
Other Casualty							
Casualty and Accident Insurance				x			
Property Owner's Liability Insurance					x		
Environmental Impairment Liability					x		
Workers Compensation			xx				
Personal Accident Insurance			x				
Motor Insurance			x				

Therefore a correct order of magnitude must suffice. Here we would assume that a deviation of $\pm20\%$ is acceptable. We suggest tackling the problem in two steps: First try to estimate the relations between the different events and the insurance policies. Then iterate the level until it falls within an acceptable range.

Run-Through Example 4. We assume a simple insurance programme with five covers. Where there is an annual aggregate limit it corresponds to the individual limit. The financial institution blended cover shall cover losses from: EPL, D&O, Fiduciary and Bond. For our model with aggregate annual losses only deductibles beyond the thresholds make sense. The interaction of loss

TABLE 3.22: Insurance programme

	Deductible	Limit	Aggregate Deductible	Aggregate Limit	Premium
FI blended cover	0.75	75	0	75	11.25
Terrorism	5	100	0	200	0.42
Liability	2.5	50	0	50	1.45
D&O	25	185	0	185	1.7
Property	5	95	0	0	0.5

TABLE 3.23: First guess of mapping

	Internal	External	Employment	Clients	Physical	Disruption	Execution	Extreme
FI blended c.	75%	75%	15%	70%	25%	45%	65%	5%
Terrorism		0.3%			0.3%			5.%
Liability		10.9%	10.9%		7.6%			5.4%
D&O	8.8%		13.2%	17.7%			13.2%	8.8%
Property					2.0%	1.2%		1.2%

modelling and insurance mapping determines the extent of mitigation by insurance. But we have to make sure that we do not "cheat." Therefore, we back-test the mapping by comparing the external real premium with our model, especially with a theoretical premium from simulation. Assume that insurance pricing is just the expected value of the losses E plus a loading (price of risk) times the standard deviation of claims paid. Then one compares these two figures in order to satisfy the regulator. In our case this means adapting the weights of the mapping table to bring the theoretical (simulated) premium in line with the real premium. While liability covers may be sufficiently close, we have to lower the coverage of the FI blended cover. We proportionally reduce or augment the weights in order to match the 20% loading. We could

TABLE 3.24: Insurance premium

	Pure Premium	Stdev	CV	Theor. Premium with Loading			Premium
				10%	15%	20%	
FI blended c.	7.19	9.65	1.34	8.16	8.64	9.12	11.30
Terrorism	0.41	5.55	13.41	0.97	1.25	1.52	0.42
Liability	0.72	4.76	6.65	1.19	1.43	1.67	1.45
D&O	0.64	7.55	11.74	1.40	1.78	2.15	1.70
Property	0.19	3.27	17.68	0.51	0.68	0.84	0.50

also in good faith have chosen different loadings if these values were available from the insurance broker or similarly well-informed institutions. To the casual reader these figures may seem rather low. Actually, the opposite is true. The Data Collection Exercise has established a ratio of 1.7% of all events having received insurance indemnification. Our figures are therefore high and for another reason as well. The mapping represents the percentage of all events qualifying for a specific insurance. This does not directly imply a payment as deductibles, etc., reduce the number of cases where a payment actually follows. With these adjusted weights the resulting theoretical premiums (with

TABLE 3.25: Next step mapping

	Internal	External	Employment	Clients	Physical	Disruption	Execution	Extreme
FI blended cover	80.%	80.%	18.%	75.%	31%	55%	75.%	5.%
Terrorism			0.1%		0.1%	1%		1.4%
Liability		8.8%	8.8%		9.4%			6.5%
D&O	7%		10%	14%			10%	7%
Property					1.5%	1%		1%

loading 20%) come quite close to the real premiums. Because there is some uncertainty about the loading and because insurance premiums are subject to changing market conditions (so-called "soft" and "hard" markets), there is no point in matching the values perfectly. In the Table 3.26 below we see the premiums and the resulting Economic Capital (CaR).

TABLE 3.26: Insurance mitigation

	CaR	Share	Theoretical Premium	Premium
FI blended cover	21.4	3.08	10.3	11.3
Terrorism	1.07	0.154	0.711	0.42
Liability	5.02	0.722	1.9	1.45
D&O	7.04	1.01	1.87	1.70
Property	1.45	0.209	0.744	0.5

3.10 Mapping Events to Lines of Business

3.10.1 The Simulation Model

We model the loss distributions by events, using both attritional losses on an aggregate basis and a compound distribution consisting of a frequency and severity distribution for the major losses. This is very much the way property & liability insurers are accustomed to proceeding. The event dimension is a complete partition of the losses.

The regulator demands that data be collected and identified by lines of business and proposes another partition. We now map the event dimension onto the LoB dimension by means of a mapping table. The elements of the mapping table are weights to be applied both to the attritional loss distribution and the frequency distribution of the compound process. In our model the loss distributions pertaining to the lines of business are constructed as a linear combination, i.e., a mix, of the loss distributions of the events by means of the mapping table.

If we denote the mapping matrix $\mathbf{S} = [s_{ij}]$ with dimensions $\Re^{N \times M}$ where N is the number of events, viz. $N = 8$ and M is the number of lines of business, here again $M = 8$. According to our construction the total loss of line of business j in any one draw of the simulation is as follows:

$$X_j = \sum_{i=1}^{N} s_{ij} \cdot l_i + \sum_{i=1}^{N} \sum_{k=1}^{K_i} J_{\{s_{ij}\}} \cdot r_{ik} \qquad (3.189)$$

s_{ij} = element of the mapping table
l_i = random attritional loss of event i
$J_{\{s_{ij}\}}$ = random indicator
$K_i \sim \mathcal{P}_i$ = random number of losses of event i
$\mathcal{P}_i$ = distribution of frequency of event i
$r_{ik} \sim F_i()$ = random severity of loss
$F_i()$ = distribution of severity of loss of event i.

The random indicator for the line of business $J_{\{s_{ij}\}}$ is either one or zero according to a random standard uniform variate u in the following sense:

$$J_{\{s_{ij}\}} = \begin{cases} 1 & \text{if} \quad \sum_{k=1}^{j} s_{ik} < u \le \sum_{k=1}^{j+1} s_{ik} \\ 0 & \text{else.} \end{cases}$$

From all these losses an empirical cumulated probability function is constructed. There is no point in trying to force it into a formula because it can be better understood by the Algorithm 10.

The elements of the mapping matrix **S** are the probability masses for the lines of business. It is used in a kind of lottery to attribute the loss by event to a line of business in a random fashion. The same holds true for the map between events and insurance policies. Note that we generate a random attri-

---

For each event $i = 1, \ldots, N$

1. draw a random aggregate loss l_i and attribute $s_{ij}l_i$ to each X_j

2. draw the random number of losses K_i

   - for each k in K_i draw a loss $r_{\cdot k}$
     - draw an index j from $j = 1, \ldots, M$ with $J_{\{s_{ij}\}}$
     - augment $X_j \leftarrow X_j + r_{jk}$

ALGORITHM 10: Generating losses by lines of business

---

bution of losses to lines of business. This is not strictly necessary as we will use expected shortfall as a risk measure which can be determined for lines of business by a linear combination of the CaR by events with the weights from the mapping matrix. But if we want to have an empirical loss distribution by lines of business this approach is useful. On the other hand, we can determine the CaR by a new objective, say countries, by defining a map $Q \in \Re^{\text{Countries} \times \text{Events}}$ where the sum of the elements of each column equals 1. Then the capital at risk by countries would be the vector $\mathbf{k}_c$ from $\mathbf{k}_c = \mathbf{Q}\mathbf{k}_e$. We give an example on page 221.

3.10.2   The Map from Optimisation

Normally we have historic loss data with classifications according to the events type i, the line of business j and the year of occurrence t. We can infer and then build histograms of the number of losses per year, the loss size and the annual losses grouped by events, lines of business or the intersection of the two.

What we have done so far is to fit the data to an aggregate distribution for attritional losses, a frequency distribution for the number of individual losses and the severity of these losses. From the simulation we have created an empirical cumulated distribution function, say $G_i(x)$ for each event $i = 1, \ldots, N$.

According to the Basel requirements each data point comes with a tag for event, line of business and the date of occurrence. In order to construct the map $\mathbf{Q}$ between events and lines of business, we propose using the "least square fit," because this procedure has previously been used for the fitting of event distributions. The elements of $\mathbf{Q}$ are q_{ji} where j indicates the line of business and i the event.

In order to keep it comprehensive we do not separate aggregate and individual losses. Compatible with the event distribution of total losses $G_i(x)$, we just take the sums of the losses grouped by year and line of business. Let us call the pertinent variable Y_j. We use a goal function based on the mean and the standard deviation. Thus we will have an array of means from $G_i(x)$, say μ and σ and from the data regarding lines of business $\mathbf{m}$ and $\mathbf{s}$.

The values of the matrix elements q_{ji} result from a constraint optimisation exercise. The constraints stem from the fact that the grand total of all losses must be the same whether analysed from an event perspective or a line of business perspective, given that all losses from events must be attributed to some line of business. This means that the sum of elements of $\mathbf{Q}$ by columns equals one, i.e.,

$$\sum_{j=1}^{M} q_{ji} = 1 \quad \text{for all } i = 1, \ldots, N. \tag{3.190}$$

Furthermore, the weights must be greater or equal to one and less than or equal to zero. Now, we will have additional information concerning the map which translates into ranges, i.e., a lower and upper bound for q_{ji} which call u_{ji} and l_{ji}. These are additional constraints for the optimisation. Formally

$$0 \leq l_{ji} \leq q_{ji} \leq u_{ji} \leq 1 \quad \text{for all} \quad j = 1, \ldots, N; k = 1, \ldots, M. \tag{3.191}$$

The sum of the weights for each event i must obey the following restrictions:

$$\sum_{j=1}^{M} u_{ji} > 1 \quad \text{and} \quad \sum_{j=1}^{M} l_{ji} < 1. \tag{3.192}$$

With respect to the objective function we may have to choose how to weight the means and the variance or standard deviation or even a derived value from them. We want to emphasize the mean because it may be more reliable a value than the dispersion. We want to minimise in the least square sense the sum of the following $2 \times M$ equations for $j = 1, \ldots, M$

$$g_j(\mathbf{Q}) = \frac{1}{m_i} \left[\sqrt{\sum_{j=1}^{N} q_{ji}^2 \sigma_i^2} - s_i \right] \overset{!}{=} 0 \tag{3.193}$$

and

$$h_j(\mathbf{Q}) = \sum_{j=1}^{N} q_{ji}\mu_j - m_i \overset{!}{=} 0. \tag{3.194}$$

The objective function is thus

$$f(\mathbf{Q}) = \sum_{j=1}^{M} h_j(\mathbf{Q})^2 + g_j(\mathbf{Q})^2. \tag{3.195}$$

The division by m_i can be seen as a weighting factor between means and standard deviations. Though we have not taken into consideration the correlations between the event losses, this can easily be done.

Example 3.31. The following example illustrates the effectiveness of this approach. There are innumerable optimisation algorithms available. This problem can be solved by expanding the classical linear optimisation algorithms when they are brought into the required form. This involves recasting the matrices $\mathbf{Q}$, $\mathbf{U}$ and $\mathbf{L}$ into a vector, say $\mathbf{q}$, $\mathbf{u}$ and $\mathbf{l}$ and supplying Jacobian and Hessian matrix of the objective function with respect to the desired variables. Now the formulation reads

$$\begin{aligned} \min_{\mathbf{q} \in R^n} \quad & f(\mathbf{q}) \\ \text{subject to} \quad & A\mathbf{q} = \mathbf{b} \\ & \mathbf{l} \le \mathbf{q} \le \mathbf{u}. \end{aligned} \tag{3.196}$$

Our problem is a non-linear, least square optimisation with linear constraints, bound both by inequalities and equalities. We choose a `Fortran` program called `DQED` from Hanson and Krogh (1992). It can be found at `http://www.netlib.org/opt/dqed.f`. First we input the means and standard deviations by events and the same variables to be fitted for the lines of business according to the table below.

Events							
σ	5.62	4.7	3.6	5.2	30.7	1.95	16	52
μ	6.24	13.4	5.88	11	12.3	2.3	25.4	11.5
Lines of Business							
$\mathbf{s}$	13.6	25.4	32.0	31.5	13.3	13.3	13.6	19.8
$\mathbf{m}$	4.2	15	29.0	20.5	3.9	3.8	3.7	8.3

In addition we formulate lower constraints for the weights of the map, as follows

$$
\mathbf{L} = \begin{pmatrix}
0.01 & 0.01 & 0.01 & 0.15 & 0.05 & 0.05 & 0.01 & 0.05 \\
0.01 & 0.01 & 0.01 & 0.15 & 0.05 & 0.10 & 0.01 & 0.20 \\
0.40 & 0.40 & 0.40 & 0.15 & 0.05 & 0.10 & 0.15 & 0.30 \\
0.01 & 0.01 & 0.01 & 0.15 & 0.20 & 0.10 & 0.20 & 0.20 \\
0.01 & 0.01 & 0.01 & 0.01 & 0.05 & 0.25 & 0.01 & 0.05 \\
0.01 & 0.01 & 0.01 & 0.01 & 0.05 & 0.15 & 0.01 & 0.05 \\
0.01 & 0.01 & 0.01 & 0.01 & 0.05 & 0.05 & 0.01 & 0.05 \\
0.01 & 0.01 & 0.01 & 0.15 & 0.20 & 0.10 & 0.01 & 0.05
\end{pmatrix}.
$$

Also upper bounds are introduced according to $\mathbf{U}$:

$$
\mathbf{U} = \begin{pmatrix}
0.70 & 0.70 & 0.70 & 0.70 & 0.70 & 0.70 & 0.70 & 0.70 \\
0.70 & 0.70 & 0.70 & 0.70 & 0.70 & 0.70 & 0.70 & 0.25 \\
0.60 & 0.70 & 0.70 & 0.30 & 0.70 & 0.70 & 0.70 & 0.40 \\
0.70 & 0.70 & 0.70 & 0.70 & 0.50 & 0.70 & 0.30 & 0.25 \\
0.70 & 0.70 & 0.70 & 0.70 & 0.70 & 0.40 & 0.70 & 0.70 \\
0.70 & 0.70 & 0.70 & 0.70 & 0.70 & 0.70 & 0.70 & 0.70 \\
0.70 & 0.70 & 0.70 & 0.70 & 0.70 & 0.70 & 0.70 & 0.70 \\
0.70 & 0.70 & 0.70 & 0.70 & 0.70 & 0.70 & 0.70 & 0.70
\end{pmatrix}.
$$

The programs presents the following optimal variables $\mathbf{Q}$:

$$
\mathbf{Q} = \begin{pmatrix}
0.010 & 0.010 & 0.010 & 0.184 & 0.075 & 0.050 & 0.010 & 0.05 \\
0.010 & 0.010 & 0.010 & 0.150 & 0.050 & 0.100 & 0.387 & 0.20 \\
0.400 & 0.594 & 0.700 & 0.282 & 0.050 & 0.200 & 0.263 & 0.30 \\
0.010 & 0.346 & 0.240 & 0.150 & 0.200 & 0.100 & 0.300 & 0.20 \\
0.010 & 0.010 & 0.010 & 0.010 & 0.182 & 0.250 & 0.010 & 0.05 \\
0.010 & 0.010 & 0.010 & 0.010 & 0.194 & 0.150 & 0.010 & 0.05 \\
0.193 & 0.010 & 0.010 & 0.010 & 0.050 & 0.050 & 0.010 & 0.10 \\
0.357 & 0.010 & 0.010 & 0.204 & 0.200 & 0.100 & 0.010 & 0.05
\end{pmatrix}.
$$

The effectiveness of the optimisation can be gauged by comparing the required values $\mathbf{m}$ and $\mathbf{s}$ with the estimated linear combinations $\hat{\mathbf{m}}$ and $\hat{\mathbf{s}}$ of the table below. $\triangle$

	Lines of Business							
$\hat{s}$	11.219	24.522	31.479	30.605	11.473	11.511	11.420	18.264
s	13.6	25.4	32.0	31.5	13.3	13.3	13.6	19.8
$\hat{m}$	3.724	13.064	25.281	22.619	4.378	3.569	3.268	8.300
m	4.2	15.0	29.0	20.5	3.9	3.8	3.7	8.3

Such algorithms extending non-linear programming may present difficulties when it comes to convergence. Within the Operational Research community there are many algorithms that solve the problem of *minimum cost network flows* (Bertsekas and Tseng, 1994; Toint and Tuyttens, 1992). Here the constraints come much more naturally as such models can be exemplified with transportation of goods (or assignment of jobs to machines, etc.). Several sources produce goods that can be conveyed through different channels with given capacity and cost function. Here the continuity – that the goods cannot disappear and are identical with Eq.(3.190) on page 211 – is always assumed. The goal is to transport all goods at the lowest cost.

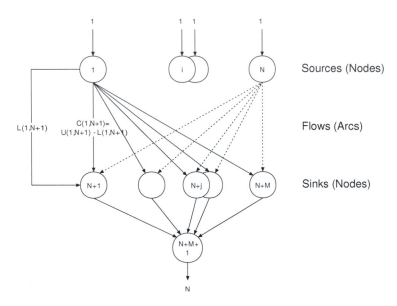

FIGURE 3.38: Mapping as Minimum Cost Flow problem.

In these models (see Figure 3.38) we have nodes n and arcs (j,i), i.e., directed edges, from node j to node i. The nodes are either sources (supplies), sinks (demands), trans-shipment nodes (depots) or combinations thereof. The goods flow through the arcs from node to node with specific unit-cost functions c_{lk} while the maximum capacity of the arc is bound by u_{lk}. In order to enforce minimum bounds, we detract this amount l_{lk} from the starting node and reduce the capacity to $u_{ji} - l_{ji}$. These fixed costs are not relevant to optimisation. After solving the problem we add the lower bounds to the solution for the flows and add the cost to the total cost. Dummies are introduced to ensure that over the whole network sinks and sources are balanced, i.e., adds up to zero. The solution consists in finding the flows q_{ji} that incur the lowest over-all cost.

One area in which network flow is widely used is in the management of foreign exchange trades of banks.

Now we have finished the journey through the input to the model. We have assessed the parameters of the loss generators and the correlations and we have assessed maps from events to insurance and lines of business. Lastly, we have outlined the insurance parameters. Now we are able to run a complete simulation.

Run-Through Example 5. For our QIS example, the mapping to Lines of Business is also very easy. We just take the share of the LoB as defined by the total loss. For the extreme events we attribute 5% to all LoBs, excepting Trading, Retail and Commercial, to which we allocate the residual 80%. This procedure yields the following Table 3.27.

TABLE 3.27: Mapping events to LoB

	Internal	External	Employment	Clients	Physical	Disruption	Execution	Extreme
Corporate	0.09	0.00	0.00	0.16	0.00	0.00	0.02	0.05
Trading	0.11	0.03	0.13	0.19	0.05	0.08	0.31	0.20
Retail	0.60	0.66	0.67	0.25	0.05	0.13	0.19	0.35
Commercial	0.04	0.27	0.04	0.15	0.57	0.09	0.27	0.20
Payment	0.04	0.02	0.02	0.01	0.01	0.37	0.04	0.05
Services	0.00	0.00	0.01	0.00	0.05	0.19	0.08	0.05
Asset	0.01	0.00	0.02	0.08	0.00	0.01	0.05	0.05
Brokerage	0.11	0.00	0.10	0.16	0.27	0.13	0.04	0.05
Total	1.00	1.00	1.00	1.00	1.00	1.00	1.00	1.00

3.11 Calculating the Economic Capital

3.11.1 Accuracy of Result

In statistical analysis, the researcher is usually interested in obtaining not only a point estimate of a statistic but also an estimate of the variation in this point estimate and a confidence interval for the true value of the parameter.

Traditionally, researchers have relied on the central limit theorem and normal approximations to obtain standard errors and confidence intervals. These techniques are valid only if the statistic, or some known transformation of it, is asymptotically normally distributed. Hence, if the normality assumption does not hold, then the traditional methods should not be used to obtain confidence intervals. A major motivation for the traditional reliance on normal-theory methods has been computational tractability. Now, with the availability of modern computing power, scientists need no longer rely on asymptotic theory to estimate the distribution of a statistic. Instead, they may use resampling methods which return inferential results for either normal or non-normal distributions.

Resampling techniques such as the bootstrap and jackknife provide estimates of the standard error, confidence intervals and distributions for any statistic. For a more rigorous treatment see Efron and Tibshirani (1986) and Efron and Tibshirani (1993).

Economic Capital EC is defined by the regulator as a quantile, i.e., $EC_p := x_p = \inf \{x \,|G(x) > p\}$. The quantile is an important performance measure. However, the estimation of quantiles is very different from the estimation of functions of means. The natural estimator for x_p is the sample quantile from the order statistics. Thus determining a point estimator for x_p is easy, viz. Eq.(3.170)

$$\hat{x}_p = \min(x_{k:N} \,|k = 1, ..., N; k/N > p) \qquad (3.197)$$

with the order statistic $x_{1:N} < x_{2:N} < \cdots < x_{k:N} < \cdots < x_{N:N}$ of the simulated losses x_i and the confidence level p.

The difficulty arises when we wish to develop a confidence interval for $\hat{x}_p$. The central limit theorem for the independent and identically distributed quantiles leads to an asymptotic result known as the Bahadur representation of the sample quantile (Haas, 2005, 2):

$$\hat{x}_p \sim \Phi(x_p, \sigma^2) \quad \text{with} \quad \sigma = \frac{1}{f(x_p)} \sqrt{\frac{p(1-p)}{N}} \qquad (3.198)$$

where $\Phi(\cdot)$ denotes the normal distribution. The standard deviation measures the potential error. This formula argues against the very high confidence value of 99.9% required by the regulators as the density at this quantile is so low that accuracy can be influenced by the number of simulations n. Due to the square root the accuracy does not even grow linearly. Of course there are special techniques to improve convergence. To estimate the density is notoriously difficult.

3.11.2   The Jackknife Estimator

This method was first used in 1949. The technique is based on removing sub-samples from a given data set D_N under analysis and recalculating the estimator $\hat{\theta}$. We want to estimate some parameter θ as a complex statistic of N data points, i.e., $\hat{\theta} = f(D_N) = f(x_1, x_2, \ldots, x_N)$. We first compute the so-called *ith jackknife replication* of $\hat{\theta}$ denoted by $\hat{\theta}_{(i)} = f(x_1, x_2, x_{i-1}, x_{i+1} \ldots, x_N)$. From this we define the *pseudo-values*

$\eta_{(i)} = N\hat{\theta} - (N-1)\hat{\theta}_{(i)}$. The jackknife estimate is given by

$$\hat{\theta}_{jack} = \frac{N-1}{N} \sum_{i=1}^{N} \eta_{(i)} = \frac{1}{N} \sum_{i=1}^{N} \left[N\hat{\theta} - (N-1)\hat{\theta}_{(i)} \right]$$
$$= N\hat{\theta} - (N-1)\hat{\theta}_{(\cdot)} \qquad (3.199)$$

with $\hat{\theta}_{(\cdot)} = \frac{1}{N} \sum_{i=1}^{N} \hat{\theta}_{(i)}$.

The variance of the estimator can be calculated as:

$$Var(\hat{\theta}_{jack}) = \frac{1}{N} \sum_{i=1}^{N} (\hat{\theta}_{(i)} - \hat{\theta}_{(\cdot)})^2. \qquad (3.200)$$

This approach is also known as "l-o-o" for leave-one-out.

Equation (3.199) can be verified for the estimate of the mean μ by the average $\hat{\mu} = \frac{1}{N} \sum_{i=1}^{N} x_i$. Here the pseudo-value $\eta_{(i)} = N\hat{\mu} - (N-1)\hat{\mu}_{(i)}$ with

$$\mu_{(i)} = \frac{1}{N-1} \sum_{j \neq i}^{N} x_j = \frac{N\hat{\mu} - x_i}{N-1} \qquad (3.201)$$

yields $\eta_{(i)} = x_i$ and thus $\hat{\mu}_{jack} = \hat{\mu}$. Similarly it can be verified that the jackknife method applied on the variance of the mean gives the correct result, viz.

$$Var(\hat{\theta}_{jack}) = Var(\hat{\mu}) = \frac{1}{N(N-1)} \sum_{i=1}^{N} (x_i - \hat{\mu})^2. \qquad (3.202)$$

For big data sets with several hundred thousand points, this method is not feasible. Fortunately there is a variant, where entire groups or sets of data are excluded in turn. A recommended number of groups is approximately $m = 20$. Instead of leaving out one, we remove $l = N/m$ points from the replicate. All else remains unchanged.

Having established the mean and the variance of the pseudo-value we can determine the confidence interval as

$$\left[\hat{\theta}_{jack} - t_{m-1} \sqrt{\frac{Var(\hat{\theta}_{jack})}{m}}, \hat{\theta}_{jack} + t_{m-1} \sqrt{\frac{Var(\hat{\theta}_{jack})}{m}} \right] \qquad (3.203)$$

where t is the $(1 - p/2)$-quantile of the Student-t distribution with $m - 1$ degrees of freedom.

Run-Through Example 6. We perform nine additional simulations of 1 million draws each by choosing different seed values. Then we determine the confidence interval with a (two-sided) t-test on the level 95%.

Run	# 1	# 2	# 3	# 4	# 5
Net CaR	665.9	673.5	678.4	679.5	657.0
Gross CaR	694.4	704.0	709.2	702.8	689.7
Insurance	4.10	4.32	4.33	3.31	4.74

Run	# 6	# 7	# 8	# 9	# 10
Net CaR	682.3	661.8	674.3	669.0	676.3
Gross CaR	712.0	694.2	707.1	698.6	707.4
Insurance	4.18	4.67	4.63	4.22	4.40

The statistics are given in the following table.

	Mean	Stdev	$t(9,0.025)$	CaR upper	CaR lower	Range	Range[%]
Net CaR	671.81	7.77	2.69	678.41	665.21	13.20	0.02
Gross CaR	701.92	7.01	2.69	707.88	695.97	11.91	0.02
Insurance	4.29	0.39	2.69	4.62	3.96	0.66	0.15

We see that the range of values is some 2%, which we deem satisfactory. We find it thus plausible that a lesser number of simulations is not very appropriate. Had the regulator opted for a more sensible confidence level, say 99%, the number of draws could be lowered sensibly.

3.11.3   Conditioning the Data

For quantiles whose confidence is close to 1, i.e., 99.9%, it is difficult to estimate x_p accurately (Haas, 2005, 7). By transforming the data points one can try to achieve a more robust procedure. We section the data into s blocks of $r = N/s$ data points and of those we take the so-called *block maxima*:

$$M_j = \max\big[X_i : (j-1)r < i \le jr\big] \tag{3.204}$$

with $j = 1, 2, \ldots, s$. Since the distribution of the maximum of r i.i.d. random variables may be computed as

$$Pr(M_j < x) = Pr(X_1 < x, X_2 < x, \ldots, X_r < x) = F_X^r(x) \tag{3.205}$$

it follows that $F_M(x_p) = F_X^r(x_p) = p^r$. This implies that estimating for F_M the p^r-quantile is equivalent to estimating the p-quantile from F_X. With $r = 50$ we have the confidence value $p^r = 0.605$ probability.

This conditioning of the data comes at the cost of searching in addition m-times the matching data sets and find the maximum.

3.12 Results of the Run-Through Example

We perform a simulation run with 1,000,000 draws. The high number of simulations is necessary because using a confidence limit of 99.9% strains the stabilty of results.

In a first step we check whether the empirical distribution of the losses is plausible. Especially the tail is of interest. This aspect is best analysed with the so-called empirical *mean excess function* $e_n(t)$:

$$e_n(t) = \frac{\sum_{i=1}^{n}(x_i - t)J_{\{x_i>t\}}}{\sum_{i=1}^{n} J_{\{x_i>t\}}}. \tag{3.206}$$

Again, $J_{\{a>b\}}$ is an indicator variable and for the threshold t we require $x_{1:n} \leq t < x_{n:n}$ for the ordered set of sample points x_i. For a heavy-tailed distribution this function has a positive slope with increasing threshold t. The analytical function for the Pareto distribution has been developed on page 116.

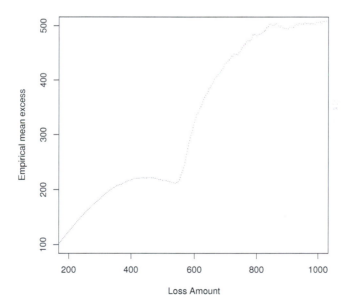

FIGURE 3.39: Empirical mean excess function of the gross total loss.

3.12.1   The Economic Capital

The Economic Capital EC is determined as a quantile from the empirical cumulated distribution function of the simulation. We show this figure both net and gross of insurance mitigation. The values are net 665.9m and 694.4 gross. The mitigation of 4.1% stems from the five insurance contracts. With our assumptions the ratio between the average loss of 87.6m and the net EC is 7.6.

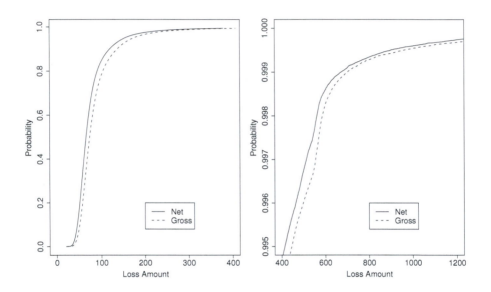

FIGURE 3.40: Total loss.

The allocation to the objectives "events" and "Lines of Business" is shown in Table 3.28 and Table 3.29. We also calculate what we shall call "leverage," i.e., the ratio between Capital at Risk (CaR) and the Expected Loss (the average).

From our very assumptions it is evident that most of the risk resides with the Extreme Event. On the other hand the "Physical" event with its overall average loss of 73m by occurrence bears a major chunk. For the other events the EC is approximately the same as the expected loss.

Because here we map these events to lines of business, the situation changes. Besides "Corporate" and "Trading" all LoBs have leverages of approx 10, that is a factor of 1.3 compared to the average. So these are relatively riskier than the two mentioned above. Commercial Banking carries the greatest risk, both because it has a high Expected Loss and a high leverage. Retail Banking, on the other hand, has a very high Expected Loss but low risk, as

TABLE 3.28: Economic capital by events

	CaR	Share	Expected loss	Leverage
Internal	5.47	0.821	5.25	1.04
External	13.3	2	13.2	1.01
Employment	5.88	0.884	5.8	1.01
Clients	9.52	1.43	9.1	1.05
Physical	230	34.6	9.2	25.00
Disruption	2.29	0.344	1.9	1.21
Execution	24.9	3.74	23.7	1.05
Extreme	374	56.2	10	37.40
Total	665.9	100	78.3	

shown by leverage. The well-known conclusion is that the Expected Loss is not necessarily a good proxy for risk.

These results rely on the mapping assumptions as it is possible to match the Expected Loss with other weights and create different leverages. In our example, Retail Banking and Commercial Banking need the most capital for their business, while Asset Management requires the least.

TABLE 3.29: Economic capital by lines of business

	CaR	Share	Expected loss	Leverage
Corporate	33.6	5.04	3.5	9.60
Trading	99.6	15	13.5	7.38
Retail	163	24.5	26.8	6.08
Commercial	188	28.2	18	10.44
Payment	37	5.56	3.1	11.94
Services	30.3	4.55	3.4	8.91
Asset	33	4.95	3.2	10.31
Brokerage	81.3	12.2	6.8	11.96
Total	665.9	100	78.3	

In the following example we allocate 3.32 the CaR to some other objectives, viz. countries.

Example 3.32. From our example we have the CaR by events (see Table 3.28), written as vector

$$\mathbf{k}_e = (5.47, 13.3, 5.88, 9.52, 230, 2.29, 24.9, 374)^T. \tag{3.207}$$

We want to allocate capital at risk by countries. As it is easier to think in lines of business categories, our first step is to generate the matching vector

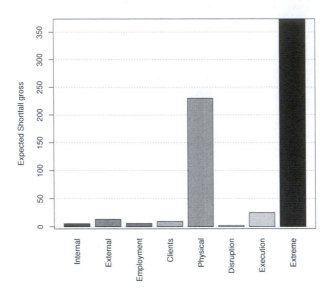

FIGURE 3.41: Gross CaR allocation.

$\mathbf{k}_l = \mathbf{Q}_{l,e}\mathbf{k}_e$ and then generate from this $\mathbf{k}_c = \mathbf{Q}_{c,l}\mathbf{k}_l = \mathbf{Q}_{c,l}\mathbf{Q}_{l,e}\mathbf{k}_e$. Now let us assume four countries, i.e., Switzerland, UK, Singapore and the USA. We have to estimate how much of our business and risks originating in our lines of business – Corporate, Trading, Retail, etc. – are done in these areas. Again the sum of the column elements must add to 1.

$$\mathbf{Q}_{c,l} = \begin{pmatrix} 0.5 & 0.2 & 0.8 & 0.55 & 0.2 & 0.25 & 0.7 & 0.25 \\ 0.15 & 0.25 & 0.05 & 0.15 & 0.25 & 0.25 & 0.1 & 0.3 \\ 0.1 & 0.25 & 0.05 & 0.1 & 0.3 & 0.25 & 0.15 & 0.1 \\ 0.25 & 0.3 & 0.1 & 0.2 & 0.25 & 0.25 & 0.05 & 0.35 \end{pmatrix}.$$

The matrix $\mathbf{Q}_{l,e}$ is from Table 3.27 on page 215. The resulting vector, in tabular form, is as follows:

Switzerland	329.7
UK	110.6
Singapore	86.0
USA	139.0
Total	665.4

This allocation is consistent as we are basing our risk measure on Expected Shortfall.   △

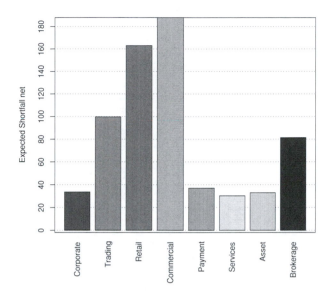

FIGURE 3.42: Net CaR by LoB.

3.12.2 Insurance Effect

In order to assess the mitigation effect of the individual insurance policies we must look at the gross situation while splitting the gross losses by events into net loss by event and claims paid by insurance. Again, the allocation method is Expected Shortfall (ES), the conditional expectation beyond a given threshold. The threshold for ES has been chosen such that ES of the total comes as close as possible to the quantile. The tiny discrepancy is then removed with the capital multiplier CM of approx. 1, i.e., $CM = VaR/ES(total)$.

From Table 3.30 above we see that the Financial Institution blended cover removes some 21m of risk capital. The question is, what is the ratio between risk mitigation and premium? This analysis yields Table 3.31: As already mentioned, the mitigation is in the order of 4%. But it seems that this effect comes at quite a high price, because the leverage here is consistently lower than for lines of business. This can be further demonstrated if we equate the internal capital model with the external. Internally we are charged the cost of capital (CoC) for risk capital needed, and compare this amount to the insurance premium: CoC × CaR = premium. We rearrange this for the CoC in the break-even assumption to yield

$$\text{CoC}_{\text{break-even}} = \frac{\text{premium}}{\text{CaR}}. \tag{3.208}$$

This is actually the reciprocal of the leverage. These values seem higher than

TABLE 3.30: Allocation gross/net

	CaR	Share
Internal	5.56	0.801
External	13.3	1.92
Employment	5.89	0.848
Clients	9.52	1.37
Physical	230	33.2
Disruption	2.31	0.332
Execution	24.9	3.59
Extreme	367	52.8
FI	21.4	3.08
Terrorism	1.07	0.154
Liability	5.02	0.722
D&O	7.04	1.01
Property	1.45	0.209
Total	694.4	100

TABLE 3.31: Insurance effect

	Theoretical Premium	CaR	Leverage	CoC(be)
FI blended cover	10.3	21.4	2.08	48.1%
Terrorism	0.711	1.07	1.50	66.4%
Liability	1.9	5.02	2.64	37.8%
D&O	1.87	7.04	3.76	26.6%
Property	0.744	1.45	1.95	51.3%

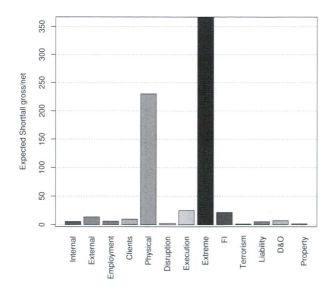

FIGURE 3.43: CaR gross/net.

what most institutions would charge internally, say 25%. But the D&O insurance is quite close. Therefore, it seems that insurance should be able to compete with self-insurance or retention with the institution.

Why is this so? Because of the difference between limit and deductible. Insurance is efficient if the deductible is very high. This means that most of the time the policy will pay no claims, but must step up to pay in case of very high losses. This may pose a psychological problem for the insurance manager.

It also seems evident that for the actual purchase of insurance it is virtually impossible to reach the 20% mitigation limit imposed by the Basel regulation. The banking and insurance industry started lobbying for a larger mitigation share before they were able to assess whether it was sensible.

3.12.3 Comparison with the Standardised Approach

In Example 1.5 on page 38 we have calculated the capital charge for the QIS data to be 1037.7m EUR. Recall that it is the sum of gross income by lines of business times given factors. Our calculation with the Loss Distribution Approach yields a gross capital charge – before insurance mitigation – of 694.4m EUR and net of insurance of 665.9m EUR. In percentage points this is 66.9% and 64.2% respectively. This is completely plausible, but the reduction might be too large for the regulators. Nonetheless, having produced a lower number than the standardised approach is an encouraging sign.

In an early publication BCBS (2001c, 18) it was proposed to require for the Advanced Measurement Approach (AMA) a floor equal to 75 percent of the Standardised Approach capital charge for operational risk.

3.13   Summary and Conclusion

In this chapter about the modelling of one possible implementation of the Advanced Measurement Approach we have described several things. Firstly, we have touched on statistical tools like Monte Carlo simulation, fitting data to a set of comprehensive distributions, assessing correlations and generating dependent random variables. Secondly, we have described data modelling with three classes of loss generators, i.e., attritional, individual and catastrophic and the mechanisms to map events, lines of business and insurance to each other. A third focus was on risk measures, their properties and how to assess related economic capital. Fourth, we have illustrated our points with examples, whose purpose is to demonstrate that the concepts presented are effective. As he capital requirement from the model turn out to equal two thirds of the Standardise Approach, it looks very promising.

To go beyond a working model we have established that the extent of the insurance mitigation may generally be over-stated, as our example with a very generous insurance programme did not reduce the required capital by more than 5%. This does not approach the permitted maximum of 20%.

Chapter 4

Managing Operational Risk

This chapter focuses on managing operational risk using the Advanced Measurement approach. In the preceding chapter we devised a quantitative method for calculating the risk capital for operational risks based on historical data. At first sight this may have seemed irrelevant as the past is just that–past. But many regulators and managers feel it necessary to relate management to actual quantification, and the past lends itself most obviously to quantification. Ideally, we could transform the actualised past information into a closed-loop feedback system. However, as that is generally impossible, the next best option is to revitalize quantification, making it more agile and sensitive to input besides historical data. This can be achieved by quantifying the "management system" itself and relating it to the quantification of the risk capital. This divorce of quantification from the passage of time is very desirable from a motivational perspective, as unit managers implementing measures for reducing operational risk need not wait for a certain period to elapse before a pattern manifests itself.

For this purpose we must describe the "management system" and how it lends itself to quantification. We favour the construction of compound indices or indicators which could be called *qualitative management factors*. Therefore, such indices will be the second focus of this chapter. The third focus is on scenarios, as they are widely believed to be useful for risk management. Lastly, we will go into some detail concerning the structure of the insurance programme, the most immediate and obvious management tool in the LDA context.

4.1 Management and Organisation

4.1.1 Management System

In order to discuss managing operational risk we have to understand what management means in its broader context, and its constituent parts. Many books on risk management pay all too little attention to management.

A management system, or "management," includes the following:

- general management principles,

- organisational structure and processes,

- goal system,

- planning system,

- control system,

- information system,

- incentive and motivation system,

- human resource development system.

All these items are interdependent. They are grounded in economic, ecological, social and technological principles. The Basel Accord implicitly defines operations with *internal processes, people* and *systems*. While processes and systems with respect to operational risk management are described in the Accord, in a service environment like banking, people and their management deserve a closer look. In business, after all, all theory must take a back seat to the reality created by those who design management systems based at least in part on their intuitive understanding of situations.

Management is a self-similar issue applicable to different levels of any firm. Therefore, these items are valid both in general and in operational risk management.

4.1.1.1 Conception of Man

"No society can exist without an implicit conception of what people are like" (Herzberg, 1966, 13). Such conceptions are culturally determined and pertain to the society as a whole, or may vary individually. That they have a great influence on how an organisation is conceived is obvious. But it has become evident that a leadership style, while grounded on such conceptions, cannot divorce itself from the context in which it operates.

Older, normative, concepts include, e.g., Old Testament's (Leviticus 19:18): "Love your neighbour as yourself," New Testament's (Matthew 7,12) "Therefore all things whatsoever ye would that men should do to you, do ye even so to them: for this is the law and the prophets," Confucius' golden rule: "Do not to others what you do not want done to yourself," Immanuel Kant's *categorical imperative*: "Act only according to that maxim by which you can at the same time will that it should become a universal law," Jeremy Bentham's *greatest happiness principle*: "the correct action in any situation is that which brings the most happiness to the most people."

Similarly, over many centuries in the Western culture there were seven *cardinal virtues*.[1] Four of them stem from Greek culture, or more precisely from Plato, and the other three ("Divine virtues") from Christian culture as

---

[1]There are also the seven deadly sins, i.e., pride, covetousness, lust, envy, gluttony, anger and sloth.

interpreted by Saint Thomas Aquinas. Again, these prescribe human conduct. They read:

1. prudence,

2. justice,

3. temperance,

4. fortitude,

5. faith,

6. hope,

7. charity.

Today it is quite difficult to convey to others that you are a virtuous and decent individual. Learning an ethical code – and then taking an oath to abide by it – can, ironically, lead to unethical actions. All too often people permit themselves to believe that whatever is not expressly forbidden is permitted. Of course, it is impossible to develop a comprehensive ethical code which will provide guidance for every single ethical dilemma.

More recently, the positive conception, at least from economists and businessmen, has changed from the so-called *economic man*, to the *social man* and the *self-actualizing* to the *complex man*. According to Schein (1980), the characteristics of these four ideas of man are:

- The rational-economic man is primarily driven by monetary incentives. He has a passive attitude and is led by the organisation. His feelings are described as irrational;

- The social man is motivated by social needs and acknowledgement within his group. This is seen as a result of the specialisation and work-sharing and must be respected by the manager;

- The self-actualizing man is mostly self-motivated. His needs can be hierarchically ordered and he aspires to independence. There is not necessarily a conflict between self-actualisation and organisational aims;

- The complex man is adaptive and flexible. His motives can change suddenly, and may vary in different situations.

In a famous paper Jensen and Meckling (1998) describe humans as follows:

> The growing body of social science research on human behavior has a common message: Whether they are politicians, managers, academics, professionals, philanthropists, or factory workers, individuals are resourceful, evaluative maximizers. They respond creatively to the opportunities the environment presents, and they work to loosen

constraints that prevent them from doing what they wish. They care about not only money, but about almost everything – respect, honor, power, love, and the welfare of others.

As conditions have changed over time, so have employees' needs and motivations. To understand and support a complex employee requires a well thought-out organisation and a mature leadership style.

Example 4.1. Gigerenzer (2007, 1) gives a nice anecdote to introduce the *intelligence of the unconscious* or the importance and efficiency of gut feelings which often transcends the analytical-rational decision making.

> A professor from Columbia University was struggling over whether to accept an offer from a rival university or to stay. His colleague took him aside and said, "just maximize your expected utility – you always write about doing this." Exasperated, the professor responded, "Come on, this is serious."

The careful weighting of utilities by their probabilities does not describe how actual people reason. △

4.1.1.2 Leadership Theory

Leaders are born and not made. This is the core assumption of this simple and androcentric *great man theory*. It implies that leadership is generated by prominent traits of the leader and that it is *independent* of the task, situation or working group. Developed by army psychologists, this theory plays into the nature versus nurture debate.

A less crude approach is to look at a set of personality characteristics. People are born with inherited traits, some of which are particularly suited to leadership. Table 4.1 presents typical leadership characteristics of early research (Stogdill, 1974). Such theories flatter top-managers and not surprisingly are often embraced by them. Whether these traits are necessary or sufficient or neither has not been, and probably cannot be, scientifically proven. Does one need those skills to get the "top job" or does one develop these skills upon acquiring the position? Does a leader only influence the team, without being influenced by the team? In short, such traits have not been shown to produce consistently positive outcomes for organisations. Moreover, if this "great man" theory were true, there would be no point in attempting to develop or create leaders. The only sensible thing to do would be to observe employees and applicants and select the natural leaders among them.

Transformational leadership is defined as a process whereby leaders and followers engage in a mutual process of "raising one another to higher levels of morality and motivation" (Burns, 1978). In so doing, individuals model values and use charisma to attract others to the values and the person modelling them. Followers often attribute heroic or extraordinary leadership abilities

TABLE 4.1: Leadership traits and skills. (From Yukl, 2006, 183, ©2006. Printed and Electronically reproduced by permission of Pearson Education, Inc., Upper Saddle River, New Jersey.)

Traits	Skills
adaptable to situations	intelligent
alert to social environment	conceptually skilled
ambitious and achievement-orientated	creative
assertive	diplomatic and tactful
cooperative	fluent in speaking
decisive	knowledgeable about group task
dependable	organised (administrative ability)
dominant (desire to influence others)	persuasive
energetic (high activity level)	socially skilled
persistent	
self-confident	
tolerant of stress	
willing to assume responsibility	

to charismatic leaders who are sensitive to their environment, and to their follower needs, who take risks confidently and sometimes indulge in unconventional behaviour. Such leaders ignore rules, are beyond the influence of colleagues, act on whims, ignore the contribution of their context to their success and may in the process create deliberate and damaging destabilisation to the organisation.

A study (McFarlin and Sweeney, 2002) suggests that there are more narcissistic leaders whose obsessive egotism and self-absorption fuel the worst excesses of leadership. Open immorality is the most visible indication of power.[2] Social psychological studies (Gruenfeld, 2006) seem to show that power leads to disinhibition, i.e., acting on his or her own desires in a social context without considering the effects of the action. Power heightens sensitivity to one's own internal states while reducing sensitivity to others' interests and may lead to using others as a means to the power holder's ends. This kind of leader is hardly concerned with what people think of them and their behaviour is less aware of what others think of them in general.

Path-Goal theory of leadership formulated by Robert House proposes that the leader can affect the performance, motivation, and satisfaction of a group in different ways (House, 1971):

- Offering rewards for achieving performance goals,

- Clarifying paths towards these goals,

---

[2] "Die offene Unmoral ist das sichtbarste Zeichen der Macht" (Noll and Bachmann, 1999).

- Removing obstacles to performance.

A person may perform these by adopting a certain leadership style, based on the situation:

- Directive leadership: Specific advice is imparted to the group and basic rules and structure are established, e.g., clarifying expectations, specifying and assigning specific tasks.

- Supportive leadership: Good relations are sought with the group and sensitivity to subordinates' needs is shown.

- Participative leadership: Decision making is based on consultation and information sharing with the group.

- Achievement-oriented leadership: Ambitious goals are set and high performance is encouraged while confiding in the group's ability.

Supportive behaviour increases the group's satisfaction, especially in stressful situations; directive behaviour on the other hand is suited to uncertain and ambiguous situations. It is also purported that leaders who can influence their superiors can increase group satisfaction and performance.

4.1.1.3 Motivation Theory

Work is the major means by which individuals develop and maintain their identity and self-esteem. Our mental self-image depends to a large extent on what other people communicate to us about ourselves, not just verbally, but also in work-related messages: salary, rank, job assignments, status in the organisation and working conditions.

Motivation is the ensemble of forces within an individual – willingness, desire, compulsion – that account for the level, direction, and persistence of effort used at work. Motivation theories, and the importance of motivation, are underlying themes in modern management thinking. Understanding what motivates an employee enables a leader to enhance the firm's performance while increasing the employee's satisfaction. Discontent, on the other hand, poses a risk for an institution. At the peril of oversimplification, most theories have at their core the old *carrot and stick* metaphor.

Abraham Maslow posited a hierarchy of human needs based on two groupings: *deficiency needs* and *growth needs* (Maslow, 1943). Within the deficiency needs, each lower need must be met before moving to the next higher level. Upon detection of a deficiency, the individual will act to remove it. The five levels, four for deficiency and one for growth, of the pyramid are:

1. Physiological: hunger, thirst, bodily comforts,
2. Safety/security: out of danger,
3. Belonging and love: affiliate with others, be accepted,

4. Esteem: to achieve, be competent, gain recognition and

5. Self-actualization: to find self-fulfilment and realize one's potential.

Maslow's later conceptualisation included three more growth factors, i.e., cognitive and aesthetic needs and finally, self-transcendence.

FIGURE 4.1: Maslow's hierarchy of needs.

In 1960 Douglas McGregor formulated two models which he called *Theory X* and *Theory Y* (Mcgregor, 1960). Theory X assumes that employees inherently dislike working and try to avoid it if possible. Therefore, they have to be forced or controlled by management and threatened so that they work hard enough. The average employee wants to be directed, because he or she does not like responsibility. This is quite a negative conception of men.

The opposing Theory Y's assumptions about workers are:

- People view working as being as natural as play and rest. Humans spend similar amounts of physical and mental effort in work as in private lives;

- Provided people are motivated, they will be self-directing to the objectives of the organisation. Control and punishment are not the only means to make people work;

- Satisfaction is key to recruit employees and ensuring their commitment;

- Average humans, under the proper conditions, will not only endorse but even seek responsibility;

- Employees are imaginative and creative, their ingenuity should be used for solving problems.

The new organisation lies or should lie firmly at the Theory Y end of the spectrum. Alderfer (1969) distinguishes three categories of human needs that influence workers' behaviour:

- **E**xistence: physiological and safety needs;

- **R**elatedness: social and external esteem, e.g., involvement with family, friends, workmates and employers;

- **G**rowth: internal esteem and self-actualization, e.g., the desire to be creative, productive and to fulfil meaningful tasks.

Contrary to Maslow's idea that progress to the higher levels of his pyramid required prior satisfaction in the lower level needs, Alderfer assumes that the three ERG areas are not traversed in any specific way. ERG Theory recognises that the importance of the three areas may vary for each individual. Managers must recognise that an employee has multiple needs to satisfy simultaneously. According to this theory, focusing exclusively on one need at a time will not effectively motivate anybody. If a higher level need remains unfulfilled, the person may resort to lower level needs that appear easier to satisfy. For example, if advancement opportunities are not provided, employees may regress to relatedness needs and socialise more with colleagues. If management can recognise these conditions early, measures can be taken to satisfy the thwarted needs until the employee is able to pursue progress again.

Frederick Herzberg's motivation theory has two parts (Herzberg, 1959). Firstly, the *hygiene theory* whose factors do not lead to higher levels of motivation, but without which there is dissatisfaction. They include, e.g., working conditions, interpersonal relations, remuneration, status and security. The second part of the motivation theory centres on what people actually do on the job. The so-called *motivators* are achievement, recognition, growth and advancement, responsibility for enlarged task and interest in the job. These factors arise from within the employees themselves, yielding motivation rather than movement. These approaches (hygiene and motivation) must be simultaneous. Treat people as best you can so they feel a minimum of dissatisfaction. Utilize people's skills and gifts so they feel a sense of achievement, receive recognition for that achievement, feel interest and assume responsibility, so that they can grow and progress in their work.

4.1.1.4 Management Styles

Management style is defined as a temporarily stable but situational, relatively consistent leadership behaviour of a superior towards his or her subordinates. There are many theories of management style. Here we touch upon a representative selection, from the very simple to the more complex. With the University of Iowa studies Kurt Lewin explored three leadership styles experimentally (Lewin et al., 1939):

- autocratic,

- democratic and

- laissez-faire.

The performance of these styles with respect to group performance, social integration and general contentment are mixed and it seems that there was a bias (by the experimenter) favouring the democratic style. Nonetheless, the democratic or co-operative style is purported to be superior.

A more differentiated approach is the so-called *LPC Contingency Theory*. Here contingency underscores the leadership style's dependence on situations or events, leading to an amalgam of leadership techniques. This theory claims that leaders choose between task-focus and people-focus. This two-dimensional focus is mapped by the so-called "Least Preferred Co-Worker scoring." It is determined by asking employees to think of the individual he would least like to work with again and then to score the person on a range of scales between positive factors and negative factors. A high LPC leader usually scores the other person as positive and a low LPC leader scores him or her as negative.

Typically, high LPC leaders tend to have close and positive relationships, act in a supportive way and grow bored with details. Low LPC leaders put the task first, are hard on those who fail and are obsessed with details.

Effective leadership is determined by three situational factors:

1. Leader's relation to group members,

2. level of structure in the task being performed and

3. leader's position within the organization.

Dichotomising these items, say by "high" and "low," there result $2^3 = 8$ distinct situations. To each of these there is an attribution either to a high LPC or a low LPC leader. Generally, a high LPC approach is ideal when leader-member relations are poor, unless the task is unstructured and the leading person is weak. In such circumstances a low LPC style is better.

This approach seeks to identify the underlying *beliefs about people*, in particular whether the leader sees others in a positive or negative light. The theory is from Fiedler (1967).

Hersey and Blanchard (1996) determine the leadership style with regard to two variables, viz. relationship behaviour (or amount of support required) and task behaviour (amount of guidance required). Again, both variables can be "high" or "low." In addition, the employee's so-called "maturity score," determined by a questionnaire, must be taken into account. In the case of the employee, limited willingness and limited ability to do a job is "telling;" limited willingness and high ability is "participating"; high motivation and limited capability is "selling" and high motivation and high capability is called "delegating."

Underlying the theories above is the assumption that there is a best way to manage people. At present we are witnessing the so-called *knowledge worker* in a networked environment. Drucker (1999, 18) defines this new class of individuals as those who consider themselves not "employees" or "subordinates" but "partners." They need to know more about their job than their boss, even more than anybody else. The boss has not previously held their position. Knowledge workers must take the responsibility to educate their superior while taking directions from them. They have to be managed as if they were volunteers with an emphasis on managing for performance.

To summarize, management is a multi-faceted topic with a tremendous impact on the performance and on the risk profile of the bank. It seems to be an invariable fact that 60 to 75 percent of employees report that their immediate superior is the worst or most stressful aspect of their work. The quality of management must be a variable in the operational risk management quantification. What evidence on management indicates, according to Pfeffer and Sutton (2006, 17) is that:

> It seems clear that leaders have some chance of making things somewhat better, but they can also make things much worse by taking actions that increase employee turnover and diminish employee motivation, as well as encourage lying and stealing and by causing numerous other organizational problems. This suggests that avoiding bad leaders may be a crucial goal, perhaps more important than getting great leaders.

4.1.2   Organisational Structures

Structure and process are siblings. Both are relevant aspects of reality and thus of organisations. The Greek philosophers were more categorical. Democritus pictured all of nature as composed of a stable material substance or atoms – particles that vary only in their position in space and time. Qualitative changes or growth are caused by new arrangements or additions to the atoms composing them. In contrast, Heraclitus, viewed reality not as a constellation of things, but as one of processes. "All is flux, nothing stands still. Nothing endures but change. You could not step twice into the same river; for other waters are ever flowing on to you."

At the beginning of the 20th century, the focus was on the efficiency of the work process, while people and their needs were considered as secondary. This period has three schools of thinking: *Bureaucratic management*, which centres on rules and procedures, hierarchy and a clear division of labour; *Scientific management*, which looks at "the best way" to do a job; and *Administrative management*, which stresses the flow of information within the organisation.

Max Weber, a founder of modern sociology, used the term *bureaucracy* to describe a particular, and in his view superior, organisational form. He defined the key elements of a bureaucracy as (Weber, 1922, 128):

- a well-defined hierarchy with a clear chain of command where higher ranks have the authority to control the lower positions,

- division of labour and specialisation of expertise, where each employee has the necessary expertise and authority to complete a particular task,

- complete and accurate rules and regulations to govern all activities, decisions and situations,

- impersonal relationships between managers and employees with explicit statements of the rights and duties of staff,

- technical competence as the basis for all decisions regarding engagement, selection and promotion.

The management of the office is founded on written documents ('the files'), which are kept in their original or draft form.

There are a number of weaknesses in the bureaucratic organisation. For example, the strong emphasis on authority hinders innovation and creativity, the strict rules make the organisation inflexible and unresponsive to change, the focus on impersonality and division of labour yields boredom and discontent of employees. Furthermore, procedures and rules may become so important in themselves that the underlying processes they are meant to advance are forgotten.

Incidentally, the idea that the bureaucratic activities of the state are intrinsically different from the management of private businesses is a Continental European conception, and totally foreign to the American perspective.

Frederick W. Taylor emphasised empirical research – time and motion studies – to increase organisational productivity by augmenting the efficiency of the production process (Taylor, 1911). Taylor's scientific management theory posits that jobs should be designed in such a way that each employee has a well-specified, well-controlled task. Specific procedures and methods for each task must be strictly followed.

This management theory rests on the fundamental belief that managers are not only intellectually superior to the average employee, but that they have an explicit duty to control staff and organise their work activities. Thus, management theory was only applied to routine and repetitive tasks that could be controlled at supervisory level. Its influence can be sensed in today's financial control and costing theories.

Henri Fayol was a French industrialist and a very influential early management thinker. In contrast to Taylor, he was more concerned with the top management and less with the blue-collar workforce. Based on his experience in management, he developed a dozen general principles of management (Fayol, 1925). Besides others we cite just six of them:

- specialisation/division of labour,

- authority with corresponding responsibility,

- unity of command: an employee receives orders from only one superior,

- unity of direction: there is only one central authority and one plan of action,

- initiative is encouraged to motivate employees,

- esprit de corps is important and teamwork is encouraged.

These three models constitute the classical school. The next approach is one example of the *Human Relations Models*.

By the 1920s, it was becoming evident that the major deficiency of classical management theory was its inability to deal with the complexity of real people working in organisations. The Human Relations School (or Behaviourist School) emerged at this time to focus on the human aspects of organisations. Followers of the *Human Relations school* believe that a co-operative work environment in which the needs and values of the employees are predominant should be encouraged by democratic consultation among those who manage.

In the 1920s and 30s Elton Mayo carried out important studies at the Hawthorne plant of the Western Electric Company in Chicago (Taylor, 1933). The Hawthorne Experiments consist of two studies that attempted to determine the effect, in turn, of various principles of scientific management on worker productivity. The results of the studies showed that productivity increased significantly irrespective of which principle was being applied. It was concluded that the principal reason for this increase in productivity was that the workers were motivated by the researchers' acute attention during the course of the study. Essentially, the experimenters were seen as a part of the management team and influenced the experiment's outcome.

The Behaviourist school lasted until the 1950s, when actual thinking shifted to more complex models that do not offer a single theory, but recognise that management practice, systems and cultures are highly variable between organisations.

The *Systems Approach* emerged during the 1960s, originating from the field of biology as a reaction to reductionism. Advocates of this approach view the organisation as a dynamic entity that encompasses many smaller sub-systems.

Johnson et al. describe the systems concept as follows (Reprinted by permission. Copyright 1964 INFORMS. Johnson et al., 1964, the Institute for Operations Research and the Management Sciences, Hanover, Maryland):

> The system concept can be a useful way of thinking about the job of managing. It provides a framework for visualizing internal and external environmental factors as an integrated whole. It allows recognition of the proper place and function of subsystems. The systems within which businessmen must operate are necessarily complex, but management, via systems concepts, can help reduce some of the complexity whilst allowing the manager to recognise the nature of the complex problems and thereby operate within the perceived environment. It is important to recognize the integrated nature of specific systems, including the fact that each system has both inputs and outputs and can be viewed as a self-contained unit. But it is also important to recognize that business systems are a part of larger sys-

tems: industry-wide, or even including several industries, or society as a whole. Further, business systems are in constant state of flux – they are created, operated, revised and often eliminated. What does the concept offer students of management and/or practising executives? (...) Fundamental ideas, such as the systems concept, are more difficult to comprehend and yet they present a greater opportunity for a large-scale pay-off.

The survival of the organisation depends upon the continuous adaptation of interactions between systems in order to meet the requirements of changing *internal* and *external* environmental processes. Here, management is understood as the process of maintaining effective relationships among the sub-systems of the organisation. Although the comparison of a firm to an organism is striking, conclusions for effective management are not always evident.

The concept of a system was a key prerequisite for the further development of *cybernetics*. It is within the study of systems that *feedback control theory* has its place in the organisation of human knowledge. Feedback control (see Figure 4.2) is the basic mechanism by which systems maintain their equilibrium or homeostasis. Feedback control may be defined as the use of difference signals or information, determined by comparing the actual values of system variables to their desired values, as a means of controlling a system (see Figure 4.3). *Control* may be defined as compelling events to conform to plans.[3] Such a *control process* has four major cyclical steps, the first pertaining to *planning*, the latter to *controlling*:

1. define target performance,

2. measure the actual level of performance,

3. compare actual and standard performance and communicate,

4. take corrective action,

5. go to step 2.

Now it is evident why it is said that planning without controlling is useless and controlling without planning is impossible. The general feedback model distinguished between the controller issuing the appropriate signals to the "plant," i.e., the unit to be controlled. The latter may be a technical system like a car or, in our context, the model of a bank. The signals may suffer interference from environmental noise. The state of the plant is continuously fed back and taken to assess the difference between actual and target values.

Controllability and *observability* are the main issues in the analysis of a

---

[3]Control can also be understood as any process in which a person, a group or an organisation determines or intentionally influences, what another person will do. Control in this sense does not require either a standard to be established or feedback information on whether the standard is achieved. It is akin to instructions, commands and orders.

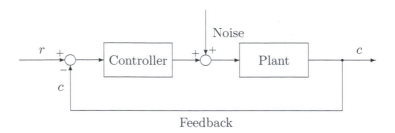

FIGURE 4.2: General feedback model.

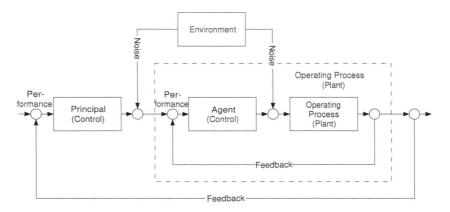

FIGURE 4.3: Basic elements of a management control system.

system before deciding which control strategy to to apply. Controllability is related to the possibility of coercing the system into a particular state by using an appropriate control input. If a state is not controllable, then no control signal will ever be able to force the system to reach a level of controllability. Observability is related to the possibility of "observing" the state of a system through output measurements. If a state is not observable, the controller will never be able to correct the behaviour. In such cases, only the addition of actuators and sensors can eventually create controllable and/or observable states and make eventual positive outcomes possible.

Example 4.2. Adam Smith made systematic use of feedback control theory in his economic arguments (Mayr, 1971). One such example is his description of how demand and supply are brought to equilibrium. Figure 4.4 shows (using a modern block diagram) the complete thought. He defines the natural price n as the cost of all input factors (Smith, 1991, 48):

> When the price of any commodity is neither more nor less than what is sufficient to pay the rent of the land, the wages of the labour and

the profits of the stock employed in raising, preparing and bringing it to market, according to their natural rates, the commodity is then sold for what may be called its natural price. (...)

The actual price at which any commodity is commonly sold is called its market price. It may either be above, or below, or exactly the same with its natural price.

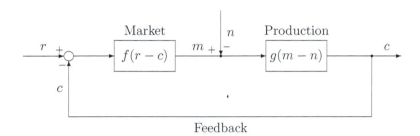

FIGURE 4.4: Adam Smith's theory on demand and supply.

Then he argues that the market price m is a function $f(r-c)$ of the demand r and the supply c (Smith, 1991, 49):

> The market price of every particular commodity is regulated by the proportion between the quantity which is actually brought to market and the demand of those who are willing to pay the natural price (...) Such people may be called the effectual demanders and their demand the effectual demand(...)

Now he explains how the market price moves depending on whether the demand is above or below the supply (Smith, 1991, 49).

> When the quantity of any commodity which is brought to market falls short of the effectual demand, all those who are willing to pay the whole value of the rent, wages, and profit, which must be paid in order to bring it thither, cannot be supplied with the quantity which they want. Rather than want it altogether, some of them will be willing to give more. A competition will immediately begin among them, and the market price will rise more or less above the natural price, according as either the greatness of the deficiency, or the wealth and wanton luxury of the competitors, happen to animate more or less the eagerness of the competition.

Then he analyses the impact of the difference between market price m and total cost n, i.e., "natural prices," on the production of the commodity c as function $g(m-n)$ (Smith, 1991, 50).

> If at any time it exceeds the effectual demand, some of the component parts of its price must be paid below their natural rate. If it is rent, the interest of the landlords will immediately prompt them to withdraw a part of their land; and if it is wages or profit, the interest of the labourers in the one case, and of their employers in the other, will prompt them to withdraw a part of their labour or stock from this employment. The quantity brought to market will soon be no more than sufficient to supply the effectual demand. All the different parts of its price will rise to their natural rate, and the whole price to its natural price.
>
> If, on the contrary, the quantity brought to market should at any time fall short of the effectual demand, some of the component parts of its price must rise above their natural rate. If it is rent, the interest of all other landlords will naturally prompt them to prepare more land for the raising of this commodity (...) The quantity brought thither will soon be sufficient to supply the effectual demand. All the different parts of its price will soon sink to their natural rate, and the whole price to its natural price.

He then concludes that the actual total cost will be the equilibrium price which clears the market – some sort of attractor. He also concedes that the input prices n may vary by accident – thus implying some stochastic disturbance to the equilibrium (Smith, 1991, 51).

> The natural price, therefore, is, as it were, the central price, to which the prices of all commodities are continually gravitating. Different accidents may sometimes keep them suspended a good deal above it, and sometimes force them down even somewhat below it. But whatever may be the obstacles which hinder them from settling in this center of repose and continuance, they are constantly tending towards it.
>
> The whole quantity of industry annually employed in order to bring any commodity to market, naturally suits itself in this manner to the effectual demand. It naturally aims at bringing always that precise quantity thither which may be sufficient to supply, and no more than supply, that demand.

This closed-loop thinking was very innovative and led to the *laissez-faire* doctrine, as it was believed that equilibrium tends to happen on its own without any outside intervention. △

Chaos in *complex systems* is not the opposite to determinism but to predictability, i.e., the notion that every event or action is an inevitable and predictable result of preceding events or actions. For decades, managers have behaved as though organisational events can always be controlled. Today, chaos theory recognises that events are rarely controlled. Many chaos theorists refer to biological systems when explaining their theory. They posit that systems naturally tend to greater complexity, and by doing so, these systems become more volatile and more susceptible to events. In addition, they must use more energy to maintain their complexity. As they do so, they seek more structure to remain stable. This process continues until the system breaks up, combines with another complex system or falls apart entirely.

Contingency theory emerged in the 1960s (Lawrence and Lorsch, 1967). As with systems theory, it does not impose the application of specific management methods. Instead it provides a *framework* for managers to develop the most suitable organisational design and management style in a given situation, where design and style depend upon combinations of variables such as the external environment, economic and technological factors and social factors like human skills and motivation. This theory recognises that there cannot be any fixed management methodology because all of these variables are in steady flux. Therefore, management's flexibility and adaptability is key.

Another radical approach, also focussed on group decision-making is the so-called *garbage can model* (Cohen et al., 1972):

> Organized anarchies are organizations characterized by problematic preferences, unclear technology, and fluid participation. Recent studies of universities, a familiar form of organized anarchy, suggest that such organizations can be viewed for some purposes as collections of choices looking for problems, issues and feelings looking for decision situations in which they might be aired, solutions looking for issues to which they might be an answer, and decision makers looking for work.

The theoretical breakthrough of this model is that it separates problems, solutions and decision makers from each other, dissimilar to classical decision theory. Actual decisions do not follow an orderly process from problem to solution, but are outcomes of several relatively independent chains of events within the organisation. Problems, solutions and participants are only connected by virtue of simultaneity. Preferences are ill-defined and the organisation's processes are not understood by members.

TABLE 4.2: Synopsis decision making

	Agree on goals	Disagree on goals
Agree on methods	Rational model	Coalition methods
Disagree on methods	Incremental model	Garbage can model

In Table 4.2 the model is arranged in a two-dimensional grid of methods and goals. The rational or classical decision-making is akin to the rational man concept, i.e., define the goals, evaluate the alternatives and decide on the most favourable cost-return characteristics. The coalition model is at work when conflicting goals are present for a decision-making situation. Different interest groups focus less on collecting information but rather form coalitions to boost their chances to overcome opposition – a rather political proceeding. The incremental model is typical for "muddling through." In order to reduce uncertainty either trial-and-error or a stepwise approach is taken. Coalition models and the garbage can model are frequently tested with mathematical game theory and simulations.

Bolman and Deal (1997) describe four "frames" for viewing the world: *structural, human resources, political* and *symbolic*. Both structural, as conceived by Weber and Fayol, and human resources models have been discussed above. Within the *symbolic* frame symbols express an organisation's culture – the interwoven fabric of beliefs, values, practices, and artefacts that define for members who they are and how they are to do things. The organisation is also seen as a theatre: actors play their roles in the organisational drama while audiences form impressions based on what they see. The four frames are not independent from each other. The core of the theory is that managers must be able to put problems in the most appropriate frame.

The political frame makes a number of assumptions about organisations and what motivates the actions of their decision-makers. Note that "political" does not have a bad connotation in this context. The political frame has the following constituents (ICAF, 2002):

1. Organisations are coalitions of individuals and interest groups, which emerge because the members need support. Through negotiation, members combine forces to attain common objectives.

2. Organisations are not rational systems. Rather, organisations are coalitions composed of varied individuals and interest groups that engage in ongoing contest for control.

3. There are enduring differences among people and groups in values, preferences, beliefs, information and recognition of reality.

4. Decisions in organisations involve the allocation of valuable resources – decisions about who gets what, when, and how much.

5. Because of scarce resources and persistent differences, conflict is key to organisational dynamics, and power is the single most important resource. Political processes resolve issues. Conflict in organisational dynamics is not negative *a priori*. Competition for scarce resources within a firm can be beneficial for the organisation.

6. Organisational goals and decisions emerge from negotiation, bargaining and competition for position among coalitions and members. The prevailing coalition defines meanings and values.

The six propositions assert that organisational variety, interdependence, re-source scarcity, and power dynamics will inevitably bring forth political forces, regardless of the players. Leaders, however, with a healthy attitude toward power and its use can learn to understand and influence political processes.

In the political context *power* is an essential. Leaders who use power in the service of their organisation are constructive. Those seeking power for their own sake and for personal satisfaction compromise their ethical position, risk the organisation's effectiveness, and perhaps even endanger the long-term viability of the organisation. Competition for power exists at two levels. Individuals compete for power within units, and units compete for power within the broader institution. In both cases, power increases when an individual or an organisation gains control of a scarce commodity that others need (ICAF, 2002).

We think that *capital allocation* within the operational or, more generally, the risk context, which is effected by committees, has this political dimension. While business units own the scarce resource "profit," the risk manager's stronghold is the infliction of cost. A risk manager at a large global bank explains some mechanics which have also lead to the sub-prime lending problems by saying (Economist, 2008b):

> Traders saw us as obstructive and a hindrance to their ability to earn higher bonuses. (. . .) Most of the time the business line would simply not take no for an answer, especially if the profits were big enough. We, of course, were suspicious, because bigger margins usually meant higher risk.

Discussions between risk management and business lines often ended in arguments. The risk manager concedes that they were bad communicators, with concern of content and manner. Pfeffer (1992, 8) who devoted his best-selling book *Managing with Power* to this topic states:

> But organizations, particularly large ones, are like governments in that they are fundamentally political entities. To understand them, one needs to understand organizational politics, just as to understand governments, one needs to understand governmental politics.

Yet another perspective on organisation starts with the needs of core people, viz. the so-called *knowledge worker*. For them, collaborative, complex problem-solving is the essence of their work, involving the exchange of information, the making of judgments, and a need to acquire various forms of knowledge in exchanges with co-workers, customers, and suppliers (Beardsley et al., 2006). To foster more interaction, innovation and collaboration, companies must become more open, steadily removing impediments such as hierarchies and organisational silos. To stimulate interactions, organisations want any information that is relevant for solving a particular problem to be shared among teams laterally and not just vertically, in real time, irrespective of reporting lines. Knowledge work is improvisational and difficult to define ahead of its solution, for it accompanies the problem being solved and the opportunity at hand. The importance of structures and labels for these people and for this type of work is diminishing. In many companies, people now are put together in project teams, resolve an issue, and then disassemble to re-start the process by joining new temporary teams. Thus, organisational charts rarely reflect what is really happening within. Although there is a need for new kinds of organisations, firms seem to be in no hurry to create them. Knowledge workers now constitute up to a quarter or more of the workforce in financial services. It is argued by some consultancy firms that although making professionals productive enables big corporations to become more competitive, most big corporations nonetheless do little to improve the productivity of their employees. Companies must change their organisational structures radically to unleash the power of their professionals and to seize the opportunities of today's economy.

4.1.3   Agency Theory

The origins of agency theory can be traced back to Adam Smith's discussion of the separation of ownership and control. He suggested that managers of other people's money cannot be expected to "watch over it with the same anxious vigilance" one would expect from owners themselves and that "negligence and profusion, therefore, must always prevail, more or less, in the management of the affairs of such a company" (Smith, 1991, 324). Other fundamental works are Berle and Means (1932) on the separation of ownership in corporations and Coase (1937) on the reasons for the existence of firms.

Agency implies one party (termed the *principal*) who seeks to achieve some outcome but requires the assistance of another (termed the *agent*) to carry out a necessary activity. This situation occurs not only in organisations but also in most contractual agreements and even whole markets. In one view agency theory is a description of the ownership and capital structure of the firm (Jensen and Meckling, 1976, 309). Other applications are with investment strategies, board relationships and even with market mechanisms in general. It is also collocated within organisational economics. For an overview see Eisenhardt (1989).

The agency conflict can be summarised as follows: When principals cannot perfectly observe, monitor, control or discipline agents, then inefficiencies may result, such as "stealing" from the firm, empire building – non value-enhancing investment activity – and managerial entrenchment, because bad agents, especially managers, are hard to remove from their positions. The ensuing costs borne by both parties are implicit and often not observable. In this context they have been classified as:

- bonding cost (paid for by the agent and possibly passed on),

- monitoring cost (paid for by the principal) and

- residual loss, i.e., any remaining inefficiency from the agency relationship.

Examples of bonding costs are malpractice insurance by the agent, escrowaccounts, non-competing clauses signed by the employee and equity stakes. They should penalise the agent in case of non-performance. Monitoring costs are inter alia frequent meetings and progress reports, hiring an external auditor, etc. Residual loss ensues when monitoring and bonding does not work. Principals have a strong incentive to minimize agency costs. What is not considered enough is that guarding against agents' opportunistic behaviour can result in suppressing initiative, creativity, entrepreneurship and innovation within firms.

Associated with monitoring is the control approach with two ramifications concerning *behaviour*, aiming at efficiency and *outcome*, focussed on effectiveness. Principals can try to collect relatively complete information about an agent's behaviour, i.e., current and past decisions and actions. But the agent can also impose high agency costs by resisting the principal's efforts to gain the information. On the other hand, principals can try to monitor the partially observable consequences of agent behaviour. However, the more uncertain the outcome – either because of complex interdependence or rapidly changing environments – the more likely that the principal will encounter ambiguity in determining how much each agent contributed to the final output.

Bonding mechanisms also frequently take the form of incentives (such as equity stakes) that align the interests of agents and principals resulting in a sharing of the benefits between principals and agents. But the welfare of the principal is affected by the choices of the agent. Policies that allow agents to be compensated in ways that are independent of the principal's interests ensure that an agent's earnings will not fluctuate with conditions outside of the agent's control. Such policies, however, seem rather ineffective.

Agency theory is largely an asocial conceptualisation because it assumes that humans are motivated solely by pecuniary self-interests and largely uninfluenced by social relations. However, it seems individuals are also motivated by status, prestigious awards, community, social relations, etc. – factors which are not taken into consideration.

4.2 Environment

Environment is what surrounds us. Often the distinction is made between internal and external environment. The internal environment is what happens within the workplace, which is considered partly controllable, whereas the external environment is both the immediate business context in which a particular organisation must function, and the greater business environment, both considered relatively immutable. Figure 4.5 shows the structure graphically. In the operational risk management context we must attempt to understand

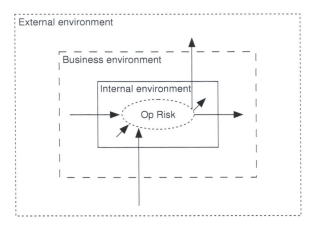

FIGURE 4.5: Concentric environments.

which influences from within and without the bank have a bearing on its risk. This will lead to a conjecture of reality – some would say a folk theory – from which we can derive a plan of action.

4.2.1   External Environment

We can stratify the the notion of external environment in many ways. In order to minimise confusion we shall stick to COSO's view. Its external environment encompasses the following dimensions:

- economic,
- natural environment,
- political,
- social and
- technological.

It has a long tradition of analysing the environment – specifically its competitiveness – by conducting so-called SWOT analysis. Here, strength, weaknesses, opportunities and threats are conjectured in order to define a winning strategy. The same can be done for the purpose of defining *risk drivers* operations. These mainly multi-dimensional terms or constructs must be operationalised. This means that the measurable attributes of the construct have to be defined. In risk management parlance the construct is called *key risk driver*, and the observable and measurable attribute is called *key risk indicator*. The construct may be defined by multiple stages of attributes, thus forming a hierarchy.

TABLE 4.3: Excerpt of the Worldwide Governance Indicators (Source: Kaufmann et al., 2006. By permission of the Worldbank)

Country	Voice and Accountability	Political Stability/ No Violence	Government Effectiveness	Regulatory Quality	Rule of Law	Control of Corruption	Mean	Min
Afghanistan	-1.28	-2.12	-1.20	-1.63	-1.68	-1.37	-1.55	-2.12
Australia	1.32	0.82	1.88	1.58	1.80	1.95	1.56	0.82
Bahrain	-0.85	-0.28	0.42	0.69	0.71	0.64	0.22	-0.85
Brazil	0.36	-0.13	-0.09	0.08	-0.41	-0.28	-0.08	-0.41
Canada	1.32	0.91	1.92	1.57	1.81	1.92	1.58	0.91
China	-1.66	-0.18	-0.11	-0.28	-0.47	-0.69	-0.57	-1.66
Cuba	-1.87	0.03	-0.94	-1.75	-1.14	-0.26	-0.99	-1.87
Denmark	1.51	0.91	2.12	1.69	1.99	2.23	1.74	0.91
Djibouti	-0.84	-0.74	-0.85	-0.86	-0.87	-0.64	-0.80	-0.87
Egypt	-1.15	-0.90	-0.35	-0.47	0.02	-0.42	-0.55	-1.15
France	1.28	0.33	1.46	1.09	1.35	1.40	1.15	0.33
Germany	1.31	0.67	1.51	1.38	1.76	1.92	1.43	0.67
Iran	-1.43	-1.14	-0.77	-1.49	-0.76	-0.47	-1.01	-1.49
Israel	0.61	-1.16	0.95	0.89	0.76	0.76	0.47	-1.16
Italy	1.00	0.21	0.60	0.94	0.51	0.41	0.61	0.21
Japan	0.94	0.94	1.16	1.17	1.33	1.24	1.13	0.94
Lebanon	-0.72	-1.14	-0.30	-0.28	-0.36	-0.39	-0.53	-1.14
Mexico	0.29	-0.29	-0.01	0.33	-0.48	-0.41	-0.10	-0.48
Netherlands	1.45	0.80	1.95	1.64	1.78	1.99	1.60	0.80
Nigeria	-0.69	-1.77	-0.92	-1.01	-1.38	-1.22	-1.17	-1.77
Russia	-0.85	-1.07	-0.45	-0.29	-0.84	-0.74	-0.71	-1.07
Saudi Arabia	-1.72	-0.70	-0.38	-0.01	0.20	0.23	-0.40	-1.72
Singapore	-0.29	1.08	2.14	1.79	1.83	2.24	1.47	-0.29
South Africa	0.82	-0.10	0.84	0.59	0.19	0.54	0.48	-0.10
Spain	1.12	0.38	1.40	1.25	1.13	1.34	1.10	0.38
Sweden	1.41	1.18	1.93	1.47	1.84	2.10	1.66	1.18
Switzerland	1.43	1.26	2.03	1.47	2.02	2.12	1.72	1.26
Syria	-1.67	-0.91	-1.23	-1.22	-0.42	-0.59	-1.01	-1.67
UK	1.30	0.34	1.70	1.53	1.69	1.94	1.42	0.34
United States	1.19	0.06	1.59	1.47	1.59	1.56	1.24	0.06
Venezuela	-0.50	-1.22	-0.83	-1.15	-1.22	-1.00	-0.99	-1.22

Alexander (2004, 24) asserts: "In operational risk, a "key risk driver" (KRD) is a variable that can be directly affected by management actions.

Examples of KRD's are product complexity, system quality and incentive schemes.

A KRI is a variable that is associated with the loss frequency and/or loss severity of an operational risk, such as the number of failed trades, system downtime and staff turnover."

It seems legitimate to question how we can directly affect "system quality," say to increase it by 20%? Yet another definition from (Pézier, 2004): " KRD – A control factor that determines a risk exposure; e.g., level of staffing, level of redundancy in systems design, KRI – A state variable that is closely related to operational losses, e.g., p.c. of unsettled transactions at days end, p.c. of system downtime, number of customer complaints."

Or we can read (Nystrom and Skoglundy, 2002): "In the operational risk management literature the concept of key risk driver is well-known. It is an observable process like employment turnover or transaction volume that influences the frequency and/or severity of operational events."

From these citations it becomes evident that there is no clear understanding of the terms KRD and KRI. We must rely, then, on the social science concept of key risk drivers as complex and multi-dimensional attributes that must be derived from observable key risk indicators.

There is a plethora of possibilities and third-party indices covering the external environment. The difficulty therefore lies with determining which ones operationalise a concept that pertains to risk drivers, possibly with respect to certain event classes. Economic indicators are legion, according to Outline 3

TABLE 4.4: Social index example

Construct	Indicator
Social cohesion	UNDP Human Poverty Index
Access to education	Education expenditure per capita
Identity, self-realisation	Unemployment rate
Security	Crime rate, corruption rate
Health of human beings	Life expectancy, infant mortality

on page 251.

Sometimes it is not so easy to make a distinction between external (or business) and internal environment. Take as an example innovation. It can be both a reaction of the bank to external innovations or a bank's internal attitude on how to promote innovation. We will go into more details concerning innovation in the next section.

4.2.2   Internal Environment

The internal environment is everything that surrounds us in the bank, i.e., people. Therefore the internal environment mirrors a good part of the external

- Output Indicator

- Growth Indicators

- Industrial Confidence Indicators

- Industrial New Orders

- Price and Cost Indicators

- Consumer Price Index

- Producer Price Index

- Employment and Wages

- Trade Indicators

- Financial Flows

- Financial Indicators

- Exchange Rates

OUTLINE 3: Economic indicators

in as far as people are members of the society. Thus, we can summarise the internal environment by the attributes of the management system (see page 227) and specific items of the wider social and political attributes, which are:

- general management principles,

- authority and responsibility,

- integrity and ethical values,

- goals,

- control,

- information,

- incentive and motivation,

- employees' competence,

- human resource development.

The human side is epitomised by motivation, competence and development, and is very important with respect to both performance and risk. The so-called internal control environment will be treated separately as it is a key notion in the Basel 2 Accord. From the list above the internal control environment covers in part control, obviously, but also ethics, goal setting and coherence, information with concern to control and reporting and incentives.

On all the above multi-dimensional constructs a quantification in the form of an index is either available or can be built. To this end, existing sub-indices may be integrated or assessed through questionnaires, interview and workshops.

Corporate values are the embodiment of what an organisation stands for. They should be the basis for its members' *behaviour*. However, what if the organisation's members do not share and have not internalised the organisation's values? Obviously, a *disconnect* between individual and organisational values will be dysfunctional. Additionally, an organisation may issue one set of values, perhaps in an attempt to generate a positive image, while maintaining a different set of internal guiding principles. When there is a disconnect between *stated* and *operating* values, it is difficult for employees to know what is acceptable. In some organisations, dissent may be rewarded, while in others dissenters may be ostracised or expelled. Group members quickly learn the operating values if they wish to survive. With a blatant disconnect, the organisation will not only suffer from performing less effectively, but also from some cynicism of its employees, who have another reason for mistrusting the leaders, or doubting their wisdom. Values determine what is right and what is wrong, and *ethics* means *choosing* what is right or wrong. People's ethical codes are generally developed before they become bank employees, but the bank defines an ethical environment and has procedures to encourage ethical behaviour. Integrity of the bank has to do with the latter two items and how it abides by the values.

Risk factors for unethical behaviour are, amongst others:

- the complexity of (strategic) issues,

- competition for scarce resources and positions,

- conflicting loyalties,

- "groupthink," i.e., a rationalised conformity,

- the presence of ideologues,

- negative response to dissent.

The first three items concern the employee as individual, while the latter three are evident at the group or collective level. Integrity and ethical values are measured with the control self-assessment, see Outline 6 on page 298 and Section 4.7.2, page 294.

A word concerning *innovation*. Technology is constantly changing. Therefore, a bank must also adapt. Hamel (2003) comes to the conclusion that:

There are two core challenges to making innovation a deep capability in any organization. First, most companies have a very narrow idea of innovation, usually focusing just on products and services. We need to enlarge our view of innovation. Second, most companies devote much more energy to optimizing what is there than to imagining what could be. We need to create constituencies for "What Could Be."

One view is that managing the legacy business is the task of a corporation to perform: they perpetuate past success by institutionalising its form. Often corporations allocate assets based on a business unit's past performance rather than on assessments of its future growth opportunities. In the short term, this may be an totally reasonable strategy but in the long run, it is likely to exacerbate corporate failure.

Välikangas and Hamel (2001) focus on three ways in which corporate hierarchies seem fail to meet their innovative potential. First, management seems to be commonly limited to an option set that is too narrow when considering investment options. Second, corporate resources appear mis-priced relative to their value-generating potential. Third, important elements like creativity and enthusiasm are not easily captured in top-down managerial processes. When new ideas surface, they are often ignored, especially if they are not aligned with existing business interests or managers' own know-how or if the organisation lacks effective means to accommodate novelty. On the other hand, it must be acknowledged that effective innovation shifts any existing power profile.

While preparing for the "year 2000" date problem, we experienced a sudden interest for such exotic skills as COBOL programming, proving that there are still very old legacy systems in production. While changing technology may foster operational risk, a refusal or reluctance to change may lead to strategic risks.

4.3 Culture

What is culture? A simple but limited answer is: "everything that is not nature." As Thomas Browne, a 17th century intellectual put it: "All things are artificial, for nature is the art of God." Culture is human made. Thus, a Corvette from 1956 is culture. In order to eliminate all hardware or material items from the definition we could say that culture is the acquired knowledge people use to interpret experience and generate behaviour. It is that complex whole which includes knowledge, belief, arts, morals, law, custom and any ̇r capabilities and habits acquired by man as a member of society. The ̇ition of knowledge is party determined by tradition. It helps to define ̇, express feelings and make judgements.

There are cultural universals, i.e., traits that are found in all human societies, other so-called "generalities" that are found in many societies and yet other traits that are present in only a few, or in one, society. Because nations and people today are better linked and more mutually dependent (due to economic and political forces, latest systems of transportation and communication) larger economic and political systems can dominate or at least complement local cultures.

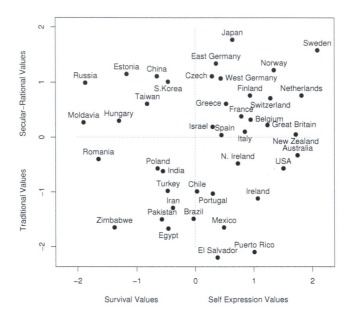

FIGURE 4.6: Inglehart-Welzel's cultural value map. (Adapted from Inglehart and Welzel, 2005, 64.)

But our values are as much a reflection of our unique selves as they are a manifestation of the social, political and cultural environments in which we live. Figure 4.6 shows the empirical findings of the World Value Survey which was first conducted in 1981 and followed by several later surveys (Inglehart, 1997).

The map represents the values and socio-economic status of many countries along two dimensions: The first dimension ranges from "traditional" to "secular-rational" values, while the second dimension goes from "survival" to "self-expression" values. More traditional societies emphasise the importance of religion and family. These traditional societies tend to be nationalisti with a high level of national pride. Wealthier countries tend to score hig on secular-rational and self-expression values than poorer countries. A tries develop, they move from the lower left to the upper right on this map (American Environics, 2006, 23). However, religion and ideology complicate the seemingly simple relationship between values and prosperity. For example, Protestant countries form a distinctive cluster separate from historically Roman Catholic societies – a trend first predicted by Max Weber and further discussed on this page.

4.3.1 America and Europe

We think it is more interesting to dwell a little bit on the differences between those nations that constitute the "Western World" than between the very obvious differences between, say Sharia-ruled countries and India, China and Japan. (By the way, we recommend Chamberlain (1902), which is still in print, for an overview on Japanese tradition.) Kagan (2004, 3) opens his best-selling essay with the following sentence: "It is time to stop pretending that Europeans and Americans share a common view of the world, or even that they occupy the same world."

According to Judt (2005) in past decades it was conventionally assumed that Europe and America were converging upon a single "Western" model of late capitalism, with the U.S. leading the way. The logic of scale and market, of efficiency and profit, would ineluctably supersede local variations and inherited cultural constraints. Americanisation was inevitable. The only hope for local products and practices was that "they would be swept up into the global vortex and repackaged as "international" commodities for universal consumption. Thus an archetypically Italian product – caffé espresso – would travel to the US, where it would metamorphose from an elite preference into a popular commodity, and then be repackaged and sold back to Europeans by an American chain store." Starbucks® is reported to have encountered some growth difficulties in Continental Europe.

Huntington (2004) asserts that "America's core culture" includes: the Christian religion, Protestant values, moralism and work ethic, the English language, British traditions of law, justice, and the limits of government power, and a legacy of European art, literature, philosophy, and music, plus the American Creed with its principles of liberty, equality, individualism, representative government, and private property. Equality is equality of opportunities and not of result. "Given the strength of the aspiration to rise, it is not surprising that Americans are more disposed to approve of high salaries and "bonuses" for "stars" in entertainment, athletics, and the market in general" (Lipset, 1996, 76).

The Protestant, or more precisely the Calvinistic, work ethic has been analysed by Max Weber (1988, 1930) and is believed to have fostered the emergence of modern capitalism. Calvinism held the dogma of predestination, i.e., either you are saved and will go to heaven or you are condemned. However, wealth was taken as a sign that you were one of God's elects, thereby encouraging people to acquire wealth. This ethic provided religious sanctions that fostered a spirit of strict discipline, promoting men to apply themselves rationally to acquire wealth. But it was considered a sin to spend this money on personal luxuries. Also, charity was generally frowned upon because a lack of worldly success was seen as the result of a combination of laziness and divine disapproval. The manner in which this paradoxon was resolved, Weber argued, was the investment of the money, which gave an extreme boost to nascent capitalism.

Freedom, security and justice are the core values that constitute the European model of society. So the differences between America and Europe have mostly to do with America's hyper-protestant moralism and work ethic in addition to individualism and pursuit of happiness. On the other hand, justice is more prominent in Europe. Additional differences are the language and the juridical system.

In the context of operational risk, cultural differences, especially with respect to morals and law, can lead to material loss and damage. Using one's own cultural context to judge people from another culture may lead to gross misunderstandings. Some years ago a magazine reported the case of an American analyst asking his British colleague about the creditworthiness of a customer in order to decide whether to extend a credit. The answer was that the customer was "on his uppers." This is a metaphor from shoemaking meaning that nothing of the shoe is left but the upper leather and thus it means being wretched. This expression was misunderstood and the customer got the credit. Language in a multi-cultural work environment is often a source of risk.

4.3.2   Gender Differences

Women are playing an increasingly important role in management, perhaps because women are attaining higher education in ever greater numbers – and often in greater numbers than men. Therefore, we speculate that it is only a matter of time before women will shatter the actual glass ceiling in business. Not surprisingly, risk management will be influenced by female attitudes and perceptions – insofar as these can be generalized. We predict that this will have profound implications for organisations.

We want to touch on gender differences in financial risk behaviour, on which rather little work has been done. It is an uncontested fact that men and women differ generally when it comes to risk taking. But it is also a fact that risk taking is contextual; an individual risk tolerance in one field (e.g., driving) does not necessarily imply the same attitude in a different context (e.g., investments). Explanations for the differences are always sought in either nature or nurture, a veritable field day for psychologists and sociologists. (A new study by Sapienza et al. (2009) suggests that testosterone has a significant effects on risk aversion regardless of gender.)

Sociological explanations rely on differential socialisation. Females are taught to be more humble, passive, obedient and caring than men. A second argument focuses on structural location, i.e., women expect to occupy the field of domestic duties and to subordinate paid employment. It is also suggested that parents, and especially fathers, influence their children in specific ways: boys are encouraged to be adventurous and bold, while girls are kept under relatively strict control. Because family structures have changed over the last decades and patriarchal structures have lost importance, these arguments are becoming less convincing.

A study by Johnson and Powell (1994) suggests that women are excluded from senior managerial positions because of stereotypes which have been formed by observing non-managerial populations. This study finds that educated managers from both sexes display similar risk propensities in business. Booth and Nolen (2009) report from controlled experiments that girls from single-gender schools choose real-stakes gamble as often as boys from either single-sex or co-educational schools, but more than co-educated girls, implying that nurture matters.

The risk differences can be analysed in a more differentiated way. An experimental investigation centres on risky decisions and tries to separate the possible loss or gain from the probabilities of occurrence. While the values of the outcome are assessed similarly by women and men, women are more pessimistic about medium or large probabilities (Fehr-Duda et al., 2004). Similarly, when the probabilities are more ambiguous, as in most real life situations, women are more ambiguity-averse than men (Gysler et al., 2002).

He et al. (2008) postulate that men and women are differentially sensitive to the extent of their perceived resources or skills to resolve an issue. Specifically, a self-assertive orientation among men makes them sensitive to accomplishment of gains (e.g., investment decisions). In contrast, a selfless orientation among women makes them susceptible to avoidance of losses (e.g., insurance decisions). This relates to our discussion of overconfidence and the illusion of superiority in Section 2.10.3.1 on page 74. In short, overconfidence is more pronounced with men.

We conclude by saying that the explanations for gender differences in risk behaviour in financial decision-making are only partially convincing but nonetheless we think it is an everyday experience that such differences exist. Therefore, for the sake of balance, we think it a good idea to encourage women to enter the field of risk analytics. Perhaps the current financial situation represents an opportunity to make this field more welcoming to women, who, one hopes, will bring a new perspective to the table.

4.3.3 Legal Systems

The *legal system* is also a cultural fact of paramount importance for risk management in banking. In the Western world there are two main juridical systems that we shall call *English* and *Roman*. Roman law developed over a thousand years and provided the basis for canon law. It also strongly influenced secular medieval law and it was blended with pre-existing law by the early enlightened monarchies of western Europe to extend a uniform code to their many provinces. Ultimately, it was re-framed by French jurists into the Code Civil (also Code Napoléon). It is at the core of legal systems in South and Central America, and in many former colonial areas of the world, as well as Continental Europe. In Table 4.5 (according to Reynolds and Flores, 1986), we see the English common law and its spread and the Roman-based systems differentiated by French, German and Scandinavian origin.

TABLE 4.5: Foundation of legal systems

Common law	Civil law		
English origin	French origin	Scandinavian origin	German origin
Australia	Belgium	Denmark	Austria
Canada	France	Finland	Germany
Ireland	Greece	Norway	Japan
United Kingdom	Italy	Sweden	Korea
USA	Luxembourg		Turkey
	Netherlands		
	Portugal		
	Spain		
	Switzerland		

In general, Roman law is considered systematic and English law is not. The former is codified according to coherent and logical categories, whereas common law is a loose body of principles that are invoked individually in the solution of a single case. English law has been particularly identified with precedents established by the various courts, while the Roman law has been developed to a far greater extent by scholars and legislators.

In 1994, a New Mexico jury ruled that McDonald's had been negligent by not warning customers that its coffee was "extremely hot," about ten degrees Fahrenheit warmer than the restaurant industry average. Although coffee is commonly known to be hot, punishment was USD 2.7 million in damages, awarded to a woman who had spilled a 49 cent cup of coffee on herself. In Figure 4.7 you can see the development of the tort cost – proxied by insurance outlays – in the USA. This is drawn from a periodic survey issued by Tillinghast (Sutter, 2005), a consultancy firm. An international comparison shows that with 2.2% of GDP the tort system in the USA is the most expensive. The European figures hover around 1%, with the exception of Italy with 1.7% (where the denominator may not include the submerged economic activities). In the US there are lawmaking initiatives to curb this inauspicious development.

US companies often settle frivolous lawsuits, because going to court and winning the case consumes too much time and money. US tort lawyers frequently work for a percentage of the money paid to settle the claim (on average almost 20%). These can be astronomical sums. In most of Europe, lawyers

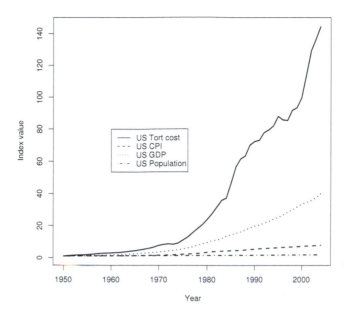

FIGURE 4.7: Tort cost in the USA. (Data from Sutter (2005). With permission.)

work for flat hourly fees, and nuisance suits are less attractive, because losers generally bear all legal costs. Another American specialty is *punitive damages* and corporate criminal responsibility. The first are available only if the defendant acted with malice or gross negligence or hurt the plaintiff in a way that expressed scorn or contempt for the plaintiff. Punitive damages greatly increase. It is argued that such damages are a means of giving satisfaction to the plaintiff.

One of the most outstanding elements of the U.S. litigation culture is the *class action*. There are signs that Europe is adopting this instrument, because it is obviously inefficient to treat all suits separately. Take a recent case in Germany involving Deutsche Telekom, which includes 2100 claims filed by 754 law firms. This case is expected to take up to 15 years to adjudicate, because German law requires judgement to be given in each individual case. Europe still lacks, and generally opposes, juries in civil trials. American and British law firms are preparing for a potential deluge of work if the floodgates to class action suits across Continental Europe are opened.

In the following dispatch of the 10th of January 2006 one can study the clash of the American and Continental legal systems (reproduced by permission of Thomson Reuters). In Germany the plaintiffs could get as much as some months' salary and a dismissal.

> FRANKFURT (AFX) - A spokesman for Allianz AG's banking unit Dresdner Bank AG said the company's investment bank Dresdner Kleinwort Wasserstein intends to "vigorously" defend itself against claims of gender discrimination. "We have a policy not to comment on pending litigation," the spokesman said. "But DrKW fully complies with all applicable employment-related laws and is confident that any claims to the contrary are without merit."
>
> A group of six currently employed female executives yesterday filed a class action gender discrimination case in the Southern District of New York against DrKW. According to the complaint, they have met a "glass ceiling," in other words they have been denied promotions, compensation on a par with male employees, and equality with respect to the terms and conditions of their employment. The complaint asserts causes of action against a number of DrKW executives who directly participated in the discriminatory conduct.
>
> The law firm representing the women – Thompson, Wigdor & Gilly LLP – said it is aiming for at least 1 bln usd in damages to be paid to at least 500 women working at DrKW.

The general assumption that one bank's big losses would foster countermeasures by peer banks can be questioned. In Table 1.4 on page 32 we can see that another bank has already incurred a loss of USD 250m on a similar case that apparently had no effect on DrKW.

Half a year later, on August 28th, 2006 came the news:

> FRANKFURT (AFX) - A Deutsche Bank AG spokesman declined to comment on a report in today's Frankfurter Allgemeine Zeitung, which said the US Equal Employment Opportunity Commission (EEOC) had presented the bank with evidence of alleged gender discrimination at one of its US locations. The bank spokesman said he cannot comment on "pending or possible litigation," adding that the bank does not tolerate discrimination of any kind. According to the report, the EEOC presented the bank with evidence of alleged discrimination against one of its female workers in its securities operations, and against a group of women in similar roles at the bank. If the bank does not reach a solution through mediation, it could be faced with a lawsuit, the newspaper said.

The U.S. tort system is seen by many as favouring extortionate behaviour by using the legal suit as a menace to force an agreement.

But globalisation is nudging both the U.S. class action suits and the tort system into Continental Europe. Not surprisingly, many banks now find legal and liability risk, which represent a different classification form the Basel 2 accord, the major threats in their operating environment.

4.3.4 Crime

Another factor contributing to risk is crime. Crime rates in the United States are for all categories approximately three times higher than in other developed countries. But types of crime are distributed similarly, i.e., violent offences (homicide, rape, robbery, and aggravated assault) comprise 10 percent of crimes perpetrated in the USA, while 90 percent are property crimes (Lipset, 1996, 269). What conclusions should one draw from this for operational risk? Should American subsidiaries take such indicators into account? Sociologists like Lipset (1996, 47) argue:

> The stress on equality and achievement in American society has meant that Americans are much more likely to be concerned with the achievement of approved *ends* than with the use of appropriate *means*. In a country that values success above all, people are led to feel that the most important thing is to win the game, irrespective of the methods employed in doing so.

Another source of the high American crime rate may be found in the strong "due process" guarantees for individual rights. In Europe the policing powers of the prosecution are much more wide-ranging. On the other hand, the draconian punishments in the USA seem not to deter offenders. The EU has 87 prisoners per 100,000 people, whereas the United States of America have 685.

What does criminology say about the motivations to get an offender? We cannot give an exhaustive account here, but would like to offer a broad *pars pro toto* summary. The classics like Jeremy Bentham posited (see page 228) that man is a calculating animal who will balance potential gains against the probable pain to be imposed. If the pain outweighs the gains, she will be deterred. This produces maximal social utility. Therefore, in a rational system, the punishment system must be graduated in such a way that the punishment closely matches the crime. This theory is problematic because it strongly assumes that the potential offender acts rationally (whereas much crime is a spontaneous reaction to a situation or opportunity). The positivist view taken by Cesare Lombroso, analogously to the "great man" theory for leadership, assumes that the biological nature of offenders fosters crime (Lombroso, 1876). He made the distinction between a *born criminal, criminal by passion, insane criminal,* and the *occasional criminal.* Observers attribute such differences as the result of nurture (i.e., society's influence on its members) rather than nature (individuals' innate predilections.) The Nobel laureate Gary S. Becker has expounded on crime and punishment, proposing a method and model to minimise the social loss in income from offences (Becker, 1968). This paper is seen as a resurrection, modernisation and improvement of the "economic" calculus already applied by Bentham.

Robert K. Merton's theory does not concentrate on crime, but instead on various acts of deviance, which may lead to criminal behaviour. He notes that there are definite goals that are strongly emphasised by society. Society accentuates certain means to reach those goals (such as education, hard work, etc.). Nevertheless, not everyone enjoys equal access to the legitimate means to attain those goals. The stage is then set for anomie (Merton, 1938).

Another stream called *social control theory* turns the question "why do people become criminals" on its head, asking instead "why do people not become criminals?" This theory assumes that all people are capable of committing criminal acts and would do so if left to their own devices. Social control theory analyses the failure of society to control criminal tendencies (Hirschi, 1969).

Walter Reckless with his *containment theory* suggested that people can be "insulated" from criminal acts (Reckless, 1961). If properly socialised by their parents and peers, the individual will control herself and contain those natural impulses which may lead to criminal behaviour. According to this theory, there are internal constraints, e.g., a favourable self-image, a high level of frustration tolerance, internalised morals and ethics and a well-developed ego. External constraints are consistent moral values, institutional support for internal values, positive role models and a sense of belonging. The probability of deviance increases as internal and external constraints weaken.

The so-called *drift theory* suggests that people live in a continuum somewhere between the extremes of total freedom and total restraint (Sykes and Matza, 1957). The associated process by which a person moves from one extreme of behaviour to another is called drift. Delinquents develop a special set of justifications – rationalisations – for their behaviour: the victim is caused no real harm, or the victim can "afford" the crime, the victim deserves the injury, or the perpetrator himself is a victim of circumstance and has no choice but to perpetrate crimes. Sometimes lawbreakers appeal to a higher loyalty – which allows them to neutralise and temporarily suspend their commitment to common values, providing them with the freedom to perpetrate criminal acts.

Here again reality is too complex for a single, simple answer that could be used to better control banks' risk for crime, especially fraud. But it cannot be denied that opportunities to perpetrate crimes must be kept to a minimum. Figure 4.8 depicts the crime environment in a simplified manner. The saying "opportunity makes the thief"[4] condenses the wisdom of history. This wisdom is not confined to theft and burglary but pertains to all crimes. Crime prevention must therefore:

- increase the perceived effort of crime,

---

[4]Also the title of a one act opera by Gioacchino Rossini, German "Gelegenheit macht Diebe," Italian "L'occasione fa il ladro."

- increase the perceived risks,

- reduce the anticipated rewards and

- remove excuses for crime.

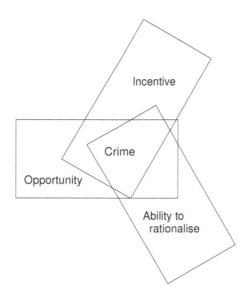

FIGURE 4.8: Crime environment.

Example 4.3. An annual survey on fraud conducted by PwC, an audit company, concludes (PricewaterhouseCoopers, 2005, 15) that the typical corporate fraudster is in 87% of cases male, between the ages of 31 and 40 and educated to degree level or higher. Most startling is the extent to which – according to this survey – members of senior management are incriminated. Overall, they are responsible for 23% of reported frauds – in smaller companies even 35%. Perpetrators within senior management appear to be driven by the incentive to maintain expensive lifestyles and once discovered, claim that they were unaware of committing any wrongdoing.

The survey shows that senior managers are also treated more leniently, because the company wants to curb possible damage to its reputation, to avoid questioning the capabilities and character of the senior management group and to prevent demoralisation of the staff. Of course, it is debatable whether it is more demoralising for an employee to know he is working for a thief who is then dismissed or for him or her to later learn the company covered up for the thief and continued to pay him his salary.   △

4.3.5   Communication

Now we shall jump from crime to *communication*, a very different, but equally important topic in the vast sea of culture. In Figure 4.9 we depict a simple model of communication. It is the model proposed by Claude Shannon (Shannon, 1948, 380) for technical systems – in the figure represented by capitalised words – with the addition of a feedback and a translation to human communication as proposed by Koontz and Weihrich (1988, 463). Encoding and decoding, i.e., writing, graphs, presentation, translation in another language, etc. and reading and listening all depend on the cultural background of the parties engaged in a dialogue. Here it is instructive and amusing to recall a story that circulated amongst students at the Swiss Federal Institute of Technology. Once there was a mechanics professor, who, during his late-in-life driving test was asked by his examiner how he would accelerate the car. Probably baffled by this obvious question he replied: "I would turn the steering wheel." How this discussion then went on is not known. The point is that, as every physics student knows, there is always a perpendicular acceleration when the curvature of a trajectory of a mass point is changed – that is, you really need not push the accelerator pedal to accelerate. But whether the driving examiner understood this – and whether the professor cared if his examiner understood or not – make this exchange a perfect example of miscommunication.

In addition to the content, the means by which information is presented affects the communication. For example, the all-pervasive use of slides both for presentations and for analysis, where they have supplanted solid reports, has a bearing on the audience's understanding. Tuft (2003) argues that slides may help speakers outline and structure their talks, but such convenience comes with a price. The standard presentation elevates format over content, hardly concealing an attitude of commercialism that turns everything into a sales pitch. Style routinely disrupts, dominates, and trivialises content. Tuft writes:

> In a business setting, a PowerPoint slide typically shows 40 words, which is about eight seconds' worth of silent reading material. With so little information per slide, many, many slides are needed. Audiences consequently endure a relentless sequentiality, one damn slide after another. (...) Often, the more intense the detail, the greater the clarity and understanding. This is especially so for statistical data, where the fundamental analytical act is to make comparisons.

Presentations largely stand or fall on the quality, relevance, and integrity of the content. If the numbers are boring, then you are presenting the wrong numbers. Using the gimmicks and gadgets of presentation software can be counter-productive. "Audience boredom is usually a content failure, not a decoration failure." Indeed, such extravagant presentations are a good means of hiding information – or the lack thereof. Related to how information is presented is content. Frankfurt (2005, 1) asserts in his best-selling essay *On*

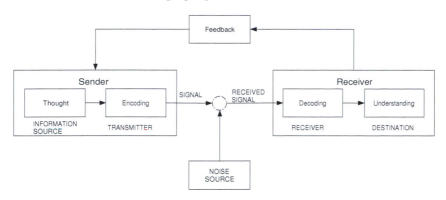

FIGURE 4.9: Communication process model. (The original model of Shannon is given by the capitalised terms.)

Bullshit: "One of the most salient features of our culture is that there is so much bullshit. (...) Most people are rather confident of their ability to recognize bullshit and to avoid being taken in by it." After attempting to define the term he sees one source of the "bullshit" proliferation as people's belief that they can no longer decode reality. It replaces an attempt to create an accurate representation of the world with a more likely to be successful attempt to create an honest representation of the bullshitter himself. In short, since it makes no sense to be true to the facts he tries to be true to himself. Refraining from making any assertion whatsoever (keeping his mouth shut) may often be seen as an impossible or unappealing option. The main manifestation of bullshit is when circumstances require someone to talk without knowing what he is talking about (Frankfurt, 2005, 63):

> Thus the production of bullshit is stimulated whenever a person's obligations or opportunities to speak about some topic exceed his knowledge of the facts that are relevant to the topic. This discrepancy is common in public life, where people are frequently impelled – whether by their own propensities or by the demand of others – to speak extensively about matters of which they are to some degree ignorant.

In risk management it seems that such obligations and opportunities occur rather often. Remember Table 4.1 on page 231, where naturally born leaders are depicted as creative, intelligent and fluent talkers. Pfeffer and Sutton (2000, 47) argue:

> For better or worse, people who want to get ahead in organizations or to achieve influence often learn that talking a lot helps them to reach such goals, perhaps even more reliably than taking action or inspiring others to act. Once people achieve high status, they are

expected to talk more than ever. Dominating the group's airtime is one way to let everyone know that they are still in charge.

Because risk management is so deeply multi-disciplinary, partial ignorance abounds, especially with respect to quantitative concepts. Chief Risk Officers often indulge in bullshit, confident that their position precludes any contradiction by subordinates. This in turn reinforces the position holder's belief of thinking that he is right. One CRO who believed he had rediscovered credit risk conducted an off-site workshop during which his staff calculated the unexpected loss as fourteen times expected loss, while still paying KMV licensing fees.

4.3.6 Innumeracy, or More on Communication

Innumeracy is the analogue to functional illiteracy. Someone who is functionally *innumerate* does not well comprehend or express logical arguments with numbers. Since the 1980s (Hofstadter, 1982; Paulos, 1988), some specialists in Western societies have been very much concerned with innumeracy, or statistical illiteracy, while society as a whole seems unconcerned.

As Sowey (2003) writes, it is not the mathematical manipulation of numbers or symbols representing numbers that is crucial to the notion of numeracy. Rather, it is the ability to draw correct meaning from a logical argument embedded in numbers. When such an argument relates to events in our uncertain world, the element of uncertainty makes it a statistical argument.

Paulos (1989) said in an article preceding his best-seller (Paulos, 1988) that the high abstraction level of mathematical models is at the root of the phenomenon:

> Whether literary types are more prone to innumeracy or not, it's certain that the abstractness of mathematics is a great obstacle for many intelligent people. Such people may readily understand narrative particulars, but strongly resist impersonal generalities.

What are the reasons for this wide-spread innumeracy, especially in the very abstract realm of statistics? Paulos (1989) examines a cultural attitude:

> The most obvious causes of innumeracy are poor education and "math anxiety" but the deeper sources are prevailing cultural attitudes, in particular misconceptions about the nature of mathematics. These attitudes and misconceptions lead to an intellectual environment that welcomes and even encourages inadequate mathematical education and pride in ignorance.

Sowey (2003) puts forward some plausible reasons based on the education system. From early school days the innumerates have retained that anything expressed in numbers is precise beyond any doubt. However, the significant issue is usually not whether a number is *precise*, but whether it is *accurate*. This miss-conception is deeply rooted. Secondly, where mathematics is concerned, many people set aside their otherwise natural scepticism in favour of an unthinking obedience to argument from authority. Not fully understanding the mathematics and numerical results, many become accustomed to just accepting figures as meaningful. And further:

> And how do strongly innumerate people respond to a statistical argument? Here again, one may hypothesise the residual influence of the school mathematics experience. Children learn mathematical theorems and, at the same time, they learn that (under given assumptions) these theorems are always true. But, surprisingly, they are rarely taught why these theorems are always true, namely because they rely on deductive logic – which is, moreover, only one of an array of logics of systematically lesser reliability (deduction, induction, analogy, intuition, ...). Not knowing that statistical arguments are inductions, statistically illiterate people assign to such arguments the same status as the deductive theorems of their school days, and thus hold the conclusions to be beyond question.

But perhaps most detrimental are those people who have had some shallow statistical education and have transformed their initial learning into a mixture of partial truths and plausible, but incorrect, beliefs. Evidently, statistical arguments judged on intuitive principles are unlikely to produce deep knowledge.

Gigerenzer and Edwards (2003, 741) maintain that statistical innumeracy is seemingly attributed to problems inside our minds. They say:

> We disagree: the problem is not simply internal but lies in the external representation of information, and hence a solution exists. Every piece of statistical information needs a representation – that is, a form. Some forms tend to cloud minds, while others foster insight.

The following examples illustrate this point.

Example 4.4. Gigerenzer and Edwards (2003) deal with three numerical representations that foster confusion: single event probabilities, conditional probabilities, and relative risks.

The weather forecast says: "There is a 30% chance of rain today." This is a probability statement about a single occurrence: it will either rain or not rain today. Most often it is left open to which class the probability refers. Some will interpret this statement as meaning that it will rain today in 30% of the area, others that it will rain 30% of the time, and a third group that it will rain on 30% of the days with similar metereological conditions. Area, time, and condition are examples of reference classes. Each class gives the probability of rain a different meaning. Gigerenzer (2008, 174) shows that children can

TABLE 4.6: The red nose problem (From Gigerenzer, 2008, 174. By permission of Oxford University Press, Inc.)

Conditional Probabilities	Natural Frequencies
Pingping goes to a small village to ask for directions. In this village, the probability that the person he meets will lie is 10 percent. If a person lies, the probability that he or she has a red nose is 80 percent. If a person doesn't lie, the probability that he or she also has a red nose is 10 percent. Imagine that Pingping meets someone in the village with a red nose. What is the probability that the person will lie?	Pingping goes to a small village to ask for directions. In this village, 10 out of every 100 people will lie. Of the 10 people who lie 8 have a red nose. Of the remaining 90 people who don't lie, 9 also have a red nose. Imagine that Pingping meets a group of people in the village with red noses. How many of those people will lie? (n out of m).

solve Bayesian problems. In Table 4.6 two different representations of the same problem are given. The description on the left-hand side follows the Bayesian formula by the letter. The right-hand side is based on enumerated groups without any re-normalisation.[5] In tests respondents are much better at answering the second problem correctly. If you transpose the problem to lying as illness and the red nose as a positive response to a clinical test, then one understands the importance of the calculation. △

When communicating an increase or a decrease of risk, whether the probabilities are presented in relative or absolute terms is critical to producing appropriate responses. To say that the probability of a loss has doubled instead of saying that the probability has increased from 2% to 4% can lead to misperception. It is better to clearify the baseline probability as a reference and thus to use absolute terms.

Framing can have a profound effect on a message's response. Whether the famous glass is half full, i.e., gains from taking measures, or half empty, i.e., losses from not acting, can have an important effect. (Gigerenzer and Edwards, 2003, 743) report from clinical studies that positive framing is much more persuasive.

In quantifying risk with the Monte Carlo approach the risk produced is an

---

[5]The problem stated as conditional probability is solved with Bayes' rule. With r for "red nose is true" and l "liar is true" and their complement denoted by $\bar{r}$ and $\bar{l}$ we calculate

$$Pr(l \mid r) = \frac{Pr(l)Pr(r \mid l)}{Pr(l)Pr(r \mid l) + Pr(\bar{l})Pr(r \mid \bar{l})} = \frac{0.1 \cdot 0.8}{0.1 \cdot 0.8 + 0.9 \cdot 0.1} = \frac{0.8}{1.7}. \qquad (4.1)$$

With frequencies it is very intuitive, because there are 8 who have a red nose and are liars and 17 who have a red nose. Thus there are 8 out of 17.

estimate and thus has some margin of error. This margin is more pronounced with very small probabilities or risks, because assessment is difficult. In other situations, like scenario analysis, uncertainties arise from different assumptions and extrapolations. The most common approach when communicating such a figure is to give the most likely value. An other approach is to give the upper bound or "worst case" and yet another way is to give the fuller picture in the form of most likely value in addition to the confidence interval. This last method has the advantage of reminding the recipient that the figures have some relative but confined uncertainty. Alas, the recipient may just mentally suppress the information.

A common tendency in attempting to elucidate risk numbers is to over-simplify. As we have seen, it must be assumed that the audience does not know too much. Clarity and *oversimplification* are obviously not the same thing. Again, the audience must and often can be sufficiently motivated to understand complex quantitative information (Hyer and Covello, 2005, 106). In presenting risk information it may be helpful to explain how they were obtained and thus demystifying the risk assessment process. Yet, there is the risk of adding too many intricacies and much too much relativism, thus hindering a clear decision making. As often, a good balance is difficult to find.

Sometimes, the risk in question, which may not so familiar, is compared with a well known everyday risk. This may prove tricky because the latter risk, like the probability of a car accident, is often assumed and accepted and therefore it can be understood that the risk communicator wants to pre-empt a decision. Now, risk perception is not a one-dimensional beast, and thus such comparisons may be unproductive. Ignoring the qualitative factors affecting the perception of risk (see Appendix 1 at page 349) are likely to backfire unless not appropriately considered.

Why is this of any concern to risk management? Puzzling enough, banks like to make the distinction between managers and experts or "quants." Only rarely does a quant, and thus a numerate person by definition, makes it to the top of risk management – a quick glance at the officers' and directors' directory of the major banks shows not one CRO with a quantitative background. Chief Risk Managers tend to be (in the best case) economists, lawyers, general managers or accountants, but also historians and masters of art are in charge. The CRO is seen as a minor senior management position with a lower salary then those of the heads of business divisions. It is also a transitional job on the way to heading a business line, as experienced by Deutsche Bank, UBS, ANZ, etc. Therefore, even in the risk management organisation there is a big communication challenge, not to mention the gap between the CRO and the board of directors.

This does not imply that senior managers are neither intelligent nor well-educated. But it can mean that those people may lack statistical insight and thus insight into the meaning of risk quantifications and abstract models in general. Even the harsh reactions in the media voiced by many banking pro-fessors and central bankers attributing the big losses of the sub-prime crisis to the flawed risk models shows the same symptoms, viz. statistical illiteracy. As with technical matters, the lay users are generally much more confident of the devices than the engineers – and much more disappointed when they fail. Other authors make similar observations, e.g., Rebonato (2007, 137) writes:

> I believe that a dangerous disconnect is forming between special-ists (statisticians, mathematicians, econometricians, etc.) on the one hand, who are undoubtedly discovering more and more powerful sta-tistical techniques, and policy makers, senior managers, and politi-cians on the other, who are ill-equipped to understand when, and how, and to what extent these sophisticated techniques should be used and relied upon. Unfortunately, the mathematicians love and know a lot about mathematics but understand the financial and so-cial topics their techniques are applied to less deeply. The senior managers, the politicians, and the policy makers know the "social" is-sues well, but are likely to understand the mathematics used to tackle them even less than the statisticians know the underlying problems.

Risk managers frequently face the problem of having to explain uncertainty to a senior management that thinks scientific findings are absolute, precise, repeatable and reliable. In addition, the audience frequently associates cor-relation and association with causality. As a result, risk managers often face the additional task of trying to explain the data's and models' limitations and uncertainties.

A key obstacle is the term "risk" itself – in this book defined probabilis-tically – how it is measured, described, and ultimately perceived. Members of the board, senior management and risk managers perceive risk differently. Less technically trained people often personalise risk with the same conviction that most risk managers and quants depersonalise it. Ultimately, the board of directors must decide how much risk is acceptable. However, its decision is based on personal factors. One goal for the risk managers should be to edu-cate the senior management and the board of directors on the extent of risk. Nonetheless, by listening to and addressing concerns, the targeted audience will be better able to understand and, possibly decide on the appropriate risk.

The communication between quants and innumerate stakeholders can be seriously impaired by the assumption that the addressee lacks enough a deep understanding to warrant a diligent effort to convey a precise message. This can lead into a vicious circle followed by questionable decisions which in turn enforce the quant's prejudices of a poor understanding by the decision maker (see Figure 4.10). In the press release guidelines for scientists of the Hubble European Space Agency Information Centre (HEIC) you can find the following items:

> *Simplify:* A fundamental rule of written science communication is to make texts as simple as possible. Nowadays people simply do not have time for lengthy explanations.

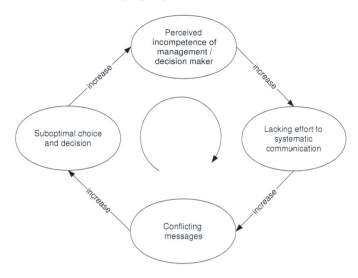

FIGURE 4.10: Vicious communication.

Explain: It is always necessary to match the writing to the level of the target group, but never more so than in science writing. Remember that "the reader knows nothing." (...) Build up your explanations from the lowest level, but try not to patronise, and avoid being overly didactic. The educational aspects go unnoticed in the best communication.

In this recipe we can see another difficulty in communication: conveying the message and educating at the same time. This is aggravated by the fact that the expert is normally in a lower hierarchical position than his or her audience.

While there is an impressive body of research and application guidelines for risk communication between government agencies and citizens, especially with respect to health care, food safety, natural catastrophes and terrorism, there is almost no information regarding communication of financial risks to ordinary citizens. In the former field, there is much effort undertaken to bridge the gap between experts, mainly engineers who dismiss their critics as being driven by emotions, and those who have to act, mainly lay stakeholders and politicians. Thus Fischhoff (2006) argues that the social sciences, especially with regards to communications and biases in judgement of those who analyse risk, have to play a paramount role in addressing the crisis in communication of financial matters.

We are not aware of a specific guideline to help communicate financial risks. The financial crisis of 2008 is also – or maybe first of all – a communication and leadership failure. The United States Government and the Federal Reserve (Fed) may have failed to explain and thus educate the general public about the crisis, its causes and the remedies taken. "'The fundamental business of the country is on a sound and prosperous basis,' said President Hoover in October 1929. The reasoning behind such piffle has been that any sign of panic from the nation's leaders would only make matters worse" (Economist, 2008a). President George W. Bush said on September 24th, 2008: "Our entire economy is in danger" adding that "our country could experience a long and painful recession." The mere talk of recession, echoed by the governor of the Fed, undermined the confidence of both business men and consumers and is thereby bringing about the recession everyone is trying to avoid. It should not be the officials making such a valuation, but rather the media. But in this crisis the media are not playing the key role of calming financial markets. Tabloid journalism has succeeded with hysterical, uninformed coverage dominating the news. There is a certainty though: we will not return to the old barter system any time in the near future. The financial crisis played itself out right in time for the 2008 U.S. elections. Ironically, the top officials of the U.S.A. do not follow the rules imposed on their subordinates. Covello and Allen (1988), working in a governmental agency, give the so-called *Five Rules for Building Trust and Credibility*:

1. Accept and involve the public as a partner. Work with and for the public to inform, dispel misinformation and, to every degree possible, allay fears and concerns.

2. Appreciate the public's specific concerns. Statistics and probabilities don't necessarily answer all questions. Be sensitive to people's fears and worries on a human level. (...)

3. Be honest and open. Once lost, trust and credibility are almost impossible to regain. Never mislead the public by lying or failing to provide information that is important to their understanding of issues.

4. Work with other credible sources. Conflicts and disagreements among organizations and credible spokespersons create confusion and breed distrust. Coordinate your information and communications efforts with those of other legitimate parties.

5. Meet the needs of the media. Never refuse to work with the media. The media's role is to inform the public, which will be done with or without your assistance. Work with the media to ensure that the information they are providing the public is as accurate and enlightening as possible.

Disseminating information without paying close attention to the complexities and uncertainties of risk is not simply effective risk communication. The risk decision makers should ask some standard questions when presented risk data and analysis. The six questions given in Outline 4 should be a mandatory selection. Paraphrasing the National Research Council (1989, 38): Well-informed

---

1. What are the weaknesses of the available data?

2. What are the assumptions and models on which the estimate is based?

3. What are the confidence limits of the estimate?

4. How sensitive are the estimates to changes in the assumptions and model properties?

5. How sensitive are potential decisions to changes of estimate's values?

6. What other risk assessments have been discarded?

OUTLINE 4: Questions of risk decision makers

choice about activities and products that present risks requires a wide range of knowledge. It depends on understanding financial products, markets, banking operations, human behaviour, legal matters to name a few and their mechanism by which they can cause loss or damage; on analysis and modelling of exposures and loss; on statistical expertise; on knowledge of the economic, social, ecological and other costs and benefits of various options; on knowledge of the constraints and responsibilities of risk managers; and on the ability to integrate these disparate kinds of knowledge, data and analysis. Needless to say, it is often hard in practice to gather all this knowledge. Nevertheless, the more complete the knowledge and the more quantitative answers are found, the better informed the ultimate decision will be.

4.4 Operational Risk Framework

Depending on the authors or institutions there may be many explicit or implicit definitions of the Operational Risk Framework. With the passage of time things have become clearer. An economic view would favour the least onerous approach if there were no benefits in choosing otherwise. A simple definition of the framework is to take the "Sound Practices for the Management and Supervision of Operational Risk," a publication of the Basel Committee (BCBS, 2003c) and to adhere to its eight principles addressing management. The eight principles are: *verbatim.*

Developing an Appropriate Risk Management Environment

1. The board of directors should be aware of the major aspects of the bank's operational risks as a distinct risk category that should be managed, and it should approve and periodically review the bank's operational risk management framework. The framework should provide a firm-wide definition of operational risk and lay down the principles of how operational risk is to be identified, assessed, monitored, and controlled/mitigated.

2. The board of directors should ensure that the bank's operational risk management framework is subject to effective and comprehensive internal audit by operationally independent, appropriately trained and competent staff. The internal audit function should not be directly responsible for operational risk management.

3. Senior management should have responsibility for implementing the operational risk management framework approved by the board of directors. The framework should be consistently implemented throughout the whole banking organisation, and all levels of staff should understand their responsibilities with respect to operational risk management. Senior management should also have responsibility for developing policies, processes and procedures for managing operational risk in all of the bank's material products, activities, processes and systems.

Risk Management: Identification, Assessment, Monitoring, and Mitigation/-Control

1. Banks should identify and assess the operational risk inherent in all material products, activities, processes and systems. Banks should also ensure that before new products, activities, processes and systems are introduced or undertaken, the operational risk inherent in them is subject to adequate assessment procedures.

2. Banks should implement a process to regularly monitor operational risk profiles and material exposures to losses. There should be regular reporting of pertinent information to senior management and the board of directors that supports the proactive management of operational risk.

3. Banks should have policies, processes and procedures to control and/or mitigate material operational risks. Banks should periodically review their risk limitation and control strategies and should adjust their operational risk profile accordingly using appropriate strategies, in light of their overall risk appetite and profile.

4. Banks should have in place contingency and business continuity plans to ensure their ability to operate on an ongoing basis and limit losses in the event of severe business disruption.

Role of Disclosure

1. *Banks should make sufficient public disclosure to allow market partic-ipants to assess their approach to operational risk management.(End verbatim.)*

After carefully reading those principles you will find many references to processes and none to structure. In this respect it is similar to the COSO framework. Nonetheless, we need a structure or organisation in order to bring all the single steps of a circular process about.

The risk process as defined in this Basel framework consists of some 5 elements, viz. identification, assessment, monitoring, and mitigation and control. In other publications of the Committee, "assessment" is substituted for "measurement." It may be assumed that reporting is dealt with as a subsection of monitoring. In Table 4.7 you can see three different structures. Semantics aside, they are very similar. In Figure 4.11 the principles are summarised.

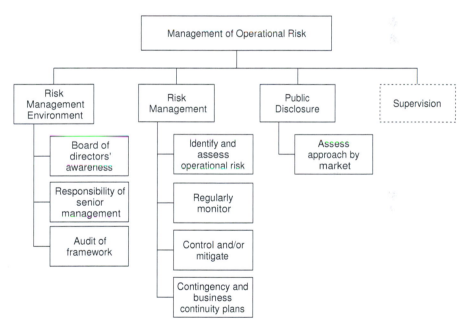

FIGURE 4.11: Basel 2 operational risk framework.

Although oversight is mainly implied in Basel 2 Accord, the AS/NZS and COSO frameworks explicitly acknowledge the importance of setting the organisational context and objectives and demanding clear communication throughout the organisation. The Basel 2 Accord requires that documentation should be available for review and consultation by external auditors and experts, similarly to the others. Thus we can conclude that there is convergence of

TABLE 4.7: Three different frameworks

AS/NZS 4360: 2004	COSO ERM	Basel 2
Establish the context	Internal environment, objective setting	(implicit)
Identify risks	Event identification	Identify
Analyse risks	Risk assessment	Assess
Evaluate risks	Risk assessment	(Measure)
Treat risks	Risk response and control activities	Control/ mitigate
Monitor and review	Monitoring	Monitor
Consult and communicate	Information and communication	(implicit)

views and that therefore not much may be gained by being fanciful and implementing a proprietary outline. We will go into more details concerning the management process after having looked at the organisational structure.

4.5   Operational Risk Structure

The structure is mainly determined by the roles and responsibilities of the framework. One starts at the top with the *board of directors* – as described in Section 4.4 – whose central supervisory task is to control the management with respect to risk. In bigger banks, a special *risk committee* of the board of directors focusing on the bank's risks and internal control system can support the monitoring of the overall approaches to implement the framework.

Generally speaking, *line management* is responsible for the operative implementation of the risk strategy and, thus, for operative risk management. The *senior management* is responsible for all risks of the bank and hence for designing and implementing its risk strategy. Often a *Chief Risk Officer* (CRO) – member of the executive board or one level below – is responsible for implementing the risk policy adopted by the entire senior management board. To what extent risk management is centralised or decentralised is discretionary and most often depends on the institution's corporate culture (Toevs et al., 2003). Of late, a matrix risk management organisational design has emerged. The Chief Risk Officer (CRO), often reporting to the CEO, is made accountable for both overseeing the risk measurement process and endorsing high-level risk governance decisions. He or she defines policies and develops tools and methods. These days, the role of the CRO is expanding.

Especially in Europe, a distinction is made between risk management and risk control. The first is understood as the comprehensive and systematic control and treatment of risks on the basis of economic and statistical knowledge. Risk management comprises identifying, measuring, assessing, managing and reporting on individual or aggregates of risk items. Risk management is effected on the appropriate organisational levels, with methods that are appropriate for and which reflect the idiosyncrasies of the institution. Risk control on the other hand is seen as an independent, controlling function, which oversees the risk profile of the institution. It prepares the risk information needed for monitoring risks and prepares the ground for the institution's risk policy, its risk appetite and tolerance as well as its risk limits, all of which must be approved by senior management or the board of directors.

Traditionally, the CRO addresses credit and market risks as well as operational risks. More specifically, these areas are covered by a Chief Market Risk Officer, Chief Credit Risk Officer and a Chief Operational Risk Officer. In leading institutions, the role often includes capital management tasks and assurance functions. This implies a close relationship with the Chief Financial Officer who also performs some global risk management duties.

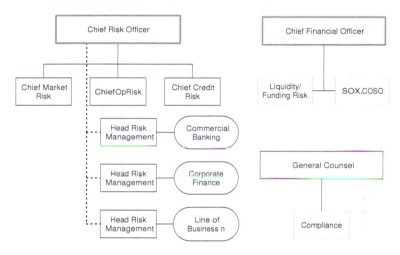

FIGURE 4.12: Frequent organisational structure.

CROs are responsible for staffing of governance committees for each of the three major types of risk, and lead a dedicated staff of specialists. Qualitative criteria for the implementation of the AM approach include (BCBS, 2005, 146):

> There must be regular reporting of operational risk exposure ... The bank must have an independent operational risk management function that is responsible for the design and implementation of the bank's operational risk management framework. The operational risk management function is responsible for codifying firm-level policies and

procedures concerning operational risk management and controls; for the design and implementation of the firm's operational risk measurement methodology; for the design and implementation of a risk-reporting system for operational risk; and for developing strategies to identify, measure, monitor and control/mitigate operational risk.

An additional level of management exists – what we term Head of Risk Management (see Figure 4.12 on page 277). These individuals carry no P&L responsibilities, but they have advisory, monitoring and control duties for all risk types within a particular line of business. Because of varying degrees of centralisation, they lack of a standard reporting hierarchy. In some cases they report directly to the CRO while in others they report to their business unit executives, with informational and reporting responsibility to the CRO.

The *matrix structure* – here implemented for the Head of Risk Management of the business units – is one of the most difficult and least successful organisational forms. Some have argued that it leads to conflict, confusion, informational logjams, and the proliferation of committees and reports; also that the overlapping responsibilities produce turf battles and a loss of accountability. When it comes to allocation of risk capital and its cost for the lines of business, then loyalty to the unit can undermine the risk organisation. As the Romans said, a slave who has three masters is a free man. It is a sound principle not to put anyone into a conflict of loyalties. While it is true that a universal bank environment with a profitable investment bank arm may lead to parallel structures, the core of this issue is the very fact that risk management is both supposed to limit risk and contribute positively to the bank's return.

As Figure 4.12 shows, not all risks are managed by the Chief Risk Officer. Often, the *general counsel* is responsible for compliance. Moreover, the Chief Financial Officer also has at least two risk issues on his desk: the extended reporting introduced by the Sarbanes-Oxley Act and the traditional liquidity risk management effected by treasury.

In many banks the insurance risk manager is still within the supply organisation or the financial division and thus not participating directly in the overall risk management effort. Increased awareness that insurance may be used as a regulatory mitigant of risk should lead those institutions to reconsider the collocation of the insurance buying department.

Not having said much about the structure in its papers, the Committee may have meant to imply that this can be left to the bank's discretion.

Besides the Operational Risk organisational unit there are other units participating in the management. We will just focus on the prerequisites necessary to the risk model's implementation. What is the role of internal and external audit? With regard to the first, there are several Basel papers pertaining to this question (BCBS, 2001b; International Auditing Practices Committee,

2001; BCBS, 2000). We shall focus narrowly on the model and its implementation. The regulatory text asserts (BCBS, 2005, 167):

> Effective control of the capital assessment process includes an independent review and, where appropriate, the involvement of internal or external audits.

Because the methodology is embodied in the computer model, the computer model itself must be included in any real review of the process and of capital assessment.

Now, how is the supervision organised? We have at least three players in the supervision of the risk capital assessment process:

- the regulatory supervisor
- external audit and
- internal audit.

In the case of well-established principles and practices, some duties can be delegated, if the delegate's qualifications meet predetermined requirements. Therefore the regulatory supervision may rely on the external or internal audit, who in turn may commission some supervisory tasks to external experts, especially where specific know-how is involved (see Figure 4.13).

It is worthwhile mentioning that these supervisory bodies have different, albeit partially overlapping, stakeholders.

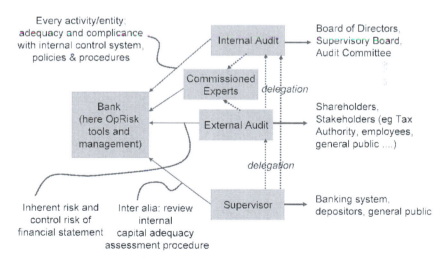

FIGURE 4.13: Roles and relations between supervising bodies.

According to the Basel Committee's papers on Internal Audit (BCBS, 2001b), the scope of internal audit includes, inter alia, the following:

- the accuracy and reliability of the accounting records and financial reports;

- the bank's system of assessing its capital in relation to its estimate of risk;

- the systems established to ensure compliance with legal and regulatory requirements;

- the review of the bank's internal capital assessment procedure;

- the testing of the reliability and timeliness of the regulatory reporting.

From this task list it is easy to infer that the internal audit plays an important supervisory role, in conjunction with the computer models that calculate regulatory capital, and of course, the required compliance checks.

Now, what is an audit trail in a computerized environment? For typical accounting programs (Boritz, 2001) states:

> When assessing the control structure to be built into a planned system, auditors can also assess the adequacy of the trail to be established to allow transactions to be traced from their source to their final destination, and vice versa. This trail may take the form of visible printouts, or may merely represent the ability to reconstruct details supporting a total figure on request.

In close analogy we therefore state the following for the quantification:

Definition 4.1. An *audit trail* within the calculation of regulatory capital, e.g., Loss Distribution Approach, means being able to reconstruct and recalculate exactly the reported figures on request. □

What does this imply for the process and the computer model? Obvious but necessary elements for successful reconstruction of data are:

- integrity and availability of input and output data,

- availability of computer model,

- verifiable set of input data and parameters,

- verifiable computer model version, configuration and implementation, and

- completeness and non-modifiability of required data.

The incorporation of such features expedite the regulatory assessment of methods and processes enormously.

Because the quantification of the required capital relies on the input of data which in turn may have been estimated in a tortuous exercise, good practice extends and reaches far up-stream in the process. The quality of the calculation cannot be better than the quality of the input.

Next we turn to the structure's partner, i.e., the process.

4.6   Operational Risk Process

The operational risk process is defined in the framework. We outline the process as a superset of the three structures from Table 4.7 on page 276:

- objectives,

- identification,

- assessment an measurement,

- monitoring and communication,

- mitigation and control.

Please note that while the term "process" implies forward motion from one item to the next, these activities are either recurrent or event driven tasks. Such steps are taken periodically and are synchronised with other reporting tasks. But, major changes in the internal or external environment must trigger the process, taking risks identification as a starting point.

Objectives include establishing risk policies to reflect the bank's risk principles, and its risk capacity and risk appetite or tolerance, consistent with evolving business requirements and best practice. In the AMA context this risk identification revolves around the continuous monitoring of existing processes and portfolios with respect to risks, assessing new business opportunities and complex or unusual transactions, and by reviewing the risks in the light of market developments and external events.

Measurement of quantifiable risks should use methodologies and models that have been independently validated and approved.

Monitoring is the act of observing the risk profile – both the internal and external variables – and recording it. It is the prerequisite for communication and interventions. There should be implemented comprehensive risk reports to stakeholders, and to management at all levels, showing the actual and historic risk profile, and all variances with respect to targets and limits etc.

Risks are controlled by monitoring and enforcing compliance with the risk principles, and through policies, limits and regulatory requirements. Mitigation regularly requires buying appropriate insurance coverage.

4.6.1   Risk Objectives

Objectives must exist before management can identify potential events affecting achievement. The framework, as defined above, ensures that management has in place a process to set objectives and that the chosen objectives support and are aligned with the bank's mission and are consistent with its *risk appetite*.

Risk appetite is the amount of risk an entity is willing to accept in pursuit of value. Recalling the Definition 1.1 on page 14, it is a random variable characterised by its distribution. Consequently, it should be determined as a pair: amount and confidence level. It reflects the bank's risk management philosophy, and in turn influences its operating style.

Risk appetite is directly related to a bank's strategy. It is considered in a strategy setting, as different bank's strategies entail different risks. Risk management helps management select a strategy that aligns anticipated value generation with the entity's risk appetite.

Moreover, risk appetite guides resource allocation. Management divides resources among business units and initiatives taking into consideration the bank's risk appetite and the unit's plan for generating desired return on invested funds. Management considers its risk appetite as it aligns its organisation, people, and processes, and plans the infrastructure necessary to effectively respond to and monitor risks.

Some authors differentiate between risk appetite and risk tolerance. But besides the understanding that appetite is more of a target and tolerance more of a limit, we do not see the necessity to treat these two terms differently.

4.6.2   Risk Identification

It must be assumed that the bank has a detailed outlay of its operational risk. This often goes beyond the Basel three-tiered classification as reported in the Appendix 5.3 on page 351. It makes perfect sense to group risks in a manner that reflects the nature of a specific organisation, in line with its unique business focus.

Example 4.5. As an example, a major Swiss bank structures its operational risks as:

- *Transaction processing risk,*

- *Compliance risk,*

- *Legal risk,*

- *Liability risk,*

- *Security risk,*

- *Tax risk.*

The bank defines as compliance risk what Basel explicitly sees as legal risk, whereas legal risk in turn is identified mainly with *ultra vires* type of risk. For transparency purposes it would be easier to stick to standards as set by the Accord.   △

Therefore, the Basel 2 Accord nonetheless requires that data must be mapped according to its classification. The main activities identified are:

- further analysis of one's own and external potential and actual risks and

- assessment of the risk implications from innovations and changes (processes, products, technologies),

- the prediction of new risks,

- an attempt to quantify perceived risks,

- and an attempt to make people aware of risks.

In Figure 4.14 (RAMP, 1998, 68) we show schematically a matrix along the two dimensions "risk exposure" (or quantification) and "risk perception." The goal must be to predict or invent new risks and then bring them into the second quadrant, i.e., to the "known-known" field. In the first quadrant we are concerned with quantification. If it is not possible to put a figure on an individual hazard then perhaps it is possible to group it and then assess the group. The third quadrant suggests a lack of awareness. Therefore it is a task of acknowledging the risk in an appropriate manner. From our discussion on probability we know that the greatest potential losses are shrouded in uncertainty and ignorance, as symbolised by quadrant four. Therefore, it is important for internal and external "seers" and "futurologists" to creatively predict or "dream up" such potential losses. The analysis of losses, incidents

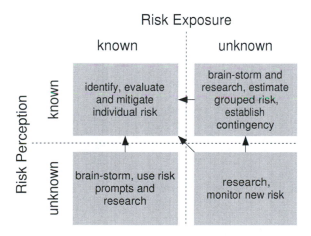

FIGURE 4.14: Degrees of risk awareness. (From RAMP, 1998, 68. By permission of Thomas Telford Ltd.)

and occurrences brings us to the causality discussion in Section 2.10 starting on page 61. It seems important to dissect one's own (or even publicly known) cases with an eye to discerning sets of sufficient and necessary conditions to imply a "causal" link. A further, useful step in such analysis would be to group cases and then assess both the costs and benefits of eliminating the original risk.

The findings of an older study (Latter, 1997, 40) shows poor asset quality and mismanagement as the major causes for problems. In general, analysis must go further than just to find that mismanagement, i.e., an uncontrollable variable, is the principal cause of problems.

Such analysis can also involve the much advertised Bayesian networks. The actual knowledge is mapped to so-called *influence diagrams*. On the most abstract level these are *directed acyclic graphs* DAG, without feedback loops. The graph joins chance nodes that represent events described by continuous or discrete probability distributions. Each node is identified with its variable. Figure 4.15 shows such a graph where in addition we have annotated the direction of the influence by plus and minus signs. A plus sign means more (or higher level) of the starting entity leads to more (higher level) of the ending entity, e.g., more incentives lead to greater motivation.

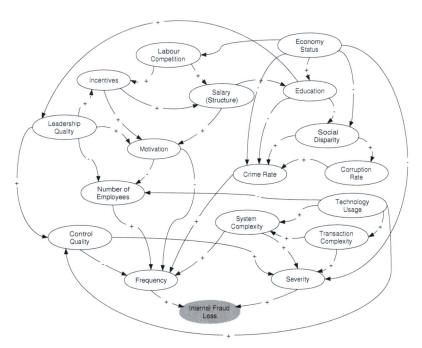

FIGURE 4.15: Example influence diagram for internal fraud.

Bayesian networks are constructed from just such graphs. For each node having one or more parents, a local probability distribution contingent on the states of the parents is defined. The distributions may be discrete (i.e., tabular) or may depend on the tools chosen.

For any given exhaustive probabilistic model, the joint probability distribution can be specified. Let us denote the set of variables $X = \{x_1, x_2 \ldots, x_n\}$. The joint probability distribution is $Pr(X) = Pr(x_1, x_2, \ldots, x_n)$. Note that the complexity grows exponentially with the number of variables. The chain rule as generalisation of the product rule, i.e., $Pr(x_1, x_2) = Pr(x_1 \mid x_2)Pr(x_2)$, leads to

$$Pr(x_1, x_2, \ldots, x_n) = \prod_{i=1}^{n} Pr(x_i \mid x_1, \ldots, x_{i-1}). \tag{4.2}$$

Bayesian networks give a compact representation of $Pr(x_1, x_2, \ldots, x_n)$ by factoring the joint distributions into local, conditional distributions for each single variable x_i given its parents. If we let $pa(x_i)$ denote a set of values for the variable's x_i parents and we assume that x_i is independent of all other, then the joint distribution is the product of all conditional probabilities according to

$$Pr(x_1, x_2, \ldots, x_n) = \prod_{i=1}^{n} Pr(x_i \mid pa(x_i)). \tag{4.3}$$

For scenario and other exploratory purposes, states can be changed and the outcome concerning the variable of interest (here loss or damage due to internal fraud) can be studied. There is a second important purpose. Any reconfiguration of the network can quite easily be translated into a new network. These changes can be identified with external interventions. The example below will illustrate this feature.

Example 4.6. This very simple example is taken from Pearl (2000, 15). We analyse the DAG consisting of five nodes $(X_1, X_2, X_3, X_4, X_5)$, representing the season of the year, whether a sprinkler is on, whether rain falls, whether the pavement got wet and whether the pavement is slippery (see Figure 4.16). The season's node can take four distinct states with appropriate probabilities, while the others are dichotomous, i.e., yes or no, on or off. According to

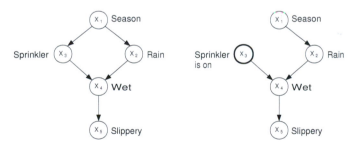

FIGURE 4.16: Two graphs representing Bayesian networks.

Eq.(4.3) and the left graph of Figure 4.16 the distribution will be

$$Pr(x_1, x_2, x_3, x_4, x_5) =$$
$$Pr(x_1)Pr(x_2 \mid x_1)Pr(x_3 \mid x_1)Pr(x_4 \mid x_2, x_3)Pr(x_5 \mid x_4).$$

Under the assumption of stable and autonomous local relationship some external intervention could change just one arrow without the need to reconfigure the whole model. For example, turning the sprinkler on results in a new DAG with the arrow from X_1 to X_3 removed as the season will no longer have any effect. The function corresponding to the right DAG of Figure 4.16 is

$$Pr(x_1, x_2, x_3 = on, x_4, x_5) =$$
$$Pr(x_1)Pr(x_2 \mid x_1)Pr(x_4 \mid x_2, x_3 = on)Pr(x_5 \mid x_4).$$

This example illustrates the difference between observing X_3 being "on" and the action of turning the sprinkler on. In the first case we may infer that the season is a dry one, that it has not rained for a while, etc. The ease of coping with interventions is a major step forward compared with the sole description of a probability function. There are additional extensions asking "do x if you see y" or even "do x with probability p if you see z" (Pearl, 2000, 107).   △

Difficulties exist both in defining a sensible model and setting appropriate probabilities. We shall not delve further into this topic as it is covered extensively by the literature on LDA's "sibling," i.e., the score-card approach. See Alexander (2002) and its references or King (2001).

4.6.3   Risk Assessment and Measurement

Assessment is a *qualitative* process. However, its result are amenable to further *qualitative* analysis, because it implies counting, classifying and estimating in a systematic way. In Section 4.7.3 on page 301 we show, for example, how to construct an index from a questionnaire.

Assessment's typical tools in social sciences are, according to Galtung (1967, 110), the following:

- observation,

- structured and unstructured interviews,

- questionnaires.

In addition, facilitated workshops are sometimes held, where groups form the units of investigation. All informal or unstructured approaches may be more exhaustive but come at a significant cost. They are very difficult to evaluate. Questionnaires, on the other hand, lend themselves at inter-temporal comparison as they can be conducted under similar environmental conditions.

If we have defined a bundle of indicators that may be calculated from third parties or internally, then this step in the Operational Risk Process means updating them regularly and contextualizing them. This is of concern especially for the assessment of the actual state of the pertinent external environment.

Physical scientists are obsessed with measuring, e.g., William Thompson of Kelvin stated:

When you can measure what you are speaking about, and express it in numbers, you know something about it; but when you cannot measure it, when you cannot express it in numbers, your knowledge is of a meagre and unsatisfactory kind: it may be the beginning of knowledge, but you have scarcely, in your thoughts, advanced to the state of science.

In our context this is not true. We prefer to look at things like Albert Einstein who allegedly said: "Not everything that can be counted counts, and not everything that counts can be counted."

This booklet is concerned with quantifying risks using the Loss Distribution approach. Therefore, we need not dwell too long on the topic of measurement in general. In short, we believe that measurement, even if it is not a perfect science as applied to risk management, is of tremendous importance to successful business.

It is often presumed that quantification implies controllability. This is not necessarily true. Moreover, the reverse is not necessarily true either: that is, controllability need not imply quantitativeness, although in most adaptive systems some sort of measurement is present in the feedback loop. These assertions are often not reflected. As we have seen earlier on control is frequently associated with over-confidence.

4.6.4 Risk Monitoring and Communication

As we already mentioned, monitoring is the act of observing the risk profile through the indicators or new facts, calculating derived risk metrics or indices, relating them to the economic capital and recording all pertinent information systematically. Then the data must be made available to the relevant stakeholders in an appropriate way. This means it must be processed to the needs and control possibilities of the units and managers.

On the one hand, ongoing monitoring should be embedded as much as possible in business processes performed by all employees performing their assigned tasks. Process and risk owners play a crucial role as does the establishment of incentive schemes motivating employees to continuously fulfil their responsibility and applying sanctions for failures. This kind of task is called *in-process monitoring*.

On the other hand, there should be separate inspections by several internal and external entities, called *process-independent monitoring*. In combination with provisions in banking supervision, the internal audit unit, supervisory board as well as independent auditors constitute essential safeguards against assuming excessive risks. Although differing in methods and tasks, all the parties involved should aim at actively cooperating, in particular, to avoid the emergence of monitoring gaps. A follow-up mechanism regularly ensures that deficiencies are eliminated and agreed-upon measures and recommendations are implemented in time.

An integral part of any risk management process is the current and comprehensive provision of information. The New Basel Accord demands of banks that intend to implement the Advanced Measurement Approach, that (BCBS, 2005, 146)

> There must be regular reporting of operational risk exposures and loss experience to business unit management, senior management, and to the board of directors. The bank must have procedures for taking appropriate action according to the information within the management reports.

In order to close the management loop the continuous provision of relevant data must potentially lead to action especially when some set thresholds are reached.

One of the most popular summaries of risk is the so-called *risk map*. It essentially consists of a two-dimensional grid with classes for probability or frequency of loss events and for severity. In Figure 4.17 we see an example. The map has three risk regions to account for gradations in severity and probability. The information consists in the placement of risk on the map. The simplest risk maps show "risks" as a function of expected number of

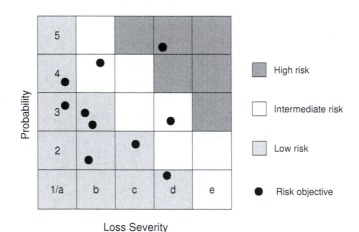

FIGURE 4.17: Typical risk map.

occurrence and expected severity of impact. The product of these two terms is taken as risk measure, being the expected annual loss or damage. (In lieu of a point estimate one could depict ranges in both dimensions or show ellipses. Nonetheless, it is not the risk measure we mean.)

The limits for each cell can be constructed as shown in the table below. The figures are taken from SAP's annual report 2005 as an example.

Probability			Severity [m EUR]		
Class	Description	Range	Class	Description	Range
1	very probable	81...100%	a	insignificant	≤0.2
2	probable	61...80%	b	small	0.2...1
3	possible	41...60%	c	moderate	1...5
4	unlikely	21...40%	d	large	5...25
5	rare	1...20%	e	very large	≥ 25

Risk maps have been a widely used representation of risk assessment for many years – not only in the financial industry. The risks are most often analysed in some kind of self-assessment of the senior management and in-house experts resulting in a consensus view. It is a typical top-down approach. Risk maps are thus also a piece of evidence that risks are identified at the top of the organisation – also for audit or regulatory purposes.

As decision aids, risk-maps should direct management to develop action plans for the most severe risks which are outside a pre-determined risk tolerance. Fervent enthusiasts argue that risk-maps can make explicit the link between risk and strategic objectives of the company. But the main problem is the actionability, especially for very rare events with a high loss potential.[6] Therefore, it is rather a tool for governance and less for internal decision making.

Communication goes beyond pure risk reporting – understood as periodically delivering a report to the main stakeholders. While the Chief Risk Office wants to demonstrate effort and competence, preserve status and influence the senior management, senior management itself is desirous of facts, reassurance and, perhaps, somebody to blame in case of failure.

The communication from the risk experts should *inter alia* (Fischhoff, 2002):

1. identify common knowledge – what goes without saying,?

2. identify critical gaps – what's worth knowing,?

3. present quantitative information – how severe and how frequent is the risk,?

4. present qualitative information – what determines the new risk,?

5. create a coherent mental model – find a "story line,"

6. evaluate success.

---

[6]Cardinal Richelieu, Prime Minister of France said about low probability events: "Un mal qui ne peut arriver que rarement doit être présumé n'arriver point. Principalement, si, pour l'éviter, on s'expose à beaucoup d'autres qui sont inévitables et de plus grande conséquence."

Items 1 and 5 deserve further comment. The specialist in charge must be able to assess the extent of the audience's knowledge, and, where necessary, to provide that knowledge required to understand the complex profile of risks. Risk officers must make certain that they provide coherent mental models for the risk profiles they pass on, lest senior management, when reviewing the risk profiles, create a "folk theory" to fill in personal knowledge gaps.

A standard risk report for operational risk for a bank using the LDA should incorporate items as mentioned in Outline 5 on page 291. It is intended as a super-set. Risk reports may and should vary in depth depending on their target audience. The content should focus rather on changes of values and time series than on exceptions. Unit managers need risk reports to discuss the allocation of capital and the attribution of insurance premiums, make comparisons with other units and determine the outcome of their risk initiatives. Senior management needs the reports for policy approval, resource allocation and strategic purposes.

The presentation, i.e., graphics, tables, text, interactive features, etc., and the quantity of information have a crucial effect on the reception of the content (see also our discussion on page 264). Although the SEC has asked reporting companies to present filings in a plain and comprehensive way (Office of Investor Education and Assistance, 1998), there is, unfortunately, still a widespread disregard for such straightforwardness. All too often, confusing presentations and reports and a flood of information are linked to obfuscation. Frees and Miller (1998) give some examples on how not to present data and Timm (1991) shows how to lie with it. Some older research suggests further that in spoken communication the non-verbal aspect is more prominent than the verbal (Mehrabian, 1968):

- 7% verbal (words used),

- 38% vocal (pitch, stress, tone, length and frequency of pauses),

- 55% facial (expression, eye contact).

An effective communicator must, in some sense, be an actor with respect to body language, i.e., posture, gesture, body movement.

We have tried in the preceding pages to make clear that with the Monte Carlo approach – but also with other estimating and testing techniques – the numbers of interest are best represented by confidence intervals, evidencing the fact that figures are ranges containing additional information, i.e., the range's spread. Nonetheless, a specific number has a vividness and simplicity that makes it an inevitable focus of debate. And often where ranges are given, the reader simplifies matters by calculating the mean himself.

The reports should also be checked with respect to all the different requirements related to stakeholders, i.e., corporate government, legal and regulatory requirements and so on.

External reporting on the bank's risk management – also concerning operational risks – is becoming increasingly important. Many banks preparing

1. Summary

2. Risk index and indicators

3. Risk capital figures by objectives (events, lines of business, etc.)

4. Internal loss events with analysis

5. External loss events

6. Stress and scenario analysis

7. Status of major initiatives

8. Issue track

9. News (from regulator, industry, etc.)

10. Conclusions and recommendations

OUTLINE 5: Items of a standard risk report

their reports according to IFRS include this information as a part of the notes on the annual report. The organisation and its integration into company-wide risk management are often described in a general, introductory section of the risk reports. In most cases, the definitions used in the bank and a concise description of the most important principles and methods are provided in a section specifically focusing on operational risk management and operational risks. Nowadays, big banks usually specify the allocation of their economic capital to individual risk categories.

4.6.5 Risk Mitigation and Control

Classically, there are four typical responses to new risks in order to control, i.e., bring to a desired level, and mitigate their impact:

1. *Avoidance*: stopping or reducing specific activities, products, markets and so forth, even selling parts of the institution and outsourcing;

2. *Reduction*: taking measures to reduce the probability of an occurrence and the severity of events achieved through everyday management decisions;

3. *Sharing*: transferring risk to a third party or pooling risks to increase diversification;

4. *Acceptance*: taking no specific action and relying on the capital cushion.

These responses are not specific to operational risk but are shared with all risk categories.

Risk mitigation involves prioritising, evaluating, and implementing the appropriate risk-reducing actions recommended by the risk assessment process. Often, the elimination of risk may be impractical or nearly impossible, because the risk is not directly controllable or cannot be effected with immediate result. But where feasible, the results of the prior steps in the risk process must be used to control the risk drivers in such a manner as to comply with the risk appetite in a cost efficient manner. Risk self-assessment is especially useful in highlighting gaps that need attention.

The tools of risk mitigation mainly include a multitude of organisational safeguards and control measures within the framework of an internal control system:

- guidelines and procedures,

- separation of functions and "four-eyes principle,"

- need-to-know and need-to-access principle,

- physical access control,

- coordination and plausibility checks,

- limit management,

- inventories and

- disaster recovery and business continuity planning.

Generally speaking, risk mitigation and control encompasses risk prevention, limitation and avoidance. Theses are all internal measures. The most effective external mitigation technique, on the other hand, is risk transfer, mainly through the purchase of insurance. Acceptance of those risks that could be otherwise mitigated must follow cost-benefit analysis, although it may admittedly be difficult to quantify the benefit side of the balance.

Not surprisingly, insurance may be the form of mitigation most amenable to direct and quantifiable analysis with the LDA approach. Unfortunately, our run-through example shows that a realistic level of insurance premium only decreases risk capital by percentage points. This holds true even if a great deal of coverage is purchased. We will cover this phenomenon in more detail in Section 4.9 starting on page 317.

Another aspect of this stage considers the improvement of controls. One main issue is with adapting, updating or introducing new *policies*. Procedures describe the way these policies are effected. Policies must be well communicated, available everywhere as comprehensive documents and judiciously and consistently implemented.

4.7   Business Environment and Internal Control Factors

Under the new Basel Accord (BCBS, 2005, 150) the following holds:

> a bank's firm-wide risk assessment methodology must capture key business environment and internal control factors that can change its operational risk profile. These factors will make a bank's risk assessments more forward-looking, more directly reflect the quality of the bank's control and operating environments, help align capital assessments with risk management objectives, and recognise both improvements and deterioration in operational risk profiles in a more immediate fashion.

According to this rule banks must assess or measure – that is, construct indices for – two areas, viz. business environment and internal controls. While the latter is sufficiently defined by the COSO framework and thus for the institutions applying the Sarbanes-Oxley Act (SOX), the first needs some further analysis.

4.7.1   Business Environment

As we try to be true to Ockham and his appreciation of simplicity, we shall refrain from reinventing the wheel and apply instead a well-known standard. This standard can be taken from Porter (1990), who describes the business environment with five items:

1. competitors,

2. new entrants,

3. substitute products,

4. suppliers and,

5. customers.

This is intended to supplement the external environment, where we would classify factors like interest rate level, stock market conditions and so forth. If you consider the external, internal and business environments as a whole we need not debate where to classify certain difficult-to-categorize items.

The likelihood of new players entering the actual banking sector is very improbable because of the overwhelming regulation. In the supplier category we have to, alas, include the regulators and other bodies of stakeholders. The business environment supplements the general environment, where especially the economic aspect encompasses many factors and indicators relevant for banks but also for other industries.

In Figure 4.18 we report some indicators from the banking statistics as compiled quarterly by the Federal Reserve Bank of New York (FRBNY, 2006). The data is based on the largest bank holding companies in the U.S. There are

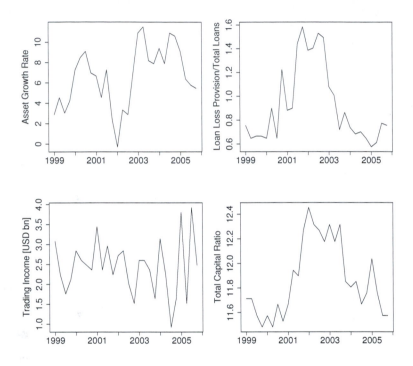

FIGURE 4.18: Some quarterly U.S. banking statistics. (Data: Federal Reserve Bank of New York.)

some regularly assessed data on banking from which indications concerning the items above can be analysed. The main issue here remains the linking of the business environment to the actual state of operational risk within the bank. Is fierce competition – perhaps operationalised by decreasing margins – dangerous? As often, here is food for thought.

4.7.2 Internal Control

The COSO's first framework (COSO, 1992) for internal controls has defined internal controls as follows (COSO, 2004, 107):

Definition 4.2. *Internal control* is a process, effected by an entity's board of directors, management and other personnel, designed to provide reasonable assurance regarding the achievement of objectives in the following categories:

- Effectiveness and efficiency of operations,

- Reliability of financial reporting,

- Compliance with applicable laws and regulations.

□

The framework consists of the three dimensions:

- components,

- objectives and,

- entity unit.

These constitute the COSO cube (see Figure 4.19). The components in turn consist of five interrelated elements:

- Control environment sets the tone of an organisation, influencing the control consciousness of its people. It is the foundation for all other components of internal control, providing discipline and structure.

- Risk assessment is the way in which an entity identifies and analyses relevant risks to achieve its objectives, forming a basis for determining how the risks should be managed.

- Control activities are the policies and procedures that help ensure that management directives are carried out.

- Information and communication systems support the identification, capture, and exchange of information in a form and temporal frame that enable people to carry out their responsibilities.

- Monitoring is a process that assesses the quality of internal control performance over time.

The objectives are three, namely

- financial reporting,

- operations and,

- compliance.

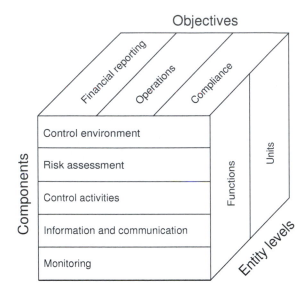

FIGURE 4.19: COSO's cube of internal control.

There is a direct relation between objectives (what an entity endeavours to achieve) and components (which constitute what is needed in order to achieve the objectives). Moreover, internal control is relevant to the entire organisation and to any of its operating units or business functions.

For the Advanced Measurement Approach, management has an opportunity to build on many of the processes required by the PCAOB Standards for SOX reporting as well as Federal Deposit Insurance Corporation Improvement Act's internal control assessment requirements. Specifically, the risk control self-assessments (RCSA) required by the AMA should ideally be integrated into similar RCSA developed for SOX and FDICIA compliance, notwithstanding the fact that AMA has a higher standard for both qualitative and quantitative risk assessments. SOX actually has the much narrower objective of attesting to internal control over *financial reporting* (ICOFR). Here are some synergies between the CRO and the CFO office – which is a rare piece of good news.

The control system is mainly organised around business processes. Davenport (1993) defines a *business process* as follows:

Definition 4.3. A process is a specific ordering of work activities across time and space, with a beginning and an end, and clearly defined inputs and outputs: a structure for action.   □

Another view of process is as the coordination of work, whereby a set of skills and work routines are exploited to create a capability which cannot be

Process XY1							
	Control Objective	Control Activity	Effectiveness	Efficiency	Traceability	Others	Score
Security							
Risk 1	CO 1	Policy 1	3	4	3	4	4
Risk 2	CO 2	Policy 2	4	3	2	2	3
...			...	...	...	...	...
Risk n	CO n	Policy n	3	3	3	3	3
Traceability							
Risk n+1	CO n+1	Policy n+1	4	3	4	3	4
Risk n+2	CO n+2	Policy n+2	5	3	4	2	4
...			...	...	...	...	...
Risk m	CO m	Policy m	3	4	4	4	4
Summary							3.67

FIGURE 4.20: Synopsis of the control process.

easily matched by others. We endorse the first description which is also a work-flow description.

For the relevant processes we assess the potential risks, define control objectives and *control activities*. Control activities are the specific policies and procedures management uses to achieve the control objectives. The most important control activities involve segregation of duties, proper authorization of transactions and activities, adequate documents and records, physical control over assets and records, and independent checks on performance. The board defines relevant quality requirements for the processes. Some auditors (PricewaterhouseCoopers, 2007) suggest a scale with five rungs according to the following table.

5	Optimised	Control activities mainly automated
4	Monitored	Regular monitoring
3	Standardised	Documented and traceable
2	Informal	Controls hardly traceable
1	Not very reliable	Few or no internal controls

Whether internal control is exclusively a process can be questioned. Of course it is also about structure. The internal control factor list is organised according to the Outline 6 on page 298.

Controls, in a narrower sense, are any action undertaken by management to increase the probability that established goals and objectives are achieved. They can be grouped into three broad categories:

- *preventive controls* deter undesirable events from occurring, e.g., restricting access to records,

- *detective controls* uncover and help correct undesirable events that have occurred, e.g., preparing reconciliations, and

- *directive controls* cause or encourage a desirable event to occur, e.g., written job descriptions.

- Control Environment

    1. Integrity and Ethical Values
    2. Commitment to Competence
    3. Management's Philosophy and Operating Style
    4. Organisational Structure
    5. Assignment of Authority and Responsibility
    6. Human Resource Policies and Practices

- Risk Assessment

    7. Goals and Objectives
    8. Risks
    9. Managing Change

- Control Activities

    10. Policies and Procedures
    11. Controls
    12. Controls over Information Systems

- Information and Communication

    13. Access to Information
    14. Communication

- Monitoring

    15. Management Supervision
    16. Outside Sources
    17. Response Mechanisms
    18. Self-Assessment Mechanisms

OUTLINE 6: Internal control factor classes

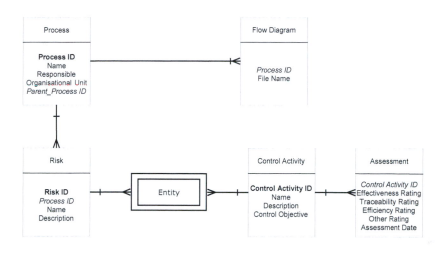

FIGURE 4.21: Control repository design.

In order to assess the quality of internal controls – not only for financial reporting – it is suggested that participants compile a questionnaire with several items to be rated or scored by values ranging from 1 to 5 (alternatively from -2 to 2). The so-called five-point scale is often used because it is a good compromise and it has a neutral or average value. According to the Outline 6 we have 18 items grouped in five classes. If we foresee five to ten questions per item, we have a total of some 90 to 180 questions. As an example we give in Outline 7 on page 300 some questions, or better assertions, to be rated. It is taken from Stanford University (1996). All the other items of the above COSO structure should be developed similarly. In Section 4.7.2 on page 294 there is a brief excerpt.

Now we have to define who should participate in this so-called *self-assessment*, how it should be carried out and how often. There are a couple of vendors offering web-based solutions for compliance that include risk self-assessment features. Such impersonal tools promise flexibility in scheduling, controlling – timeliness, completeness, etc. – and evaluating the assessment. On the other hand, such automated methods lead easily to boredom and their usefulness is diminishing. Thus the quality diminishes over time. Therefore, the methods for questioning must be varied. As risk and control are ubiquitous in the line organisation and pervasive to all hierarchies, a full count sample could be envisaged. On the other hand, standing panels of selected or randomly chosen assessors could ensure a more diligent assessment.

Rate each question with 5, if you strongly agree, 4=agree, 3=uncertain, 2=disagree, 1=strongly disagree.

1. A process exists to identify and consider the implications of external risk factors concurrent with establishing unit-wide objectives and plans.

2. A process exists to identify and consider the implications of internal risk factors concurrent with establishing unit-wide objectives and plans.

3. The likelihood of occurrence and potential monetary impact have been evaluated. Risks have been categorized as tolerable or requiring action.

4. In-depth, cost benefit studies are performed before committing the unit.

5. A risk management programme is in place to monitor and minimise exposures.

6. External advisers are consulted as needed to supplement internal expertise.

7. Internal Audit's assistance is requested whenever internal control issues surface.

OUTLINE 7: Questions concerning risks in risk assessment

The resulting *data matrix* $\mathbf{R} = [r_{ij}]$ of the questioning – according to social science practice – is the Cartesian product of the questions $\mathbf{Q}$ times the response unit $\mathbf{U}$ on the response $\mathbf{R}$. The element r_{ij} of the data matrix is the response the unit U_j gives to the question Q_i. The response unit is the employee who actually belongs to a line of business, is located in a certain country, etc. Therefore, you may construct several conditional data matrices by just choosing those columns of response units that have the attribute. If data collection is repeated, the matrix must also take into account time. This data matrix will be used to construct indices.

In a first step we would aggregate the questions to the 18 items, e.g., by equally weighting the answers. Then we could further construct 5 indices, one for each class. And lastly, we could compound these to yield only one index standing representing quality of the internal control. The same obviously applies to indices for the business environment and the like.

4.7.3   Factors and Indices

Definition 4.4. *Measurement* of some attributes of a set of objects or variables is the systematic process of assigning numbers or other symbols to the object or variable in such a way that relationships of the numbers or symbols reflect relationships of the attributes of the object or variable being measured. □

A symbol can be a letter, a word, etc. Thus, measuring the length of some sticks leads to the correspondence of "greater than" for the numbers with "longer than" for the sticks. But 40 degrees Celsius is not twice as warm as 20 degrees. With numbers you can do much more transformations than with the objects. Obviously, what is meaningful depends on the measurement scale (see page 171).

The measurement of an index is the assignment of numbers (or symbols) to an attribute A of a variable in such a way that the scale value $S(A)$ is a function $f[.]$ of n measurements $g_i(A)$ of A on the indicators i:

$$S(A) = f[g_1(A), g_2(A), \ldots, g_n(A)]. \tag{4.4}$$

From this definition the programme to be pursued should become evident. We have to choose the attributes ("quality of risk management," "effectiveness of controls," "external environment" etc.), find appropriate measurable indices and create a functional relationship between them to yield a compound index. A *compound index* (or composite index) is an index constructed from other indices (called sub-indices). Index construction has a long-standing tradition – see the classic book by Irving Fisher (1927).

In the oft-used *additive index* the index is constructed as a weighted sum of the indicator values, i.e.

$$S(A) = w_1 \cdot g_1(A) + w_2 \cdot g_2(A) + \ldots + w_n \cdot g_n(A). \tag{4.5}$$

Two underlying assumptions are noteworthy. Firstly, it is implied that all indices are mutually independent as a low value of indicator i can be compensated by a high value of another indicator j. Secondly, the indicators must be measured at least on an interval scale. The weights can be chosen with recourse to an underlying hypothesis, or at the assessor's discretion. Furthermore, they have to make the product $w_i \cdot g_i$ dimensionless. With regard to scores this is a non-issue.

Formula 4.5 looks quite familiar as it is related to the linear regression model, viz.

$$E(Y) = \beta_0 + \beta_1 \cdot X_1 + \beta_2 \cdot X_2 + \ldots + \beta_n \cdot X_n. \tag{4.6}$$

Y is assumed as normally distributed. Given a sample of values for Y and X_i, estimates of $\beta_0, \beta_1, \ldots$ are often obtained by the least squares method. One extension of this model is the so-called *generalised additive model* (Hastie and

Tibshirani, 1990; Venables and Ripley, 1999, 281). The expected value of Y is modelled as

$$E(Y) = s_0 + s_1(X_1) + \ldots + s_n(X_n), \qquad (4.7)$$

in accordance with Eq.(4.5), where $s_i(X_i)$ for $i = 1, \ldots, n$ are smooth functions. These functions are not given a parametric form but instead are estimated in a non-parametric fashion. This amounts to allowing for an alternative distribution for the underlying random variation other than just the normal. Although Gaussian models can be used in many statistical applications, there are classes of problems for which they are not appropriate. The normal distribution may not be adequate for modelling discrete responses such as counts or proportions.

These regression models only work if there are observations Y, i.e., index values. In our setting, these index values cannot be measured. Therefore, regression cannot be applied. However, there are methods focused on the covariance structure that can be used to determine the "optimal" weights to be given to indicators. The method of choice is *factor analysis* which has many variations. It is a standard in the empirical social sciences. According to this method, weighting is used to correct for the redundant information of two or more correlated indicators, and it is not a measure of importance of the associated indicator. Generally speaking, factor analysis in its exploratory form seeks to uncover the underlying structure of a relatively large set of variables. The *a priori* assumption is that any indicator may be associated with any factor. The derivation of these methods involves some linear algebra.

The first step is to check whether the indicators are substantially correlated. The Bartlett's test of sphericity is used having the null hypothesis that the correlation matrix $\mathbf{C} = [c_{ij}]$ is an identity matrix $\mathbf{I} \in \Re^{n \times n}$. The statistic is based on a chi-squared transformation of the determinant of the correlation matrix. An alternative test for the same null-hypothesis is from Steiger (1988, 249). Its test statistics is

$$S = (n - 3) \sum_{i=1}^{n} \sum_{j=i+1}^{n} z_{ij}^2, \qquad (4.8)$$

where $z_{ij} = Z(c_{ij})$ according to Eq.(3.124) of page 172. The rejection region is $S \geq \chi_{1-\alpha, df}$. The degree of freedom of the χ^2-distribution is $df = n(n-1)/2$.

If the indicators are only slightly co-varying then there is only a small probability that they share any common factor. This means that equal weights are appropriate.

The principal components decomposition is *not* scale-invariant. This means that you obtain different decompositions depending on whether you calculate them for the covariance or the correlation. In general, you use the covariance matrix when the original observations are on a common scale as, for example, in our data set. You use the correlation matrix when you have observations of different types. This amounts to z-transforming the responses by using $z_i = (x_i - \mu_x)/\sigma_x$, where μ_x is the mean and σ_i the standard deviation of X.

Many criteria have been suggested for deciding how many principal components and factors to retain. One graphical method is the so-called *screeplot*. It shows the eigenvalues against their indices, and shows visually a steady downward slope and a gradual tailing away (see Figure 4.22 on page 305). The fracture from the steady downward slope indicates the break between the principal components under consideration and the remaining ones which make up the scree.

A second approach is to include as many components to explain some arbitrary degree of variance, typically 90%. A third method is to exclude those principal components with eigenvalues smaller than the average. For principal components computed from a correlation matrix, this criterion excludes components with eigenvalues less than 1. The first criterion typically results in too many included components, while the third criterion typically includes too few. The second method is often a useful compromise.

The model posits the following relation between the data matrix $\mathbf{R} \in \Re^{n \times p}$ and some latent variables $\mathbf{F} \in \Re^{n \times p}$, the *factor loadings* $A \in \Re^{n \times n}$ and a residual u called *uniqueness*:

$$\mathbf{R} = \mathbf{AF} + \mathbf{U}. \tag{4.9}$$

The covariance of a matrix $\mathbf{X}$ is given by $Cov(\mathbf{X}) = \frac{1}{n}[\mathbf{X}^T\mathbf{X} - \overline{\mathbf{X}}^T\overline{\mathbf{X}}]$. Here $\overline{\mathbf{X}}$ is the mean. $\mathbf{F}$ is chosen as with uncorrelated columns, thus $\mathbf{F}^T\mathbf{F} = n\mathbf{I}$ with $\mathbf{I}$ the identity matrix.

$$Cov(\mathbf{R}) = Cov(\mathbf{AF}+\mathbf{U}) = \mathbf{A}^T Cov(\mathbf{F})\mathbf{A}+\Psi = \mathbf{A}^T\mathbf{IA}+\Psi = \mathbf{A}^T\mathbf{A}+\Psi \tag{4.10}$$

Now this result is not unique as every $\mathbf{VA}$, with $\mathbf{V}$ an orthogonal (unitarian) matrix, can replace $\mathbf{A}$. This multiplication is also called "rotation" (analogous to the co-ordinates transformation in linear algebra). Now these loading matrices or rotations are chosen such as to successively maximise the variance. This method is called *varimax*. See Lawley and Maxwell (1971) or Bortz (1999) for more information.

The final step involves the determination of the weights used to construct the composite indicator "Internal Control." The approach followed is to weight each indicator with respect to the proportion of its variance explained by the factor with which it is associated – i.e., the normalised squared loading – while each factor is weighted according to its contribution to the portion of the explained variance in the dataset – which is the normalised sum of squared loadings. In Nicoletti et al. (2000) you can find more details.

Of course, there are also some objections to this method. The weights can be changed with every survey or assessment of the indicators. In order to guarantee a minimum of inter-temporal stability the weights should change only smoothly or after some periods of constant application.

Example 4.7. Suppose some 16 experts have filed out the questionnaire. The questions have been aggregated to five classes, i.e.,

> 1. Test for significant correlation
>
> 2. Determine number of factors with screeplot on eigen-values
>
> 3. Perform Factor Analysis
>
> 4. Calculate weights from normalised squared loadings

ALGORITHM 11: Weights calculation from factor analysis

1. Control Environment,

2. Risk Assessment,

3. Control Activities,

4. Information and Communication,

5. Monitoring.

The data matrix **R** thus has five rows and 16 respondents. First we construct the data matrix and then we calculate the correlation matrix in order to get a feeling for the structure. If there is no significant correlation then the analysis need not be performed as the indicators are already the factors sought after.

```
# The data responses (per question)
ContrEnv   <-c(3.29,1.91,3.62,3.58,2.07,3.25,2.28,3.70,4.00,
               2.85,2.55,1.97,3.55,1.40,1.55)
RiskAssess<-c(3.69,3.16,3.62,4.45,1.03,3.80,1.99,4.11,3.71,
               2.74,2.99,2.16,3.89,0.877,1.14)
ContrActiv<-c(3.47,2.72,3.61,4.04,1.99,3.17,1.64,4.10,3.11,
               1.78,2.59,1.82,4.53,1.74,1.19)
InC        <-c(2.75,1.58,4.22,2.66,1.81,2.71,1.92,2.97,3.82,
               2.63,2.25,1.62,3.29,2.44,1.75)
Monitor    <-c(2.64,1.68,3.30,1.96,2.02,2.48,2.71,2.10,4.37,
               1.29,2.12,2.16,3.45,2.590,1.58)
# The data matrix R
R <- cbind(ContrEnv,RiskAssess,ContrActiv,InC,Monitor)

# Show correlation
cor(R)
            ContrEnv RiskAssess ContrActiv       InC   Monitor
ContrEnv   1.0000000  0.8832741  0.8345453 0.8225994 0.5358630
RiskAssess 0.8832741  1.0000000  0.8826130 0.6069726 0.3117740
ContrActiv 0.8345453  0.8826130  1.0000000 0.6596226 0.4248429
InC        0.8225994  0.6069726  0.6596226 1.0000000 0.7135326
Monitor    0.5358630  0.3117740  0.4248429 0.7135326 1.0000000
```

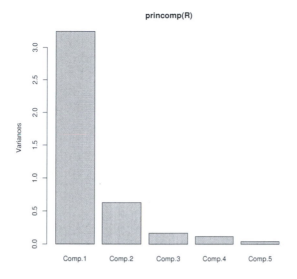

FIGURE 4.22: Screeplot.

The data shows some high correlations, as one might expect. The data matrix **R** alone is the input to the factor analysis performed by an R function `factanal`. The number of factors is set to two and the standard rotation algorithm "varimax" is chosen.

```
# screeplot
screeplot(princomp(R))
```

The screeplot (see Figure 4.22) suggests using two factors in order to capture a sufficient degree of variance.

```
# factor analysis, rotation="varimax" is the default
factanal(R, factors=2)

Call:
factanal(x = R, factors = 2)

Uniquenesses:
  ContrEnv RiskAssess ContrActiv        InC    Monitor
     0.076      0.005      0.192      0.058      0.420

Loadings:
          Factor1 Factor2
ContrEnv    0.794   0.543
RiskAssess  0.978   0.198
ContrActiv  0.835   0.334
```

```
InC        0.447    0.861
Monitor    0.169    0.742
```

```
              Factor1 Factor2
SS loadings      2.511   1.739
Proportion Var   0.502   0.348
Cumulative Var   0.502   0.850
```

```
Testing the hypothesis proves that 2 factors are sufficient.
The chi square statistic is 0.23 on 1 degree of freedom.
The p-value is 0.634.
```

The "SS loadings" above stands for the sum of squares of the loadings on each factor, here 2.511, respectively 1.739, or 59.1% and 41.9% respectively of the variance. The two factors contain approximately 85% of the variation. Now we select the higher loadings per index r_i and square them. This yields the weights as shown in the last column of the table below.

	Loading 1	Weight 1	Loading 2	Weight 1	Weight
Control Environment	**0.794**	27.6%	0.543		0.176
Risk Assessment	**0.978**	41.9%	0.198		0.267
Control Activities	**0.835**	30.5%	0.334		0.195
Information & Comm.	0.447		**0.861**	57.4%	0.207
Monitoring	0.169		**0.742**	42.6%	0.154

In this fictitious example the final index for the Internal Control Environment would thus be

$$S(A) = 0.176 \cdot g_1 + 0.267 \cdot g_2 + 0.195 \cdot g_3 + 0.207 \cdot g_4 + 0.154 \cdot g_5, \quad (4.11)$$

where $S(A)$ is the summary index and g_j the sub-indices per class $j = 1, \ldots, 5$. We can see a certain deviation from the equal weights of 0.2 in this case.

Now we push our investigation a little bit further by applying another multi-variate technique known as *cluster analysis* (Venables and Ripley, 1999, 336; Bortz, 1999, 547). The aim of this method is to group objects in such a manner as to minimize the differences within the cluster of objects, while maximizing the differences between the clusters themselves. There are several different distance measures, the simplest of which is the euclidean metric. There are also different criteria applied to the metric. Now we want to graphically depict the similarities among the questions classes and the respondents. This is accomplished by the so-called dendrograms or trees. Again, this can be done easily with the following command:

```
par(mfrow=c(2,1), mex=1, oma=c(1,0,2,0), mex=.6, mar=c(5,5,3,1))
```

```
plclust( hclust(dist(t(R),method="euclidean")), xlab="")
plclust( hclust(dist(R,method="euclidean")), xlab="")
```

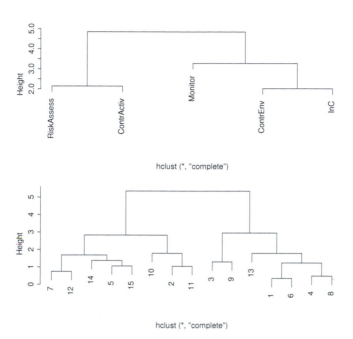

hclust (*, "complete")

FIGURE 4.23: Cluster analysis and dendrograms.

This yields the Figure 4.23. It nicely shows that there are several groupings depending on the height at which the tree is cut. With respect to the classes we understand that "Risk Assessment" and "Control Activity" are related, while the others form a second cluster. If we identify the respondents with lines of business their affinity is revealed, a fact that can be used for further activities and measures. △

While statistical methods assign weights in the additive model, more often equal weights are assigned, as an expression of indifference or ignorance. Yet another method is to ask for expert opinion.

An alternative to the additive model is to employ the multiplicative relationship between indicators according to the following formula (see Ebert and Welsch, 2004):

$$S(A) = \prod_{j=1}^{n} [w_j \cdot g_j(A)]^{\frac{1}{n}}. \tag{4.12}$$

For comprehensiveness, the most simple construction seems to be the best bet. After all, we have seen that not only the selection and rationalisation of the index choice is difficult, but that when building an index it is not good enough to just add the results.

4.7.4   Implications for the AMA

Operational risk assessments should capture the context of the business environment and internal control factors relevant to the operational risk profile. Having said that, a bank's idiosyncratic qualitative factors can be introduced in a number of ways:

- adjustments to the empirical estimates of operational risk capital,

- refinement of the scenario analysis and

- development of score-card approachs.

The qualitative adjustments related to self assessments obviously require a well-documented justification as the Basel documents states (BCBS, 2005, 150): "the supporting rationale for any adjustments to empirical estimates, must be documented and subject to independent review within the bank and by supervisors" (BCBS, 2005, 150).

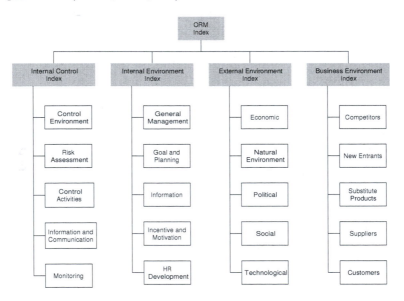

FIGURE 4.24: ORM index and sub-indices.

In brief, we want to create a programme that makes the transition from the historical data $\mathbf{H}$ driven economic capital $EC(\mathbf{H}(t))$ to a risk quantification that is contingent upon additional information as evidenced by indices and indicators $\mathbf{g}(t)$, i.e.,

$$EC(\mathbf{H}(t)) \rightsquigarrow EC(\mathbf{H}(t) \mid \mathbf{g}(t)). \tag{4.13}$$

TABLE 4.8: Examples risk drivers by events

Event	Risk Driver	Indicators
Internal	Management and supervision	Time stamp delays
	Security	Crime rate, corruption rate
	Recruitment policy	Index
	Control quality	Index
	Ethical conduct	Index
External	System quality	Number of unauthorised transactions
	Security	Crime rate, corruption rate
Clients	Product complexity	Number of client complaints
	Legal advise	Cost
	Training of sales staff	Fines for improper practice
	Tort	Tort cost index
Employment	Recruitment policy	Index
	Pay structure	Gini index
	Safety measures	Time off work, number of injuries
	Gender discrimination	Ratio of advancements, pay increase
	Leadership quality	Index
	Health	Days of vacation
Physical	Location	CRESTA zone, nat-cat premium
	Terrorism	Warning level, insurance premium
	Social unrest	Poverty index
	Theft	Insurance premium
Disruption	System quality	Down-time, systems' age
	Data integrity	Error rates
	Business continuity	Test results
	Technology	Index
Execution	Supervision	Number of failed trades
	Qualifications	Settlement delays, average employment
	Transaction quality	Number of errors
	Training of back office staff	Average hours spent in training
Extreme	Terrorism	Warning level, insurance premium
	Competition	Index of net income
	Events	Index

By summarising the preceding sections, we may be able to construct an overall index from the elements of Figure 4.24 representing the quality of the operational risk management. The underlying sub-indices and indicators may also serve to build specific indicators for the event dimension and its loss generators. In a purely additive model we can think of the basic indicators or sub-indices as a vector $\mathbf{g}$. The index vector for "operational risk" $\mathbf{s}_e$ is the multiplication of a weight matrix $\mathbf{W}$ by the indicator vector, i.e.,

$$\mathbf{s}_e = \mathbf{W}\mathbf{g}. \tag{4.14}$$

Analogously, the overall index S is the product of yet another weighting vector, say $\mathbf{u}$, and thus $S = \mathbf{u}^T \mathbf{s} = \mathbf{u}^T \mathbf{W}\mathbf{g}$. Remember that the mapping matrix $\mathbf{Q} \in \Re^{n \times m}$ is also an additive model. Thus we could also calculate the index for each line of business with a simple multiplication

$$\mathbf{s}_l = \mathbf{Q}\mathbf{s}_e. \tag{4.15}$$

The model should allow for the integration of bank-specific factors into the calculation of the economic capital requirement for operational risk. The business environment and internal control factors reflect changes in the operational risk profile of the bank. These changes stem from different sources, such as improved or deteriorated risk management processes, new product launches, access to new markets, implementation of new procedures and guidelines. In addition, changes in the business environment may affect the bank's operational risk profile.

All these changes directly affect the shape and the parameters of both the frequency distribution and the severity distributions. They might also influence the shape of the distribution of aggregate losses. This model allows to analyse the effect of variations of the loss distributions. Thus, the risk manager is able to both analyse the impact of such changes on the required economic capital for operational risk and to assess the risk sensitivity of a single change to the overall capital requirement and earning situation of the bank. The model therefore supports risk managers in constantly monitoring and analysing the economic capital impact of changes in banks' own risk management processes and in the market environment in which the bank operates.

As we can see from Figure 4.25 there are two possibilities to bring the influence of the indicators to fruition, viz. by applying them to:

- the economic capital outcome of the model or

- to the distribution parameters on input.

Technically, the index will play the role of a *fudge factor*. This is a variable factor used in calculations that produces a desired result. The slightly derogatory connotation stems from the fact that it is not buttressed by much theory. Nonetheless, it introduces some flexibility and helps the chief risk manager

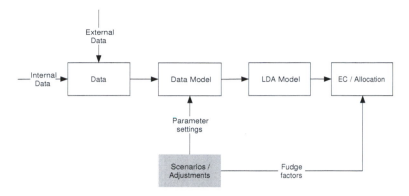

FIGURE 4.25: Synopsis of the model.

reduce the risk capital burden to those line managers who have undertaken initiatives to reduce operational risk.

To distort the output with a "fudge factor" has the drawback that the whole allocation becomes incoherent. Therefore, we think it more appropriate to apply the fudge factors to the model input. Now we profit from a sensible choice of distributions. Recall that most of the severity distributions presented have a scale parameter with the role of stretching the independent values. Often the scale is identical with a measure of dispersion. We would suggest multiplying the scale factor $b_i(t_0)$ of the distribution pertaining to event i (or the standard deviation) with the fudge factor ϕ_i centred around 1 according to

$$b_i^{(f)}(t) = b_i(t_0) \cdot \phi(t). \tag{4.16}$$

The function $\phi(t)$ will depend on the index value at times t_0 of the input data sample and t, the moment when the indices were calculated, thus $\phi(t) = f(s_i(t_0), s_i(t))$. At the outset $\phi(t_0)$ should evidently be 1.

Analogously, some similar device can be envisaged for the frequency of events. This procedure is aligned with Basel's view on scenarios, which conforms to: "expert assessments could be expressed as parameters of an assumed statistical loss distribution."

4.8 Scenario Analysis

What is a scenario? Opera buffs might say that it's a preliminary sketch of the plot, or main incidents, of an opera. Unfortunately, this description is not very relevant for our understanding of scenario analysis. Rather,

Definition 4.5. A *scenario* is a coherently structured descriptions of alter-

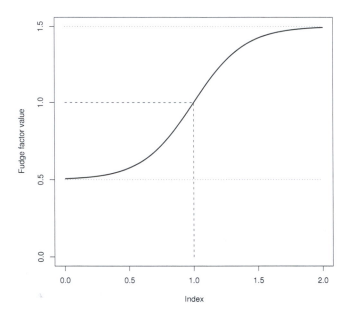

FIGURE 4.26: Fudge factor and indices.

native hypothetical or conjectured states of the world. They reflect different perspectives on past, present, and future developments. *Scenario analysis* is the investigation of consequences of scenarios. □

The "state of the world" concerns both exogenous, i.e., external and given, and endogenous, i.e., self-made, states or conditions. The analysis may also consist of several stages, where the "world" interacts with our own strategies step by step. In this sense a scenario is seen to be a two-person game with a passive environmental player and ourselves as decision-maker effecting our pre-defined strategies.

In the narrower context of LDA with our quantification model, scenarios are mainly concerned with the input to a highly complex calculation where the output cannot be guessed from the outset. Now we can define it as a set of N input states generating N output states. It makes little sense to jump directly to the output and ask: "What if we lost 100m?" In a formal shorthand, we will assess risk qua economic capital as contingent on some states $\mathbf{C}_k$ for some different scenarios k as

$$EC(\mathbf{H}(t) \mid \mathbf{C}_k). \tag{4.17}$$

The main drawback in scenario analysis is the assessment of its plausibility, which translates to probability. Actually, it would be rather startling to use the indifference argument here. Monte Carlo simulation is, in a sense, the limiting case of scenario analysis because each draw of a set of input states is an unnamed scenario. But here such states are not conjectured but randomly selected and thus they are amenable to the equally weighting in sampling.

Scenario analysis is often subject to the availability bias. Plausible and representative details, causal explanations and typical events combine to create more plausible scenarios than any subset of those events. Add details, pile up predictions, construct an elaborate narrative and the story will sell (Gardner, 2008, 61). Narratives about the future are an extremely powerful means of engaging the imagination. From earliest times, story-telling was the principal vehicle for developing and communicating explanations of the way things were and how they came to be (Lempert et al., 2003, 12). The obvious problem with using narratives about the future is that while these stories may offer compelling, insightful commentary about current options they are usually wrong in many crucial details about the future. Logic and probability theory says that the further you venture in this direction, the less likely it is that your scenario will materialise.

In a predictable world the more information the better. But in a uncertain environment, sometimes less information is more. Historic data contains two ingredients: useful information for predicting the future and arbitrary information or noise which is useless (Gigerenzer, 2008, 27). There is the well-known problem of over-fitting which arises by considering too much data. The hard work is with finding out which category the historic data belongs. Thus a complex scenario can fail because it tries to explain too much in hindsight.

While the Basel Committee recommends generating and implementing scenarios, specific how-to is limited. It says (BCBS, 2005, 150):

> A bank must use scenario analysis of expert opinion in conjunction with external data to evaluate its exposure to high-severity events. This approach draws on the knowledge of experienced business managers and risk management experts to derive reasoned assessments of plausible severe losses. For instance, these expert assessments could be expressed as parameters of an assumed statistical loss distribution. In addition, scenario analysis should be used to assess the impact of deviations from the correlation assumptions embedded in the bank's operational risk measurement framework, in particular, to evaluate potential losses arising from multiple simultaneous operational risk loss events. Over time, such assessments need to be validated and reassessed through comparison to actual loss experience to ensure their reasonableness.

Unfortunately, other than specify that scenario analysis must be robust and methodical, the Basel Committee provided few clues as to what scenario analysis should cover in practice.

We think it appropriate to assume that with banks using the AM approach, specifically the loss distribution approach, scenario should be applied to the inputs of the simulation model. The inputs are either parameters to distributions, correlations, mapping tables or eventually insurance policy parameters.

Closely related to scenario analysis is *sensitivity analysis*. Here the input scenarios are generated by perturbing the input parameters and variables ζ_i by a small amount in order to assess the impact on the output. Mathematically speaking, we want to numerically assess the first derivative of the output EC with respect to an input factor change $\Delta\zeta_i$, assuming EC to be a smooth function. This can be represented by the total differential:

$$\Delta EC = \frac{\partial EC}{\partial \zeta_1}\Delta\zeta_1 + \ldots + \frac{\partial EC}{\partial \zeta_N}\Delta\zeta_N. \qquad (4.18)$$

This is a bit like exploring one's surroundings by only taking one step in any direction in order to find out where the landscape is flat, rangy or has some precipitous ravines. However, such exploration is meant to minimize the wasting of resources by allowing for an intelligent guess as to which direction merits further exploration, i.e., where to test with more complex scenarios.

Run-Through Example 7. In the following scenarios we shall keep the average overall loss constant in order to better understand the influence of the changes introduced and thus the structure of the model at this given base case. We shall also start all simulations with the same random generator seed value. Thus, we can rule out additional statistical errors.

Higher correlations

We impose a new correlation matrix. We just double the values where they are non-zero and replace the zero correlations with 0.1. Thus the following matrix (see Table 4.9) will take effect. We expect that the Economic capital will rise. The question is, by how much will it rise?

TABLE 4.9: New correlation matrix

1							
0.6	1						
0.2	0.2	1					
0.2	0.4	0.2	1				
0.2	0.2	0.2	0.2	1			
0.2	0.2	0.2	0.2	0.4	1		
0.4	0.2	0.2	0.2	0.2	0.2	1	
0.1	0.1	0.1	0.2	0.2	0.2	0.1	1

The new values are net 681m as opposed to 665.9m and 710.7m versus 694.4 gross. The increase is in the order of magnitude of 2%. This is not very dramatic.

Higher correlation and higher tail dependency

Now let us increase the tail dependency by choosing a Student's t-distribution with degree of freedom of 5 instead of a normal distribution. This parameter pushes higher correlation into the tail, whereas normal distributions tend to zero. The effect should also be a higher EC. This simulation yields 683.1 and 716.1 for net and gross EC, respectively. This corresponds to an additional 0.3% and 0.8%.

Uniformly high correlation

As the correlation increase so far has not given spectacular results let us attempt another trial. We shall set all correlations to 0.5 just to rule out the conjecture that (realistic) correlations have a massive impact. The outcome as reckoned by the EC is as follows. The net EC is 698.6, an increase of 5%, and a gross EC of 734.3 with 6% more capital needed. The insurance mitigation is now 4.9%.

Now let us take a look at the correlations between Lines of Business induced by the frequency correlations. The latter are unrealistically high. Nonetheless, the resulting correlations are tiny, in the order of magnitude of some percentage points. The highest correlation is 6.5% between Commercial Banking and Retail Banking. This effect can be calculated analytically. The decrease bears an inverse relation to the coefficient of variation.

TABLE 4.10: Correlations between lines of business

	Corporate	Trading	Retail	Commercial	Payment	Services	Asset	Brokerage
Corporate	1							
Trading	2.3%	1						
Retail	3.0%	5.2%	1					
Commercial	2.8%	6.0%	6.5%	1				
Payment	1.2%	2.6%	3.1%	2.8%	1			
Services	1.3%	3.2%	3.1%	3.2%	1.3%	1		
Asset	1.2%	2.5%	2.8%	2.7%	1.2%	1.8%	1	
Brokerage	2.1%	3.8%	5.0%	3.4%	2.1%	2.2%	1.9%	1

This shows that dependencies do not change the results much.

Zero correlation

Although this case should not be allowed by the regulator it puts the correlation effect into perspective. Let us once again contemplate the EC which take the values 595.4 for net and 626.5 for gross. The comparison with our base case reveals an overall decrease of 10%. The uniformly high correlation has approximately 15% more in EC requirement.

Higher frequency of extremes

Now let us increase the frequency of the extreme event by 10% from 0.1124 to 0.1236. At the same time, let's alter the shape parameter of its truncated Pareto severity distribution in order to keep the average loss constant. This implies changing the parameter from 1.92 to 2.15. These changes mean that although the frequency increases, the severity has less mass in the tail. Therefore it seems difficult to predict the outcome.

The result makes the case clear: the net EC has decreased dramatically from 665.9 to 601.2 and the gross from 694.4 to 633.2. Expressed in percentage points, this is a decrease of 10%. At the same time the insurance mitigation has increased to 5.05%.

Higher threshold of extremes

The extreme events' severity is a truncated Pareto distribution whose losses start at the location parameter value and reach the maximum. The shape parameter determines how the mass is distributed between those boundaries. We have seen the results above. Now let us increase the threshold from 50m to 55m. Again, we adapt the shape parameter in order to maintain the expected loss. The new shape value is 2.14.

As in the previous case, the EC lowers by approximately 8% (7% gross) to yield the values 611.1 and 644.1 respectively. The insurance mitigation becomes 5.12%. Again, the shape parameter plays a dominant role. Its increase induces much more mass in the lower part and thus a much thinner tail that over-compensates the increase by 10% of the lower bound (location). Thus, an increase of the maximum loss or upper truncation will not have too much of an impact when the shape parameter is greater than 2 and the severity distribution has a relatively thin tail. As before, insurance profits from generally lower losses and makes it more efficient.

Higher insurance layer

We change the FI blended cover by increasing the deductible by 10% from 0.5 to 0.55. At the same time we also increase the limit in order to keep the premium constant. We know that there is little mass in the upper end of the cover. Therefore we increase the limit from 75 to 125.

The first guess turns out to be quite lucky: the premium is now 10.5 versus 10.3 from the base case. What has changed? First of all, the gross EC must have stayed the same. The net EC has lowered to 663.6 or, in terms of absolute value, 1.8m. If we gain this amount in saved capital at no additional cost, this must imply a gain in efficiency by the insurance policy.

Wrap-up of the scenarios

TABLE 4.11: Scenario synopsis

	CaR total Gross	CaR total Net	Insurance mitigation
Base case	694.4	665.9	4.10
Higher correlations	710.7	681.0	4.18
Higher correlation, higher tail dependency	716.1	683.1	4.61
Uniformly high correlation	734.3	698.6	4.86
Zero correlation	626.5	595.4	4.97
Higher frequency of extremes	633.2	601.2	5.05
Higher threshold of extremes	644.1	611.1	5.12
Higher insurance layer	694.4	663.6	4.43

In the following Table 4.11 we summarise the different scenarios' outcomes This should give a first indication about the sensitivity of the over-all model to parameter changes.

With these scenarios we have just perturbed the parameters of the given structure but not changed the latter, i.e., taken different distributions, changed the maps, etc.

With the model described herein you can quantify the Economic Capital or regulatory capital. But more importantly, this model's complex and intricate interdependencies lend themselves to much further exploration and examination and the creation of practically infinite scenarios, depending on the user's needs.

4.9 Optimising the Insurance Programme

Now we turn our attention from making the risk calculus more agile and responsive to the actual state, to the most obvious means of managing operational risk, viz. transferring some of it to another entity.

4.9.1 What Is Optimisation?

Optimum and its derivatives are ubiquitous in our every-day talk. Hirshleifer (1988, 24) states that "it is a remarkable fact that practically all the analysis (...) throughout economics generally, makes use of only two analytical techniques: (1) finding an optimum and (2) finding an equilibrium." In other words, does one ask: "Would I do better to buy a new car, or keep my old one for a while yet?" or "Are new car prices likely to be lower next year?"

Optimum means the "best" in Latin. Best relies on the notion of "good," or most desirable. Of course, not everything that is desired can be achieved, due to the imposition of restrictions. The most general class of problems associated with optimisation is efficient allocation. However, if there are several "best" situations, all equally appealing, then we have so-called Pareto-optimality. This essentially means that we have found the biggest pie but it can be sliced in different ways.

In a strict mathematical sense, optimisation is associated with a so-called programme formulated as follows: find a solution $\mathbf{x}$ that maximises (or minimises) a so-called objective function $f(\mathbf{x})$ under the constraints that some elements of $\mathbf{x}$ obey $0 \le h(\mathbf{x})$ and $0 = g(\mathbf{x})$.

The problem that concerns us – buying insurance – can be thought of as allocating a budgeted amount of money for several specific insurance policies while trying to realise the most appealing overall programme. In the above mathematical formulation the hardest thing to do is to define the objective function, i.e., what is good and what is even better. We can use this formulation to structure our problem by asking what the objective function looks like and what restrictions are to be observed. One restriction has already been imposed by the regulators (BCBS, 2005, 151), viz.

> Under the AMA, a bank will be allowed to recognise the risk mitigating impact of insurance in the measures of operational risk used for regulatory minimum capital requirements. The recognition of insurance mitigation will be limited to 20% of the total operational risk capital charge.

This rule is quite generous because it is hardy conceivable that coverage in such large extent is available.

4.9.2   The Variables

What are the parameters we can choose from? First, there are the different insurance policies. e.g., fidelity bond, Directors & Officers, etc. Each of these possesses (in our model) four parameters, viz. a (straight) deductible, a limit, an annual aggregate deductible and an annual aggregate limit. The cover – meaning the difference between limit and deductible – may have several contiguous layers with their specific premium.

These variables define the premium and thus the allocation of financial resources. The premium, however, cannot be calculated precisely as it is influenced by many factors in addition to the obvious actuarial ones. Premium may be quoted by the broker. Over a broader time horizon the establishment of a captive could also be interpreted as a possible variable.

4.9.3   The Objective Function

The objective function in our context is a kind of measure of utility of a given insurance programme. Insurance prices, differentials of price to the bank's own risk assessment, contribution to risk mitigation, etc., are all important "ingredients" to be kept in mind.

Let X_i, $i = 1, 2, \ldots, N$ denote the random claims to the portfolio covered by the programme and N the stochastic number of such claims. In this model, this random claim is generated from a linear combination of the event loss generators through the use of the map $\mathbf{Q}$. In simulation X_i is drawn according to a loss generator for losses by events and then randomly attributed to an insurance contract according to the "probabilities" from the map. In the following, the insurance contracts will be priced by the standard deviation principle. According to this principle the total premium for the contract is given by

$$\pi\left(Z, \mathbf{H}\right) = E\left[Z\right] + a\sqrt{Var\left[Z\right]}, \qquad (4.19)$$

where the loading factor a is determined by the insurance company and thus to be estimated by the bank. It can be in the range between 8% and 15% for example.

There are some rules of thumb for choosing insurance. They may be boiled down to the following (Vaughan and Vaughan, 1995, 34):

- Don't risk more than you can afford to lose,

- Consider the odds and,

- Don't risk much for little.

The first of these simple rules is especially important for the private citizen. In a corporate setting, loss is borne by the company. It is therefore often suggested that the company estimates the maximum uninsured loss it could absorb.

Insurance theory differentiates among three kinds of insurance expenditures (Vaughan and Vaughan, 1995, 39; Rejda, 1995, 47):

- Essential insurance: protection against catastrophic loss or losses threatening the survival of the company (and insurance required by law),

- Desirable insurance: protection against financial difficulties but not bankruptcy,

- Optional insurance: covering losses that can be met out of income.

What is not considered is the fact that Basel 2 with its capital charge skews this conventional wisdom. With the quantile as measure for the required capital it is evident that for one monetary unit spent you will get the highest mitigation by buying high excess layers of insurance. This means high limits with high deductibles. From a psychological point of view there is the inconvenience that the insured will very rarely experience losses that are covered by such a policy, although there is considerable expenditure for such protection. But for catastrophic losses there will be some protection.

Potential losses with high severity are differentiated by the frequency. For relatively low frequencies the deductible feature may be appropriate for structuring the policy. But for not so infrequent claims number distributions, large per-loss deductibles pose the threat of unacceptable loss assumption if several large losses or an unusually large number of small losses occur in one year. A more logical approach is to determine the organisation's annual loss assumption capability, then buy insurance that pays claims above the annual aggregate deductible. One has to be aware that from the insurance point of view if a form of deductible makes sense, then an according limit for the insurance company makes sense.

On the other hand very frequent losses with low severity pose the problem of claims handling and processing. Although these risks may easily retained and paid out of the current income the focus is on administration and resources. Therefore for high frequency/low severity losses buying insurance means self-retention with administration out-sourcing.

An additional point to mention is the fact that insurance expenditure are recognised in the income statement and have in the case of profit a tax shield effect compared with the situation where it may not be possible to set provisions aside. These considerations are summarised in the the Table 4.12.

TABLE 4.12: Insurance guiding rules

Severity	Frequency	Insured	Insurance
Low	Low	Do not insure	
Low	High	Self-insure, buy insurance in order to outsource claims handling	
High	Low	Per occurrence deductible	Per occurrence limit
High	High	Annual aggregate deductible, with or without per occurrence deductible	Annual aggregate limit

The bank actually does not start from scratch as there might be long-dated experience with insurance buying. Therefore a sensible starting point for the insurance programme should be not too difficult to find.

4.9.4 Cost and Benefit of Risk Mitigation

The premium can be thought as the cost of risk mitigation. It is a fixed and certain outlay cost when the insurance contract is stipulated. Because it reduced the net income by an amount equal to the premium times the income tax there is an inherent benefit from the tax shield.

In the context of Basel 2 where insurance mitigates or can mitigate the capital charge there is an additional benefit in the cost of capital for the saved portion. If EC denotes the economic capital, identical to the risk capital, then by ΔEC_i we define the capital saved by insurance i. By applying the cost of equity r_E – as proxy for the cost of Economic Capital – the saving of total cost ϵ is

$$\Delta \epsilon = r_E \times \Delta EC_i. \tag{4.20}$$

The *cost of equity* is the expected return required by investors to induce them to hold the equity (Grinblatt and Titman, 1998, 371; Brealey and Myers, 1996, 534). The cost of equity is most often assessed simply by arguing that the return from the assets A minus return to the debt D equals the return of equity E where $A - D = E$ holds:

$$r_A A - r_D D = r_E E \quad \Leftrightarrow \quad r_E = r_A + (r_A - r_D)\frac{D}{E}. \tag{4.21}$$

Although one should consider also the tax shield provided for debt as interest on debt is a pre-tax income item and therefore calculate the cost of equity with the corporate tax rate t:

$$r_E = r_A + (r_A - r_D(1 - t))\frac{D}{E}. \tag{4.22}$$

Nonetheless, formula Eq.(4.21) is used as approximation for the cost ratio if it is not even announced to shareholders with a view on the industry average or the closest competitors.

Now the risk capital should be less than regulatory capital where the latter is approximately shareholders' equity. Therefore, the cost for the risk capital could be taken from the Eq.(4.21) or Eq.(4.22) as a basis or defined in a discretionary fashion by the management.

It is tempting to think that r_E is a linear function of the debt ratio D/E. But this is not necessarily true as the return of debt r_D is itself a function of the debt ratio. The higher the leverage the higher the possibility of default and according to the option model of equity in conjunction with the option pricing model. If we split the asset value A in the following way

$$A = \min[0, A - D] + \max[0, A - D] + D \tag{4.23}$$

we can deduce from the intrinsic values that the value of the assets and thus of the company equals a short call plus a long call on the assets with exercise price identical to the nominal debt plus the nominal debt. Now the debt

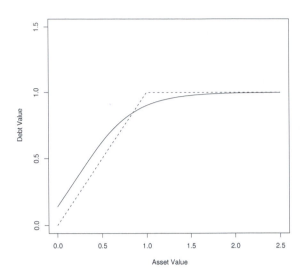

FIGURE 4.27: Debt value as option on the asset value.

holders possess the short call and the nominal debt. This corresponds to the
so-called *covered-call* strategy. The interpretation is that the debt-holders are
the owner of the company but have sold an option to the shareholders to
buy the company at the price of the nominal debt. With a high debt ratio
the probability that the asset value approaches the debt value is bigger and
thus the possibility that the option gets into the money also, the debt-holders
will require a higher return for this contingency. Besides the debt ratio also
the volatility of the assets enters the valuation of the debt. In Figure 4.27 we
see that above of the default, i.e., where $A > 1$ the option value is less than
the nominal value. Therefore debt-holders will require a premium in the form
of additional return. Where $A \gg 1$ we have the linear relationship between
return on equity and the debt ratio. Yet another model is to use the Capital
Asset Pricing Model. Here the ingrediences are the so-called risk-free rate
r_f, the return of the whole market r_M and the relation between the stock's
volatility and the market volatility expressed as β_E. Now this model claims
the following relationship:

$$r_E = r_f + \beta(r_M - r_f). \tag{4.24}$$

Example 4.8. Assume a return on assets of 9%, an interest on debt of 7%
and a tax rate of 25%. Furthermore, the ratio D/E be 7. Remember that
the capital base must be at least $1/8$. This leaves us with a value for r_E
of approximately 30%. Without tax consideration the value drops to 25%, a
figure often heard in the market for the targetted ROE.   △

4.10 Audit, Reporting and Disclosure of Operational Risk

As we have seen in Section 4.7.2 at page 294, internal control is a major component of the operational risk framework. It is closely scrutinized and critical for the independent audit of the financial statement. This overlap has led some regulators to use the result of the financial audit as part of the regulatory audit. In Switzerland regulators require three reports to be produced by the independent auditor, viz. "Risk analysis and audit strategy," "Report on the financial audit" and "Report on the regulatory audit." This close link makes it worthwhile to take a look at the principles used in the accounting profession for auditing the financial statements. The review of internal controls is relevant to operational risk's causes:

- internal fraud,

- processes and,

- systems.

In Figure 4.28, depicting the relationship between the financial and the regulatory audit, we try to show that the planning of the audit, consisting mainly of the risk assessment and the determination of its strategy, can be carried out simultaneously for both audits. An overlap also exists with the information technology and the internal control system, as auditors are faced with the challenge of understanding an entity's information technology (IT) processing and control environment (O'Donnell and Rechtman, 2005, 64). The auditors' need for an IT specialist is not surprising, given the impact of IT on the financial statement processes and the level of reliance on IT controls.

4.10.1 Financial Audit

The goal of an independent or external audit of a company's financial statements is to express and disclose an independent opinion on whether or not the financial statements are relevant, accurate, complete, and fairly presented in conformity with generally accepted accounting principles. The fact that it is an opinion, and not a certification, is meant to indicate to users of the financial statement that the independent auditor is providing *reasonable assurance*, and not complete assurance, as to whether the financial statements are materially misstated. This is reminiscent of Popper's falsification theory. Alas, the term "reasonable assurance" leaves much open to interpretation. Since the 1980s the profession has striven to close the so-called "expectation gap" – the gap between the assurance that financial statement users expect of an audit opinion and the assurance that it actually provides. For certain, auditors would like to raise the assurance to almost 100 percent, if only to avoid

FIGURE 4.28: Relationship between regulatory and financial audit.

potential litigation and the disciplinary actions that are likely in the wake of a disclosure of an audit failure. Because a failure is a career-ending event for the auditor in charge, it is in the auditor's own best interest to provide as diligent and accurate an audit as possible. In the USA, the audit failure rate for the year 2004 was less than 2%.

In their engagements, the auditors must consider the risk of material misstatements resulting from fraud or error in the financial statements. The term "fraud" refers to an intentional act by one or more individuals among management, those charged with governance, employees, or third parties, involving the use of deception to obtain an unjust or illegal advantage. But, auditors are only concerned with fraudulent acts that cause a material misstatement in the financial statements. Actually, there are two types of fraud: the first stems from fraudulent financial reporting, while the second results from the misappropriation of assets. The term "error" refers to an unintentional misstatement in financial statements, including the omission of an entry or a disclosure, such as a mistake in gathering or processing data, an incorrect accounting estimate or a mistake in the application of accounting principles.

In general, the financial statements of public companies consist of

1. an income statement,

2. a balance sheet,

3. a statement of changes in equity,

4. a cash flow statement and

5. notes to the financial statements.

The auditor expresses an *unqualified audit opinion* as to whether or not there are any restricting or limiting circumstances. Otherwise, depending on the nature and materiality of the qualification, a *qualified opinion* may be an exception opinion, an adverse opinion or an inability opinion.

Audit materiality is both a quantitative and qualitative measure. The overriding objective is to maintain the true and fair view presented by the financial statements. Audit differences, individually and cumulatively, are considered in that context.

Auditors can commit two types of errors when issuing an opinion. Firstly, the auditor may issue an unqualified audit opinion, when a qualified opinion is appropriate. This is referred to as a Type I error,[7] i.e.,

$$Pr(\text{unqualified audit opinion} \mid \text{misstatement}) \leq \alpha. \qquad (4.25)$$

The complementary Type II error, i.e.,

$$Pr(\text{qualified audit opinion} \mid \text{no misstatement}) \leq \beta, \qquad (4.26)$$

is seen as having more significant consequences. These potential errors are called *audit risk*.

In theory, overall audit risk is the product of three factors:

- the *inherent risk*, that the entity's financial statements will be misstated assuming that there were no related internal controls,

- the *control risk*, that the entity's internal controls will not deter or detect and correct material misstatements on a timely basis and

- *detection risk*, the risk that any remaining material misstatements will not be discovered by audit procedures.

Furthermore, detection risk can be subdivided into sampling and non-sampling risk. Sampling risk refers to the possibility that the sample selected is deliberately not representative of the population in order to draw false conclusions. It is both pertinent to controls testing and substantive testing. When performing tests of controls (called attribute sampling), the purpose is to determine the frequency of errors occurring. In substantive testing (called variables sampling), the purpose is to measure errors of total monetary amounts in particular account balances.

Non-sampling risk relates to errors due to the human factor, including failure to recognise errors, to analyse results properly, to identify the population and errors properly and failure to use the proper audit procedure for a given audit objective.

Thus, audit risks can be reduced to the following stylized equation:

$$\text{Audit Risk} = \text{Inherent Risk} \circ \text{Control Risk} \circ \text{Detection Risk}.$$

If we think of these risks as probabilities, then we assume mutual independence

---

[7]Whether an error is of type I or II depends actually on the choice of the null-hypothesis.

of these items. Risk as probability actually differs from our definition of risk 1.1 on page 14. The concatenation of inherent risk and control risk is also called *risk of a material misstatement.*

The audit of financial statements by the independent auditor consists of four distinct phases, shown in Table 4.13, namely

1. audit planning,

2. controls testing,

3. substantive procedures and

4. the conclusion.

The *audit planning* rests on the risk assessment, which in turn determines which strategy and approach are implemented. The goal is to obtain an understanding of the entity and its environment, including its internal controls, to appraise the risks of material misstatement at the financial statement and relevant assertion levels.

TABLE 4.13: Typical audit phases

Phase	Activities
Planning	Perform risk assessment procedures and identify audit risks Determine audit strategy Determine audit approach
Controls Testing	Understand accounting and reporting activities Evaluate design and implementation of selected controls Test operating effectiveness of selected controls Assess control risk and risk of significant misstatement
Substantive Procedures	Plan substantive procedures Perform substantive procedures Consider if audit evidence is sufficient and appropriate
Completion	Perform finalisation procedures Perform overall evaluation Form an audit opinion

In the second phase, called *controls testing*, when the auditor deems it necessary, she tests the *operating effectiveness of controls* in preventing or detecting material misstatements. When such controls are effective, the auditor has greater confidence in the reliability of the information. Because the risk of material misstatement takes into account internal control, the extent of substantive procedures may be increased as a result of unsatisfactory results from tests of the operating effectiveness of controls. It must be mentioned that for large complex banks, the number of critical controls could be thousands.

Executing *substantive procedures* aims at detecting material misstatements, i.e., mainly the inherent risk. Audit procedures performed for this purpose include *tests of details* of classes of transactions, account balances, and disclosures, and *substantive analytical procedures*.

Tests of details are generally more appropriate for obtaining audit evidence regarding certain assertions about account balances, including existence and valuation. In some situations, the auditor may determine that performing only substantive analytical procedures may be sufficient to reduce the risk of material misstatement to an acceptably low level. For example, the auditor may determine that performing only substantive analytical procedures is responsive to the assessed risk of material misstatement for a class of transactions where the auditor's assessment of risk is supported by the evidence from performance of tests of the operating effectiveness of controls. In other situations, the auditor may determine that only tests of details are appropriate, or that a combination of substantive analytical procedures and tests of details are most responsive to the assessed risks.

In designing tests of details, the extent of testing is ordinarily thought of in terms of the sample size in order to minimise sampling risks. Several methods of choosing a sample are used, e.g., fully randomized sampling, monetary unit sampling, stratified sampling or judgemental sampling. However, the auditor also considers whether it is more effective to use other means of testing, such as selecting large or unusual items from a population as opposed to performing representative sampling.

The main tools of the auditor are called procedures. They are summarised in Table 4.14. These procedures, or combinations thereof, may be used as risk assessment procedures, for tests of controls, or as substantive procedures, depending on the context in which they are applied by the auditor.

In the last phase of the audit, the auditor summarises the findings and forms his or her audit opinion. The board of directors receives a management letter which goes into greater detail than the attestation that will be published.

4.10.2   Controls Testing

According to the SEC rules in the USA and the Sarbanes-Oxley Act of 2002 two types of controls are to be assessed: "disclosure controls and procedures" (DCP) and "internal controls over financial reporting" (ICOFR).

Disclosure controls and procedures aim at ensuring that material information is identified and conveyed to the CEO and CFO so that disclosure decisions can be made, that all disclosures made by the company to its investors are accurate, complete, timely, and fairly present the company's condition, and to minimise the risks of selective disclosure. The SEC advises all public companies to create a disclosure committee. It sets parameters for disclosure and determines the appropriateness of disclosures in all publicly disseminated

TABLE 4.14: Audit procedures

Procedure	Example
Inspection of Records or Documents	Test of controls for producing evidence of authorisation.
Inspection of Tangible Assets	For example, when observing an inventory count, the auditor may inspect individual inventory items.
Observation	Looking at a process or procedure being performed by others.
Inquiry	From formal written inquiries to informal oral interviews.
Confirmation	Obtaining a representation of information or of an existing condition directly from a third party.
Recalculation	Checking the numerical accuracy of documents or records.
Re-performance	Independent execution of procedures or controls that were originally performed as part of the entity's internal control.
Analytical Procedures	Study of plausible relationships among financial and non-financial data; investigation of identified fluctuations.

information. The formation and activities of such a committee are thought to represent one of the most effective controls that a company can implement to ensure that its financial statements are fair, accurate, timely, and complete.

ICOFR is a subset of internal control specific to financial reporting objectives. Even though internal control systems review has always been an important element of an external audit, the requirement of formal reports by both senior management and the independent auditor is relatively new, introduced with the above mentioned act (see page 6). It does not encompass the elements that relate to the effectiveness and efficiency of a company's operations or a company's compliance with applicable laws and regulations, with the exception of compliance with the applicable laws and regulations directly related to the preparation of financial statements (Agami, 2006).

Under most accounting standards, auditors are responsible for acquiring an understanding of the control environment as part of the normal financial audit. As IT controls are a material component of the control environment, they must also be understood. The auditor's ability to gain the required level of understanding may depend on the type of controls involved in the engagement.

General IT controls include the procedures and processes that support the general usage of business applications of an entity. These controls cover areas such as access to programs, data and transactions, data centre operations, program development, program changes, IT disaster recovery plans and the proper segregation of duties of personnel involved with information systems. Computerised application controls include the controls involving the processing and storing of transactions. They ensure the completeness, accuracy, authorisation, and validity of processed business transactions (O'Donnell and Rechtman, 2005, 64).

Application controls include application security, controls for input, rejected-transaction, transaction-processing and output. Both types of controls are needed to ensure accurate information processing and the integrity of the resulting information required to manage and report on the organisation.

Auditors should perform a walk-through of the information system to be satisfied with the design, implementation and operation of the applicable controls. The walk-through should cover the entire process of initiating, authorising, recording, processing, and reporting individual transactions and controls for each of the significant processes identified, including controls meant to address the risk of fraud. During the walk-through, at each point at which important controls occur, the auditor should question the company's personnel about its understanding of what is required by the entity's procedures and controls and determine whether the processing is performed as originally intended and on a timely basis (PCAOB, 2007).

When trying to determine deficiencies in the quality of control, factors such as organisation size, the quantitative and qualitative features of the risk factors that the control activity was intended to mitigate, and complexity of operations need to be taken into account. Both likelihood and magnitude of the potential misstatement, as depicted in Figure 4.29 from Agami (2006), are relevant.

A *control deficiency* indicates a flaw in the design, the implementation, and/or the operating effectiveness of a control activity. Such defects could adversely affect the company's ability to initiate, record, process, summarise and report accurate financial and non-financial data. A *significant deficiency* is a control deficiency which, in the independent auditor's opinion, should be communicated to the company's audit committee. Lastly, a *material weakness* is a deficiency in control such that the design or operation of one or more of the internal control components does not reduce the risk of material misstatements caused by error or fraud to a relatively low level nor makes it possible for such an error to be detected within a timely period by employees in the normal course of performing their assigned functions. The presence of one or several material weaknesses may indicate that internal controls system is not effective. Such weaknesses need not be viewed overly negatively, as long as they are evidenced by the entity itself and remedial action is being taken.

While internal control can help mitigating risks, it does not eliminate risk

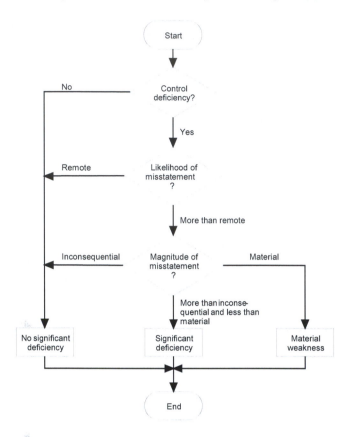

FIGURE 4.29: Audit relationship. (From Agami, 2006. Reprinted from The CPA Journal, November, 2006, copyright 2010, with permission from the New York State Society of Certified Public Accountants.)

altogether. Controls are, after all, built on processes involving people, and, therefore, susceptible to all the limitations of human endeavour. Internal control can be circumvented deliberately through fraud by individuals or collusion between employees. It can be undermined inadvertently through poor judgment, carelessness, or other breakdowns of processes and procedures. Moreover, it may be weakened by resource constraints.

There is a plethora of textbooks in the field of assurance and auditing, e.g., Arens et al. (2008).

4.10.3 Regulatory Audit

In addition to annual reports and accounts of financial institutions, the banking supervisors' principal sources of information include the reports which independent auditors produce as part of their auditing of the annual financial statements.

The national supervisory environment determines the independent auditor's role in the regulatory process. If the banking supervisor follows an active approach of conducting frequent and rigorous inspections, the independent auditor's contribution is normally low. On the other hand, if there is a tradition of supervision based on the analysis of reported information provided by bank's management or if supervisory resources are scarce, the supervisor benefits from the assurances provided by the auditor. Currently, many countries practice an approach which contains elements of both *inspection* and *analysis of reported information* (BCBS, 2001b, 16).

However, the supervisor cannot assume that the independent auditor's evaluation of internal control for the purposes of the audit will necessarily be adequate for the purposes for which the supervisor needs an evaluation. Therefore, the supervisor needs to extend the scope for which internal controls are evaluated and tested by the supervisor and the auditor. We will give several brief examples of these different approaches.

Example 4.9. The objective of the Swiss regulatory audit is to express an audit opinion on the adherence by the audited entity to the licensing requirements and other applicable provisions (see Outline 8 for the full list of mandatory audits). This opinion is based on audit standards and the provisions of the supervisor, i.e., the Swiss Federal Banking Commission (SFBC, 2005). Wherever possible and reasonable, professional standards originally conceptualised for financial audits are also to be adopted in regulatory audits. Auditors must take due account of supervisory practice with respect to the due diligence to be exercised by a responsible and adequately qualified auditor.

The regulatory audit and audit strategy comprises risk-based audits and mandatory audits. The auditor determines the strategy for the risk-based audit guided by the key audit risks. The principal business lines are listed, i.e., interest business, commission business, trading, etc. The auditor can add further business lines or subcategories of the main business lines which are of importance to the institution. The level of detail given must be appropriate to the institution's business activities, size and risk situation.

The regulatory auditor first determines the inherent risk and the control risk for both key audit risks and mandatory audit risks. The auditor can tag the inherent risk as "higher" or "lower." The control risk in turn can be categorised as "higher," "medium" or "lower." For "higher" or "lower" risk rationales must be given. The combination of inherent and control risk yields what is called the combined risk (or previously defined as risk of material misstatement in the financial audit). The combined risk determines the depth of the audit to be applied, i.e., "audit," "review," "plausibility check" or "no investigations." The pertinent analysis leads to a definitive assessment of the key audit risk (see Figure 4.31). △

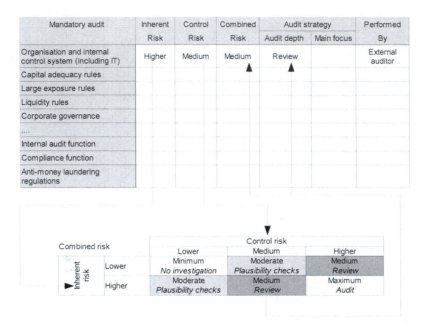

Mandatory audit	Inherent Risk	Control Risk	Combined Risk	Audit strategy		Performed By
				Audit depth	Main focus	
Organisation and internal control system (including IT)	Higher	Medium	Medium	Review		External auditor
Capital adequacy rules						
Large exposure rules						
Liquidity rules						
Corporate governance						
....						
Internal audit function						
Compliance function						
Anti-money laundering regulations						

Combined risk		Control risk		
		Lower	Medium	Higher
Inherent risk	Lower	Minimum *No investigation*	Moderate *Plausibility checks*	Medium *Review*
	Higher	Moderate *Plausibility checks*	Medium *Review*	Maximum *Audit*

FIGURE 4.30: Risk analysis and audit strategy. (See SFBC, 2005.)

Example 4.10. The Financial Services Authority (FSA), the British regulator, uses a variety of tools to monitor whether an authorised bank remains in compliance with regulatory requirements. These tools include:

1. desk-based reviews;

2. liaison with other agencies or regulators;

3. meetings with management and other representatives of a firm;

4. on-site inspections;

5. reviews and analysis of periodic returns, notifications and past business;

6. transaction monitoring;

7. use of auditors and skilled persons.

$\triangle$

- Audit of adherence to licensing requirements;

- Audit of adherence to capital adequacy, large exposure and liquidity rules;

- Other mandatory audits:

  - adequacy of corporate governance, including personnel segregation of executive management and Board of Directors;

  - adherence to generally accepted banking principles for transactions by members of governing bodies and others with a qualified participation;

  - proper conduct of the business operations by those entrusted with the administration and executive management and by those with a qualified participation;

  - adequacy of organisation and internal control system (including information technology);

  - adequacy of risk identification, measurement, management and monitoring;

  - adequacy of the internal audit function;

  - adequacy of the compliance function;

  - adherence to anti-money laundering regulations;

  - and adherence to rules on consolidated supervision.

OUTLINE 8: Mandatory audits of the Swiss Federal Banking Commission

Example 4.11. Within the Supervisory Review Process of the European Union, as described in the framework for consolidated financial reporting (see CEBS, 2005), the underlying aim of the Pillar 2 processes is to enhance the link between an institution's risk profile, its risk management and risk mitigation systems, and its capital.

Institutions themselves should develop sound risk management processes that identify, measure, aggregate and monitor their risks to an adequate degree. Institutions are expected to have an appropriate assessment process that cover all the key elements of capital planning and management and brings forth an adequate amount of capital to set against those risks.

Regulated institutions should "own," develop and manage their risk management processes; the so-called Internal Capital Adequacy Assessment Process (ICAAP) belongs to the institution and supervisors should not prescribe how it is applied. The task of the supervisory authority is to review and evaluate the ICAAP and the soundness of the internal governance processes within which it is used. The *dialogue* between an institution and its supervi-

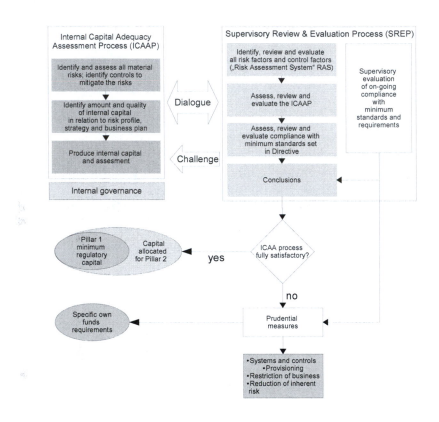

FIGURE 4.31: Supervisory review process of the European Union. (From CEBS, 2006b, 8.)

sor is a key part of the supervisory review process. The regulation highlights the respective involvement of supervisory authorities and institutions and the interaction between them, with the intention of making this dialogue clear and consistent. The dialogue should cover all aspects of business risk and internal governance (including also risk control, compliance and internal audit). To ensure transparency and consistency in the dialogue, and to advance the convergence of practices, the supervisory processes have been laid out in detail. The intensity and depth of the dialogue should be proportionate to the nature, size, complexity and systemic importance of the institution. For example, a small, non-complex institution would not be expected to have a sophisticated ICAAP, and its supervisor need not to subject it to an intense and comprehensive dialogue.

The Capital Requirements Directive (CRD) provides supervisors with several tools, including setting a Pillar 2 capital requirement. However, supervisors fully recognise that while capital has an important role to play in the mitigation of risks it needs not to be always the sole or best solution. Depending on the concrete circumstances, it may therefore be used on its own, in combination with other measures, or not at all, if other measures are more appropriate. Accordingly, supervisors will give due weight to the application of qualitative measures of an institution within its ICAAP. FINREP represents a common standardised reporting framework intended to improve comparability of financial information produced by credit institutions for their respective national supervisory authorities. Based on the international accounting standards IAS/IFRS it ensures that the regulatory capital, the denominator in all risk ratios, is assessed uniformly.

The Guidelines on Common Reporting (COREP) (CEBS, 2006a) are intended to be used by credit institutions and investment firms when preparing prudential reports to be sent to any European Union Supervisory Authority, according to the new framework established in the new capital regulation. Its principal purpose is to assess the "total own funds for solvency purposes" and the "capital requirements." △

Example 4.12. In the USA there are various federal bank supervisory agencies: the Office of the Comptroller of the Currency (OCC) for national banks, the Federal Deposit Insurance Corporation (FDIC) for state banks that are not members of the Federal Reserve System (the "Fed"), and the Federal Reserve for state member banks. Moreover, the Office of Thrift Supervision (OTS) is the primary regulator of federal savings associations, also known as thrifts.

On-site supervisory visits produce a detailed understanding of an institution's financial condition and risk profile, but they are conducted only about once a year.

In a full-scope examination many procedures are necessary to complete the mandatory components of the *CAMELS* rating system. The acronym "CAMELS" refers to the six components of a bank's condition that are assessed:

- Capital adequacy,

- Asset quality,

- Management,

- Earnings,

- Liquidity and,

- Sensitivity to market risk.

Scores are assigned for each component in addition to the overall rating of a bank's financial condition. The scores are assigned on a scale from 1 to 5. Banks with ratings of 1 or 2 are considered to present few, if any, supervisory concerns, while banks with ratings of 3, 4, or 5 present a moderate to extreme degree of supervisory concern.

An institution's risk profile may change rapidly and substantially, so that the regulator complements on-site examinations with off-site monitoring based on quarterly supervisory data reports. Off-site monitoring ensures a timelier supervisory analysis and hence a potentially more efficient allocation of scarce investigatory resources. An important part of off-site monitoring is based on econometric models (Lopez, 1999). A new econometric framework referred to as the Supervision and Regulation Statistical Assessment of Bank Risk model or SR-SABR is being implemented.

Three separate econometric modules contribute to SR-SABR surveillance ratings. Depending on the CAMELS rating a first module contributes an estimate of the probability of an adverse change in the supervisory rating of a bank, if it were examined within the next quarter. The other module estimates the probability that a bank will fail or become critically under-capitalised within the next two years. This module is referred to as the "viability" model (FED, 2006).

Up to now, there has been no official infrastructure in place. Given the contingent commitment to adhere to the Basel Accord with respect to operational risk this is not surprising. $\triangle$

4.11   Risk Management versus Internal Control

Over the last twenty years "risk management" has become a very important topic due to pressures for change in organisational practices dealing with risk. It has evolved into an organisational concept merging ideas about corporate governance and responsibility and focussing on accountability and performance. Not last, it is an appetising multi-billion industry for different constituencies with very different ideological backgrounds and aspirations. Operational risk is a natural battle ground.

How did this come to be so? The last two decades have been paved with corporate scandals which triggered regulatory action in a given political environment. One driving force was the success of the neo-liberal theory of self-regulation. It serves the interests of state agencies and regulators (more efficient meta-regulation), the regulated (more flexibility and influence) and their institutional representatives (gain in control, power, resources and prestige)(Spira and Page, 2003, 641). As an example, in the early 1990s several events, e.g., collapse of Barings Bank, led to a re-definition of corporate governance, making senior management responsible for the maintenance of an internal control system which should not just cover the traditional accounting concern with financial reports but also regulatory compliance and operations in general. In the USA a congressional inquiry triggered the first COSO document on internal control, written by audit professionals. In the same period audit had assumed the status of an all-purpose solution to problems of administrative control. Particularly in the wake of corporate scandals, auditing remains powerful (Power, 1994, 38). Power (1994, 39) defined audit in the following manner:

> Instead of involving direct observation, audit is largely an indirect form of "control of control" which acts on systems whose function is to provide observable traces. In a number of areas this results in a preoccupation with the auditable process rather than the substance of activities. This consequently burdens the auditee with the need to invest in mechanisms of compliance, a fact which apparently has produced a consistent stream of complaint.

The primary risks of many industries, especially the financial industry with credit and market risk, cannot be audited or directly inspected. But it is assumed that control can be exerted indirectly by auditing the management system of control. This may be a bold assertion but a powerful sales pitch. Typically, in 1996 Deloitte & Touche, an audit firm, proposed a study titled "Internal Control Issues in Derivatives Usage: An Information Tool for Considering the COSO Internal Control – Integrated Framework in Derivatives Applications." Overall the attempt to incorporate risk management under the internal control umbrella with its accounting perspective was so successful that its representatives could propagate the enterprise-wide risk management (ERM) with the late COSO model of 2004. If we take a quick glance at auditors' interpretation of risk management, then we see that they sell risk management advice as internal controls shaped by some variation of the COSO template (e.g., KPMG, 2008; Ernst & Young, 2008). It goes even farther.
  Inherent in the their strategy is (Power, 2005, 595):

> ... the construction of operational risk as a higher order category in relation to market and credit risk; in this way accountants as low level quantifiers can stand in a hierarchical relation to mathematical modeling expertise. This, of course, is not how the non-accountant specialists see the terrain; so, operational risk exposes tensions between different forms of expertise in organizations and becomes a stake in the competition for pre-eminence in management hierarchies.

In Section 1.4.2 on page 16 we defined risk management as a rather technical discipline concerned mainly with quantifying risk for better decision-making and trying to permeate the organisation with concepts of risk-adjusted returns and risk capital calculation and allocation. The Loss Distribution Approach for operational risk is some sort of last stand for interpreting risk management. Leitch (2008, 20) says:

> Internal control has been expanding and has claimed to include risk management. Ironically, risk management has also been expanding and claims to include internal controls.

Risk assessment is one component of the COSO cube and the so-called internal control environment is explicitly mentioned in the operational risk framework of Basel 2. The regulatory framework gives the choice to either apply the quantitative loss distribution approach or the score-card approach akin to the accounting universe.

All substantive changes within a company shift power bases and create pressure on individuals and departments to re-invent themselves. With the increasing focus on internal control the internal audit function has managed to expand from a low-key and subservient role into a consultant role with a direct link to the board of directors, aspiring to provide assurance with respect to effectiveness and efficiency of operations. Formerly, its main competitive edge was not its proven added value but its independence from business divisions. But now independence and the advisory role are at odds.

The operational risk is according to Power (2004, 39) a "boundary object" whose broad definition means that the concept of operational risk can appeal to a wide range of sub-groups within organisations and serves as an umbrella for many different interests, e.g., influence, status, career advancement, etc. Because of its fuzzy definition, the question of what it is, or could be, cannot be divided from the question of who is re-inventing themselves in the name of operational risk (Power, 2005, 601). Smart credit risk officers were quick at depicting operational risk as a special case of or at least highly pertinent to credit risk. Sceptical commentators like Power (2004, 31) contest the ability to control operational risk:

> The suspicion is that, while operational risk facilitates a greater "managerialisation of risk" via new organisational processes, and extends the scope of the risk manager's and the regulator's work into more corners of organisational and social life, it also reinforces myths of controllability in areas where this is at best limited – for example, the senior management culture and the often discussed "tone at the top" of organisations. Very much the same can be said of reputational risk management.

Is there something wrong with internal control as risk management? The main problem is that it is obsessed with processes and procedures and thus assumes that all relevant risks can be cast into this form. It is an internal view of what is and not of what could be. Procedural risk management is onerously bureaucratic and fails to add value unless we place value on alleviating the fear of responsibility. It is also a mechanism for shifting blame to the "system" as long as procedures have been followed. Power (2004, 14) writes about the *secondary risk*, i.e., because of the increased accountability of risk experts, they have become very much concerned with the risk posed to their reputation and less with their primary risk for which they are employed. There is an emergent culture of defensiveness.

This kind of risk management dwarfs the experts' judgement and his "honest professional opinion" and forces them to trade their knowledge for a process. On the other hand supplanting the direct observation of risk, say in a derivative trading desk with market and counter-party risk, by reviewing and auditing the compliance and monitoring of a policy (a second-order observation), creates awareness of the self-production of risks. Both auditors and regulators need not possess the inherent knowledge pertaining to the most difficult technical areas. Auditing makes regulation in these areas look possible and links the aspirations of a remote regulator with those of a profit-driven trader (Power, 1997, 15).

From all this we see that risk management has developed into an organising template from two distinct cultures, i.e., a quantitative, model-based culture and an accounting culture. The author's preference for the interpretation of risk management can be inferred from this book's content. Obviously, the merging of the more technical risk management with internal control will take place. But as of now we have still the following situation (Leitch, 2008):

> However, risk managers and internal controls managers tend to have different backgrounds and preoccupations. Risk managers tend to be concerned with big, non-recurring risk events and often have insurance or engineering backgrounds. Internal controls people are more concerned with smaller, recurring, internal risk events and tend to have audit or accounting backgrounds.
>
> It's ironic that internal controls thinking, despite being a movement led by the big audit firms (of accountants), has paid almost no attention to quantifying risks or the benefits of controls in a credible, mathematically competent, and data-supported way. Most assessments don't get past "high-medium-low."

The difference between internal controllers and risk managers may by summarised as in Table 4.15. The ideological differences between quants and non-quants (discussed in Section 4.3.6 on page 266), called by Mikes (2008c; 2008b) "quantitative enthusiasts" and "quantitative sceptics" or idealist and pragmatists, paired with the internal power of the different camps, decides on the specific risk management style of the institution. Mikes (2008a) gives some good empirical evidence and lucid insight on this issue.

TABLE 4.15: Quantitative risk management and internal control

	Quantitative Approach	Internal Control
Context, culture	Financial institutions and corporate treasuries	All organisations
Subject	Market, credit and operational risk	Processes and controls on entity-level, processes and general IT
Prerequisite	Quantifyability	Accountability
General principle	risk-adjusted return, cost of risk capital, risk allocation	(Cybernetic) variance and performance analysis
Source	Actuarial mathematics	Accounting, audit profession
Method	Models	Self-assessment, observation

4.12   Summary and Conclusion

In this chapter we have tried to link the quantitative model with the risk management. In a first step we have introduced traditional and some less well known management concepts. To be retained is the fact that there is no best way of management, leadership or organisational structure. Politics and power are dominant themes. Management is embedded with culture and surrounded with its environment. Both have a deep influence on risk management and risks because the first may clash with other cultures leading to misunderstandings, discontent and frustrations, and the latter may change rapidly and meet the organisation ill-prepared. Basel 2 puts special emphasis on both business environment – a sub-domain of the general environment – and internal control factors which in turn are over-lapping with the activities to securing financial reporting. These two multi-dimensional domains must also be operationalised in such a manner as to be reflected in the quantification process. The regulators' requirement reads as follows (BCBS, 2005, 150): "These factors will make a bank's risk assessments more forward-looking, more directly reflect the quality of the bank's control and operating environments, help align capital assessments with risk management objectives, and recognise both improvements and deterioration in operational risk profiles in a more immediate fashion."

We have chosen the easiest way, as always, by proposing to construct indices from either measurable sub-indices or the extraction of indicators from questionnaires. These indices enter the calculation by means of fudge factors to be applied to the parameters of the various distributions. The analysis of the bank and its processes is the same as would be needed for rather a "causal" model of the institution. This approach is akin to scenario analysis where we do also change parameters on input in order to assess the convoluted effect on the economic capital.

Somewhat more specific we suggested to use the loss distribution approach with its capability to include rather a realistic modelling of insurance to play with the insurance parametrisation. Optimisation in a narrow sense will not be possible, but a substantial improvement of the insurance programme is realistic.

Risk management – as an activity – is an adaptive feed-back process. The most critical step is to provide information that readily leads to consequential action. With the loss distribution approach, its one year horizon and the precarious loss data history, this kind of information cannot be produced. Therefore, besides the insurance programme optimisation where we have a feed-back loop, the LDA does not lend itself naturally to operational risk management. The thought that measurement is a sufficient condition for management is an innocuous wish. But, as Julius Caesar said: "in most cases men willingly believe what they wish."[8]

But, the effect of second order, that is the fact that many people must think about operational risk and raise this topic above the psychological attention level, may be the most tangible result of this exercise. But bullshit looms everywhere.

---

[8] De bello gallico, 3.18: "fere libenter homines id quod volunt credunt."

Chapter 5

Conclusions

A book on the Advanced Measurement Approach, and especially on the Loss Distribution method, rests on the fundamental assumption that the approach can be implemented in a meaningful way. This is not generally accepted. For example, Ong (1998) was, and may still be, convinced that: "the only buffer against operational risk is not some quantified measure but educated people who are careful, technologically aware, informed and compliant with internal policies." We, on the other hand, think otherwise. We believe that the rationale "capital against risk" is sound. Since the advent of public companies, shareholders' equity has been seen to fulfil exactly this role. That being so, the accurate quantification of such capital is a desirable and useful measure. We grant that it cannot be purely scientifically determined, due to the vagaries of human nature and the uncertain outcomes of human endeavours. However, within such parameters, the assessment of risk capital must nonetheless strive to be comprehensive, traceable, transparent, sensible and commensurate with the potential risk involved. As the figure is added to the other risk capital figures, it should be on an equal footing with those methodologies and their ranges of error. We think that we have argued sufficiently for this cause.

We see the Loss Distribution Approach, and especially its quantification, as consisting of two main parts: firstly, the statistical procedure to use the given data best in order to yield a risk capital quantity. This has been described mainly in Chapter 3 of this book. But this would only take into account the history as depicted by the gathered data. Unfortunately, we cannot assume a temporal structural stability of the banking industry. Furthermore, the pace of new data reflecting management and industry changes is much too slow to yield a responsive and actual model. Regulators want the measurement to be integrated in the day-to-day risk management. Therefore, a second part, primarily concerned with *adjustments* and discussed in Chapter 4, complements the LD approach. The definition of operational risk contains four main risk sources requiring coverage, i.e., processes, people, systems and external events. The first three items are included in the traditional understanding of management. This is the reason why we have – for some perhaps at excessive length – tried to understand what management is: its relationship to human behaviour and motivation, culture and so forth. This "soft underbelly" of quantification, however, demands a quantification of its own. The methods of choice stem mainly from the social sciences, with their long-standing tradition of measuring multi-dimensional concepts using indicators. Here, we have opted for the construction of several compound indices related to the parameters of the LDA model. Indices, based on Factor Analysis, can be seen as an alternative to "causal" models, based either on conditional probabilities or correlations – like Bayesian nets.

This latter quantification bears a close affinity to the other AMA method, i.e., the score-card approach. Taking into account both approaches, with their respective benefits, we have tried to create a combination approach that minimizes their differences (see also page 40).

The presented model's outstanding characteristic is the modelling of the losses. We discern three levels of losses by the given event categories, according to their frequency and severity. The lowest and most frequent losses, called "attritional," are treated as an aggregate lump sum. The second level treats losses individually by combining frequency of occurrence with severity of loss. Such losses are amenable to insurance coverage. And finally, we must take into account catastrophic losses from all event categories. We assume that quantity becomes a quality. The losses by lines of business are derived by a mapping table. The correlation between losses is introduced by a dependency model, i.e., a copula, affecting the loss frequencies. Insurance is modelled as sensitively as possible. Of course, policies must be mapped to events. Such mapping must be back-tested, lest an unreliable mitigation result. The parameters for these inputs are derived from the data available. The calculation of the risk capital is done using the Monte Carlo simulation technique, which means that the number of draws is crucial to the accuracy of the result, and therefore also needs estimating.

We propose an adjustment model based on indices using the four compounds: internal control, internal environment, external environment and business environment. For each severity of the events such an index should be constructed as to be multiplied as a fudge factor to the dispersion parameter of the severity distribution. Favourable developments should diminish the index, while unfavourable elements should augment it. This mechanism is also used for scenario analysis, – where, in addition to severity, the frequency may also be varied in order to analyse the model's sensitivity to such parameter estimates.

The "adjustment model" makes the LDA approach more respondent to exogenous occurrences. But for management – in a feedback control sense – the risk quantification is not a direct help. You cannot play with the model, find some undesired risk area and take immediate action. Firstly, because you may not know the appropriate action or because the pertinent variable is not controllable, i.e., there is no course of action that can steer it to a better outcome. Of course, insurance constitutes an exception. Managers may tweak insurance policies' parameters until they are happy with the coverage for which they are paying. In our run-through example we have tried to demonstrate that insurance, as it is conceived today, is being under-utilized as a means of risk mitigation. We find a mitigation of approximately 4%, significantly below the maximum of 20%. We believe it ought to be possible to bring insurance premiums close to the other capital costs of the bank. The bottom line is that insurance is a valuable alternative to a bank's carrying risk on its own balance sheet.

Our knowledge is admittedly confined to those risk situation where statistical models are appropriate. In those cases where statistical models cannot be employed, i.e., in the "uncertainty situation," guesswork reigns. Yet every risk manager has a personal take on risk and ranks risks differently. Our own view is that legal action, especially with concern to tort and its rampant costs, is one of the major foreseeable threats to the banking industry.

What can be done to manage operational risk with greater success? We think that more careful hiring and promotion practices, more and better training, the creation of a culture of openness that tolerates, or even encourages, dissent, the streamlining and simplifying of complicated processes, the more efficient use of technology and the exercise of basic caution must all play a critical role in refining, or even redefining, risk management practices.

Although strategic risk may pose the biggest threat to a bank's survival it is also true that in the shorter term an institution should not fail due to market, credit or operational risk. While the stability of the banking system does not presume a zero-bankruptcy rate among its constituents, bankruptcy is obviously not a desirable outcome. Given that the banking system as a whole is not a controllable entity, any regulation must apply to all institutions alike, thus strengthening the second goal, i.e., creating a level playing field. Such regulation may be cynically perceived as a tax in kind required for compliance. Whether this is indeed so is a matter of individual reflection which we shall leave to the reader. And on that rather open-ended and perhaps provocative note, we conclude our treatise on an altogether fascinating and controversial subject.

Appendix

1 Factors Influencing Risk Perception

TABLE 5.1: Risk perception.(From Covello et al., 1988, 54. By permission of the American Chemistry Council)

Factor	Conditions Associated with Increased Public Concern	Conditions Associated with Decreased Public Concern
Catastrophic Potential	Fatalities and Injuries Grouped in Time and Space	Fatalities and Injuries Scattered and Random
Familiarity	Unfamiliar	Familiar
Understanding	Mechanisms or Process Not Understood	Mechanisms or Process Understood
Uncertainty	Risks Scientifically Unknown or Uncertain	Risks Known to Science
Controllability (Personal)	Uncontrollable	Controllable
Voluntariness of Exposure	Involuntary	Voluntary
Effects on Children	Children Specifically at Risk	Children Not Specifically at Risk
Effects Manifestation	Delayed Effects	Immediate Effects
Effects on Future Generations	Risk to Future Generations	No Risk to Future Generations
Victim Identity	Identifiable Victims	Statistical Victims
Dread	Effects Dreaded	Effects Not Dreaded
Trust in Institutions	Lack of Trust in Responsible Institutions	Trust in Responsible Institutions
Media Attention	Much Media Attention	Little Media Attention
Accident History	Major and Sometimes Minor Accidents	No Major or Minor Accidents
Equity	Inequitable Distribution of Risks and Benefits	Equitable Distribution of Risks and Benefits
Benefits	Unclear Benefits	Clear Benefits
Reversibility	Effects Irreversible	Effects Reversible
Origin	Caused by Human Actions or Failures	Caused by Acts of Nature or God

2 Various Death or Harm Frequencies

TABLE 5.2: Various death or harm frequencies. (The source of this data is based on U.S. population. Excerpt from Ropeik and Gray, 2002. Copyright (c) 2002 by David Ropeik and George Gray. Reprinted by permission of Houghton Mifflin Company. All rights reserved)

Cause of Death or Harm	Annual Frequency	Lifetime Frequency
Heart disease	1 in 300	1 in 4
Cancer (all forms)	1 in 510	1 in 7
Stroke	1 in 1,800	1 in 23
Diabetes	1 in 4,100	1 in 53
Pneumonia	1 in 4,300	1 in 57
Alzheimer's disease	1 in 5,700	1 in 75
Suicide	1 in 9,200	1 in 120
Motor vehicle accident	1 in 6,700	1 in 88
Criminal homicide	1 in 18,000	1 in 240
Occupational accident	1 in 48,000	1 in 620
Flu	1 in 130,000	1 in 1,700
Struck by a falling object	1 in 390,000	1 in 5,100
Heat	1 in 740,000	1 in 9,600
Drowning in the bathtub	1 in 840,000	1 in 11,000
Lightning	1 in 3,000,000	1 in 39,000
Commercial aircraft accident	1 in 3,100,000	1 in 40,000
Hornet, wasp, bee	1 in 6,100,000	1 in 80,000
Skydiving	1 in 9,100,000	1 in 120,000
Hurricane	1 in 17,000,000	1 in 220,000
Dog bite	1 in 19,000,000	1 in 240,000
Anthrax (as of 2001)	1 in 56,000,000	1 in 730,000
Passenger train accident	1 in 70,000,000	1 in 920,000
Shark attack	1 in 280,000,000	1 in 3,700,000

3 Basel 2's Loss Event Type Classification

TABLE 5.3: Event-type category definitions

Event Type	Definition
Internal Fraud	Losses due to acts of a type intended to defraud, misappropriate property or circumvent regulations, the law or company policy, excluding diversity/discrimination events, which involves at least one internal party.
External Fraud	Losses due to acts of a type intended to defraud, misappropriate property or circumvent the law, by a third party.
Employment Practices and Workplace Safety	Losses arising from acts inconsistent with employment, health or safety laws or agreements, from payment of personal injury claims, or from diversity/discrimination events.
Clients, Products & Business Practices	Losses arising from an unintentional or negligent failure to meet a professional obligation to specific clients (including fiduciary and suitability requirements), or from the nature or design of a product.
Damage to Physical Assets	Losses arising from loss or damage to physical assets from natural disaster or other events.
Business Disruption and System Failures	Losses arising from disruption of business or system failures.
Execution, Delivery & Process Management	Losses from failed transaction processing or process management, from relations with trade counterparties and vendors.

TABLE 5.4: Basel 2 loss event type classification

Event-Type	Categories (Level 2)	Activity Examples (Level 3)
Internal Fraud	Unauthorised Activity	Transactions not reported (intentional)
		Transaction type unauthorised (no loss)
		Intentional mismarking of position ()
	Theft and Fraud	Fraud/credit fraud/worthless deposits
		Theft/extortion/embezzlement/robbery
		Misappropriation of assets
		Malicious destruction of assets
		Forgery
		Check kiting
		Smuggling
		Account take-over/impersonation/etc.
		Tax non-compliance/evasion (wilful)
		Bribes/kickbacks
		Insider trading (not on firm's account)
External Fraud	Theft and Fraud	Theft/robbery
		Forgery
		Check kiting
	Systems Security	Hacking damage
		Theft of information (without monetary loss)

TABLE 5.4: Basel 2 loss event type classification (continued)

Event-Type	Categories (Level 2)	Activity Examples (Level 3)
Employment Practices and Workplace Safety	Employee Relations	Compensation, benefit, termination issues Organised labour activity
	Safe Environment	General liability (slip and fall, etc.) Employee health & safety rules events Workers compensation
	Diversity & Discrimination	All discrimination types
Clients, Products & Business Practices	Suitability, Disclosure & Fiduciary	Fiduciary breaches/guideline violations Suitability/disclosure issues (KYC, etc.) Retail customer disclosure violations Breach of privacy Aggressive sales Account churning Misuse of confidential information Lender liability
	Improper Business or Market Practices	Antitrust Improper trade/market practices Market manipulation Insider trading (on firm's account) Unlicensed activity Money laundering

TABLE 5.4: Basel 2 loss event type classification (continued)

Event-Type	Categories (Level 2)	Activity Examples (Level 3)
Clients, Products & Business Practices (cont.)	Product Flaws	Product defects (unauthorised, etc.)
		Model errors
	Selection, Sponsorship & Exposure	Failure to investigate client per guidelines
		Exceeding client exposure limits
	Advisory Activities	Disputes over performance of advisory activities
Damage to Physical Assets	Disasters and Other Events	Natural disaster losses
		Human losses from external sources (terrorism, vandalism)
Business Disruption and System Failures	Systems	Hardware
		Software
		Telecommunications
		Utility outage/disruptions
Execution, Delivery & Process Management	Transaction Capture, Execution & Maintenance	Miscommunication
		Data entry, maintenance or loading error
		Missed deadline or responsibility
		Model/system misoperation
		Accounting error/entity attribution error
		Other task misperformance
		Delivery failure
		Collateral management failure
		Reference Data Maintenance

TABLE 5.4: Basel 2 loss event type classification (continued)

Event-Type	Categories (Level 2)	Activity Examples (Level 3)
Execution, Delivery & Process Management (cont.)	Monitoring and Reporting	Failed mandatory reporting obligation Inaccurate external report (loss incurred)
	Customer Intake and Documentation	Client permissions/disclaimers missing Legal documents missing/incomplete
	Customer/Client Account Management	Unapproved access given to accounts Incorrect client records (loss incurred) Negligent loss or damage of client assets
	Trade Counterparties	Non-client counterparty mis-performance Misc. non-client counterparty disputes
	Vendor Disputes	Outsourcing Vendors & Suppliers

4   Some Regulatory Report Templates

TABLE 5.5: OPR LOSS details

1	Internal reference number
2	Gross loss amount
3	Of which: unrealized
4	Status: ended?: Yes/No
5	Loss already directly recovered
6	Loss already recovered from risk transfer mechanisms
7	Loss potentially to be recovered directly or from risk transfer mechanisms
8	Related to credit risk or market risk
9 – 16	Breakdown of gross loss (%) by business lines (CF, TS, RBr, CB, RB, PS, AS, AM)
17	Risk event type (number)
18	Date of occurrence
19	Date of recognition
20	Date of the first payment from risk transfer mechanisms
21	Date of the latest payment from risk transfer mechanisms

This template (Table 5.5) provides information on the major operational risk losses recorded in the last year on a gross and net basis along with the status of whether they are ended or still open. It allows a monitoring of those losses above a threshold designated by the competent authorities, and provides information on the nature of the operational losses (event types) and on their location by business lines, as well as effectiveness of the hedging techniques used.

TABLE 5.6: Reporting items under the EU's COREP guidelines for AMA

Capital requirements
of which: due to an allocation mechanism
Memorandum items:
Capital requirements before alleviation due to expected loss
(-) Alleviation of capital requirements due to the expected loss captured in business practice
(-) Alleviation of capital requirements due to risk transfer mechanisms
of which: due to insurance
Excess on limit for capital alleviation of risk transfer mechanism

This template (Table 5.6) provides information on the capital requirements for Operational Risk under the Basic Indicator Approach (BIA), the Standardised Approach (STA), the Alternative Standardised Approach (ASA) and the Advanced Measurement Approaches (AMA).

5   Monte Carlo Example (continued from Example 3.3)

Note that here the matrix **p** holds the cumulated values of the original matrix.

```fortran
program mceqn
implicit double precision (a-h,o-z)
dimension v(2,2),p(2,2),b(2),istart(2)
data v/1.,1.,1.,-1./ b/.7,1.1/
data p/.1,.2,.3,.5/
data istart/1,2/

idum=-1
niter=25
do i=1,2
  sum=0.d0
  do n=1,niter
    istate=istart(i)
    x=b(istate)
    fac=1.d0
    do k=1,100000
      u=ran1(idum)
      if (u .lt. p(istate,1)) then
        iprev=istate
        istate=1
      else if (u .lt. p(istate,2)) then
        iprev=istate
        istate=2
      else
        goto 100
      end if
      fac=fac*v(iprev,istate)
      x=x+fac*b(istate)
    end do
100     continue
    sum=sum+x
  end do
  ex=sum/dble(niter)
  write(*,'(a,i1,a,f14.3)') 'E(x_',i,')=',ex
end do
stop
end
```

In the Table 5.7 below we see the first 25 paths and values of the simulation. The trajectory is marked by the succession of the states, i.e. either 1 or 2. The first column always starts in state 1 for estimating x_1, respectively 2 for x_2.

TABLE 5.7: Simulation outcome

Trajectory	X_1	Trajectory	X_2
1	0.7	22	0
11	1.4	2	1.1
1	0.7	2,	1.1
1	0.7	22	0
1	0.7	21	1.8
1	0.7	22	0
11	1.4	22	0
1	0.7	21	1.8
1	0.7	22	0
1	0.7	2	1.1
1	0.7	221	-0.7
1	0.7	221	-0.7
1	0.7	2	1.1
122	0.7	222	1.1
1	0.7	2	1.1
1	0.7	2	1.1
112	2.5	2	1.1
12	1.8	2	1.1
1	0.7	2	1.1
122	0.7	2	1.1
1	0.7	21	1.8
11	1.4	2	1.1
1	0.7	2	1.1
1	0.7	2	1.1
11	1.4	212	2.9
Sample mean	0.928	Sample mean	0.892

Bibliography

Acerbi, C. and Tasche, D. (2001). On the coherence of expected shortfall.

ACFE (2008). 2008 report to the nation on occupational fraud & abuse. Discussion papers, Association of Certified Fraud Examiners, Austin, TX.

AeA (2005). Sarbanes-Oxley Section 404: The "section" of unintended consequences and its impact on small business. Technical report, American Electronics Association.

Agami, A. M. (2006). Reporting on internal control over financial reporting. *CPA Journal online*.

Alderfer, C. P. (1969). An empirical test of a new theory of human needs. *Organizational Behavior and Human Decision Processes*, 4(2), 142–75.

Alexander, C. (2002). Managing operational risks with Bayesian networks. In C. Alexander, editor, *Operational Risk; Regulation, Analysis and Management*, pages 285–295. Prentice Hall, London.

Alexander, C. (2004). The present and future of financial risk management. Technical report, The University of Reading, ISMA Centre Discussion Papers in Finance 2003-12.

American Environics (2006). The evolution on global values. Technical report, American Environics, Oakland, CA.

Arens, A. A., Elder, R. J., and Beasley, M. S. (2008). *Auditing and Assurance Services. An Integrated Approach*. Prentice Hall, Upper Saddle River, NJ, 12 edition.

Artzner, P., Delbaen, F., Eber, J., and Heath, D. (1998). Coherent measures of risk. http://www.math.ethz.ch/~delbaen/ftp/preprints/CoherentMF.pdf.

Aumann, R. and Shapley, L. (1974). *Values of Non-Atomic Games*. Princeton University Press, Princeton, NJ.

Baddeley, M., Curtis, A., and Wood, R. A. (2004). An introduction to prior information derived from probabilistic judgments: Elicitation of knowledge, cognitive bias and herding. *Special Publication of the Geological Society of London*, **239**, 15–27.

Balzer, L. (1994). Measuring investment risk: A review. *Journal of Investing*, (Fall).

Barth, J. R., Caprio, G., and Levine, R. (2006). *Rethinking Bank Regulation:*

Till Angels Govern. Cambridge University Press, Cambridge.

Bazerman, M. H. and Neale, M. A. (1992). *Negotiating Rationally.* The Free Press, New York.

BCBS (1988). International convergence of capital measurement and capital standards. Technical Report Basel Committee Publications No. 4, Bank for International Settlements, Basel. http://www.bis.org/publ/bcbs04.pdf.

BCBS (1996). Overview of the amendment to the capital accord to incorporate market risk. Technical Report Basel Committee Publications No. 23, Bank for International Settlements, Basel. http://www.bis.org/publ/bcbs23.pdf.

BCBS (2000). Internal audit in banking organisations and the relationship of the supervisory authorities with internal and external auditors. Technical report, Bank for International Settlements, Basel. http://www.bis.org/publ/bcbs72.pdf.

BCBS (2001a). Customer due diligence for banks. Technical report, Bank for International Settlements, Basel. http://www.bis.org/publ/bcbs85.pdf.

BCBS (2001b). Internal audit in banks and the supervisor's relationship with auditors. Technical report, Bank for International Settlements, Basel. http://www.bis.org/publ/bcbs84.pdf.

BCBS (2001c). Working paper on the regulatory treatment of operational risk. Technical report, Bank for International Settlements, Basel. http://www.bis.org/publ/bcbs_wp8.pdf.

BCBS (2003a). The 2002 loss data collection exercise for operational risk: Summary of the data collected. Technical report, Bank for International Settlements, Basel.

BCBS (2003b). Consultative document. The new capital accord. Technical report, Bank for International Settlements, Basel.

BCBS (2003c). Sound practices for the management and supervision of operational risk. Technical report, Bank for International Settlements, Basel.

BCBS (2004a). Basel II: International convergence of capital measurement and capital standards: A revised framework. Technical Report Basel Committee Publications No. 107, Bank for International Settlements, Basel. http://www.bis.org/publ/bcbs107.pdf.

BCBS (2004b). Principles for the home-host recognition of AMA operational risk capital. Technical report, Bank for International Settlements, Basel. http://www.bis.org/publ/bcbs106.pdf.

BCBS (2005). International convergence of capital measurement and capital standards: A revised framework. Technical report, Bank for International Settlements, Basel. http://www.bis.org/publ/bcbs118.pdf.

BCBS (2006). Core principles for effective banking supervision. Technical report, Bank for International Settlements, Basel. http://www.bis.org/-publ/bcbs129.pdf.

BCBS (2008a). Liquidity risk: Management and supervisory challenges. Technical report, Bank for International Settlements, Basel. http://www.bis.-org/publ/bcbs136.pdf.

BCBS (2008b). Principles for sound liquidity risk management and supervision. Technical report, Bank for International Settlements, Basel. http://www.bis.org/publ/bcbs138.pdf.

Beardsley, S. C., Johnson, B. C., and Manyika, J. M. (2006). Competitive advantage from better interactions. *McKinsey Quarterly*, (2).

Beaver, W. (1966). Financial ratios as predictors of failure. *Journal of Accounting Research*, pages 77–111.

Beck, U. (1999). *World Risk Society*. Polity Press, Cambridge.

Becker, G. S. (1968). Crime and punishment: An economic approach. *The Journal of Political Economy*, **76**(2), 169–217.

Benston, G. and Kaufman, G. (1996). The appropriate role of bank regulation. *The Economic Journal*, **106**, 688–97.

Bentham, J. (1907). *An Introduction to the Principles of Morals and Legislation*. Clarendon Press, Oxford. Available at http://www.econlib.org/-library/Bentham/bnthPML.html.

Berle, A. A. and Means, G. (1932). *The Modern Corporation and Private Property*. Macmillan, New York.

Bernoulli, D. (1954). Exposition of a new theory of the measurement of risk. *Econometrica*, **22**, 123–136. Translation of "Specimen theoriae novae de mensura sortis," appeared 1738.

Bernstein, P. L. (1996). *Against the Gods. The Remarkable Story of Risk*. John Wiley, New York.

Bertsekas, D. P. and Tseng, P. (1994). RELAX-IV: A faster version of the RELAX code for solving minimum cost flow problems. Technical report, Massachusetts Institute of Technology, Cambridge, MA.

BITS (2005). Reconciliation of regulatory overlap for the management and supervision of operational risk in U.S. financial institutions. Technical report, BITS Operational Risk Management Working Group – The Financial Services Roundtable.

Bluhm, C., Overbeck, L., and Wagner, C. (2003). *An Introduction to Credit Risk Modeling*. Chapman & Hall/CRC, London.

Bolman, L. G. and Deal, T. E. (1997). *Reframing Organizations: Artistry, Choice, and Leadership*. Jossey-Bass, San Francisco, 2nd edition.

Booth, A. L. and Nolen, P. J. (2009). Gender Differences in Risk Behaviour: Does Nurture Matter? *SSRN eLibrary*.

Boritz, J. E. (2001). *Computer Control and Audit Guide*. University of Waterloo, New York, 3rd edition. http://www.arts.uwaterloo.ca/ACCT/ccag/-COVERPAGE.htm.

Bortz, J. (1999). *Statistik. Für Sozialwissenschafter*. Springer Verlag, Berlin, 5th edition.

Brealey, R. and Myers, S. (1996). *Principles of Corporate Finance*. McGraw-Hill, New York, 5th edition.

Burns, J. (1978). *Leadership*. Harper & Row, New York.

Bury, K. (1999). *Statistical Distributions in Engineering*. Cambridge University Press, Cambridge.

Cario, M. C. and Nelson, B. L. (1997). Numerical methods for fitting and simulating autoregressive-to-anything processes. *INFORMS Journal on Computing*, **10**, 72–81.

Carnap, R. (1950). *Logical Foundations of Probability*. University of Chicago Press, Chicago.

CEBS (2005). Framework for consolidated financial reporting. Technical report, Committee of European Banking Supervisors, Bruxelles.

CEBS (2006a). Guidelines for the implementation of the framework for consolidated financial reporting FINREP. Technical report, Committee of European Banking Supervisors, Bruxelles.

CEBS (2006b). Guidelines on the application of the supervisory review process under pillar 2. Technical report, Committee of European Banking Supervisors, Bruxelles.

Chamberlain, B. H. (1902). *Things Japanese. Being Notes on Various Subjects Connected with Japan for the Use of Travellers and Others*. John Murray, London.

Chernobai, A. and Rachev, S. T. (2004). Stable modeling of operational risk. In M. Cruz, editor, *Operational Risk Modelling and Analysis*, pages 139–169. Risk Books, London.

Chernobai, A. S., Rachev, S. T., and Fabozzi, F. J. (2007). *Operational Risk: A Guide to Basel II Capital Requirements, Models, and Analysis*. Wiley Publishing, Hoboken, NJ.

Clemen, R. T. and Winkler, R. L. (1999). Combining probability distributions from experts in risk ananlysis. *Risk Analysis*, **19**(2), 187–203.

Clemen, R. T., Fischer, G. W., and Winkler, R. L. (2000). Assessing dependence: Some experimental results. *Management Science*, **46**(8), 1100–1115.

Coase, R. (1937). The nature of the firm. *Economica*, **4**, 386–405.

Cohen, M. D., March, J. G., and Olsen, J. P. (1972). A garbage can model of organizational choice. *Administrative Science Quarterly*, **17**(1), 1–25.

Coles, S. (2001). *An Introduction to Statistical Modeling of Extreme Values*. Springer Verlag, London.

COSO (1992). Internal control - integrated framework. Technical report, The Committee of Sponsoring Organizations of the Treadway Commission, New York.

COSO (2004). Enterprise risk management – integrated framework: Framework. Technical report, The Committee of Sponsoring Organizations of the Treadway Commission, Jersey City.

Covello, V. and Allen, F. (1988). Seven cardinal rules of risk communication. Technical report, U.S. Environmental Protection Agency, Office of Policy Analysis, Washington, DC.

Covello, V. T., Sandman, P. M., and Slovic, P. (1988). *Risk Communication, Risk Statistics, and Risk Comparisons: A Manual for Plant Managers*. Chemical Manufacturers Association, Washington, DC.

Crow, E. L. and Shimizu, K., editors (1988). *Lognormal Distributions: Theory and Applications*. Marcel Dekker, New York.

Davenport, T. (1993). *Process Innovation: Reengineering work through information technology*. Harvard Business School Press, Boston.

Dawkins, R. (1988). *The Blind Watchmaker*. Penguin Books, London.

Daykin, C., Pentikainen, T., and Pesonen, M. (1994). *Practical Risk Theory for Actuaries*. Chapman & Hall, London.

Delbaen, F. and Denault, M. (1999). Coherent allocation of risk capital. Preprint, RiskLab Zurich, http://www.risklab.ch/Papers.html.

Demidovic, B. P. and Maron, I. A. (1981). *Fondamenti di calcolo numerico*. Edizioni Mir, Moscow.

Demsetz, R. S., Saidenberg, M. R., and Strahan, P. E. (1996). Banks with something to lose: The disciplinary role of franchise value. *Federal Reserve Bank of New York Economic Policy Review*, (2), 1–14.

Denault, M. (1999). Coherent allocation of risk capital. Preprint, RiskLab Zurich, http://www.risklab.ch/Papers.html.

Devroye, L. (1986). *Non-uniform random variate generation*. Springer Verlag, New York.

Di Clemente, A. and Romano, C. (2004). A copula-exteme value theory for modelling operational risk. In M. Cruz, editor, *Operational Risk Modelling and Analysis*, pages 189–208. Risk Books, London.

Douglas, M. and Wildavsky, A. (1982). *Risk and Culture: An Essay on the Selection of Technical and Environmental Dangers*. University of California Press, Berkeley, CA.

Dowd, K. (1998). *Beyond Value At Risk. The New Science of Risk Management*. John Wiley, New York.

Drucker, P. F. (1999). *Mangement Challenges for the 21st Century*. Harper Collins, New York.

Durbin, A., Herz, S., Hunter, D., and Peck, J. (2005). Shaping the future of sustainable finance: Moving the banking sector from promises to performance. Technical report, World Wildlife Fund in association with Bank-Track.

Dutta, K. and Perry, J. (2006). A tale of tails: An empirical analysis of loss distribution models for estimating operational risk capital. Working Papers 06-13, Federal Reserve Bank of Boston.

Ebert, U. and Welsch, H. (2004). Meaningful environmental indices: A social choice approach. *Journal of Environmental Economics and Management*, **47**, 270–283.

Economist (2008a). Buttonwood: Weaken the sinews. *The Economist*. August 2nd, 2008.

Economist (2008b). Confessions of a risk manager. *The Economist*. August 7, 2008.

Efron, B. and Tibshirani, R. J. (1986). Bootstrap methods for standard errors, confidence intervals, and other measures of statistical accuracy. *Statistical Science*, **1**(1), 54–77.

Efron, B. and Tibshirani, R. J. (1993). *An Introduction to the Bootstrap*. Chapman & Hall, New York.

Eisenhardt, K. M. (1989). Agency theory: An assessment and review. *Academy of Management Review*, **14**(1), 57–74.

Ekeland, I. (1993). *The Broken Dice and Other Mathematical Tales of Chance*. Chicago University Press, Chicago.

Ellsberg, D. (1961). Risk, ambiguity, and the Savage axioms. *The Quarterly Journal of Economics*, pages 449–470.

Embrechts, P., Klüppelberg, C., and Mikosch, T. (1997). *Modelling Extremal Events*. Springer Verlag, Berlin.

Embrechts, P., McNeil, A., and Straumann, D. (1999). Correlation: Pitfalls and alternatives. *RISK*, pages 69–71.

Ernst & Young (2008). The future of risk management and internal control. Brochure. http://www.ey.com/Global/assets.nsf/International/-AABS_-_RAS_-_The_Future_of_Risk_Management_and_Internal_Controls/-$file/AABS_RAS_The_Future_of_RM_and_IC.pdf.

EUC (2001). Promoting an European framework for corporate social responsibility: Green paper. Technical report, European Commission: Directorate-General for Employment and Social Affairs, Brussels.

EUC (2002). Comparative study of corporate governance codes relevant to the European Union and its member states. Technical report, European

Commission: Internal Market Directorate General, Brussels.

Evans, M., Hastings, N., and Peacock, B. (1993). *Statistical Distributions*. John Wiley, New York, 2nd edition.

Fayol, H. (1925). La gestion des entreprises et l'outillage administratif. *Chronique sociale de France*, pages 10–26.

FED (2006). Enhancements to the system's off-site bank surveillance program. Supervisory Letter SR 06-2, Board of Governors of the Federal Reserve System, Washington, DC.

FED (2008). Commercial bank examination manual. Technical report, Board of Governors of the Federal Reserve System, Division of Banking Supervision and Regulation, Washington, DC.

Fehr-Duda, H., de Gennaro, M., and Schubert, R. (2004). Gender, financial risk, and probability weights. Economics working paper series 04/31, CER-ETH - Center of Economic Research (CER-ETH) at ETH Zurich.

Feller, W. (1957). *An Introduction to Probability Theory and Its Applications*, volume 1. John Wiley, New York, 2 edition.

Fiedler, F. (1967). *A Theory of Leadership Effectiveness*. McGraw-Hill, New York.

Field, C. A. (2004). Using the gh distribution to model extreme wind speeds. *Journal of Statistical Planning and Inference*, **122**(1-2), 15 – 22.

Fischer, M. (2006). Generalized tukey-type distributions with application to financial and teletraffic data. Discussion Papers 72/2006, Friedrich-Alexander-University Erlangen-Nuremberg, Chair of Statistics and Econometrics.

Fischhoff, B. (2002). What's worth knowing – and saying – about the risks of terrorism? Presentation. U.S. Congressional Briefing: The Human Response to Disaster.

Fischhoff, B. (2006). Behaviorally realistic risk management. In R. Daniels, H. Kunreuther, and D. Kettl, editors, *On Risk and Disaster: Lessons from Hurricane Katrina*, pages 77–88. University of Pennsylvania Press, Philadelphia.

Fisher, I. (1927). *The Making of Index Numbers*. Houghton Mifflin, Boston, 3rd edition.

Fisher, R. A. (1915). Frequency distribution of the value of the correlation coefficient in samples from a infinitely large population. *Biometrika*, **10**, 507–521.

Fisher, R. A. and Tippett, L. H. C. (1928). Limiting forms of the frequency distribution of the largest or smallest member of a sample. *Proceedings of the Cambridge Philosophical Society*, **24**, 180–190.

Fishman, G. (1978). *Principles of Discrete Simulation*. John Wiley, New

York.

Folz, J.-M., Azema, J., and Jeancourt-Galignani, A. (2008). Mission Green. Summary report, Interim conclusions as of February 20, 2008. Technical report, Société Générale, General Inspection Department, Paris.

Frachot, A., Roncalli, T., and Salomon, E. (2004). The correlation problem in operational risk. Working papers, Groupe de Recherche Opérationnelle, Crédit Agricole SA. http://gro.creditlyonnais.fr/content/-wp/lda-correlations.pdf.

Frankfurt, H. G. (2005). *On Bullshit*. Princeton University Press, Princeton, NJ.

FRBNY (2006). Quarterly summary of banking statistics. Forth quarter, 2005. Technical report, Federal Reserve Bank of New York, Banking Studies Function.

Frees, E. and Miller, R. B. (1998). Designing effective graphs. *North American Actuarial Journal*, **2**(2), 53–76.

Frees, E. and Valdez, E. (1998). Understanding relationships using copulas. *North American Actuarial Journal*, **2**(1), 1–25.

Freixas, X. and Parigi, B. M. (2007). Banking regulation and prompt corrective action. CESifo Working Paper Series CESifo Working Paper No., CESifo GmbH.

FSF (2009). Report of the financial stability forum on enhancing market and institutional resilience update on implementation. Technical report, Financial Stability Forum, Basel. http://www.financialstabilityboard.org/-publications/r_0904d.pdf.

Fumerton, R. and Kress, K. (2001). Causation and the law: Preemption, lawful sufficiency, and causal sufficiency. *Law and Contemporary Problems*, **64**(4), 83–105.

Galtung, J. (1967). *Theory and Methods of Social Research*. Georg Allen & Unwin, London.

Gano, D. L. (2008). *Apollo Root Cause Analysis: A New Way of Thinking*. Apollonian Publications, Yakima, WA, 3 edition.

Gardner, D. (2008). *Risk: The Science and Politics of Fear*. Vergin Books, London.

Genz, A. (1992). Numerical computation of multivariate normal probabilities. *Journal of Computational and Graphical Statistics*, **1**, 141–149.

Gerber, H. U. and Pafumi, G. (1998). Utility functions: From risk theory to finance. *North America Actuarial Journal*, **2**(3), 74–100.

Ghosh, S. and Henderson, S. G. (2000). Chessboard distributions and random vectors with specified marginals and covariance matrix. Technical report, Department of Industrial and Operations Engineering, University of Michigan, Ann Arbor.

Ghosh, S. and Henderson, S. G. (2002). Properties of the NORTA method in higher dimensions. In E. Yucesan, C.-H. Chen, J. L. Snowdon, and J. M. Charnes, editors, *Proceedings of the 2002 Winter Simulation Conference*, pages 263–269, San Diego, CA.

Gigerenzer, G. (2007). *Gut Feelings: The Intelligence of the Unconscious*. Penguin, London, 1 edition.

Gigerenzer, G. (2008). *Rationality for Mortals. How people cope with uncertainty*. Oxford University Press, New York.

Gigerenzer, G. and Edwards, A. (2003). Simple tools for understanding risks: From innumeracy to insight. *BMJ (British Medical Journal)*, **327**, 741–744.

Goodman, L. A. and Kruskal, W. H. (1954). Measures of association for cross-classifications. *Journal of the American Statistical Association*, **49**, 732–764.

Goodstein, D. (2000). How science works. In Federal Judicial Center, editor, *Reference Manual on Scientific Evidence*, pages 67–82. Federal Judicial Center, Washington, DC, 2nd edition.

Grabosky, P. N. and Braithwaite, J. (1993). *Business Regulation and Australia's Future*. Australian Studies in Law No. 1. Australian Institute of Criminology, Canberra. http://aic.gov.au/publications/previous%20series/lcj/1-20/business.aspx.

Griffiths, T., Griffiths, T. L., and Tenenbaum, J. B. (2006). Optimal predictions in everyday cognition. *Psychological Science*, **17**, 767–773.

Grinblatt, M. and Titman, S. (1998). *Financial Markets and Corporate Strategy*. Irwine/McGraw-Hill, New York.

Gruenfeld, D. (2006). Behaving badly may be natural at the top. *Stanford Business*, **74**(3), 20–21.

Gumbel, E. J. (1968). Ladislaus von Bortkiewicz. *International Encyclopedia of the Social Sciences*, **2**, 128–131.

Gysler, M., Kruse, J., and Schubert, R. (2002). Ambiguity and gender differences in financial decision making: An experimental examination of competence and confidence effects. Economics working paper series 02/23, CER-ETH - Center of Economic Research (CER-ETH) at ETH Zurich.

Haas, P. J. (2005). Simulation, quantile estimation. Hand-out. http://www.stanford.edu/class/msande223/handouts/lecturenotes09.pdf.

Hájek, A. (2003). Interpretations of probability. In E. N. Zalta, editor, *The Stanford Encyclopedia of Philosophy*.

Hamel, G. (2003). Innovation as a deep capability. *Leader to Leader*, (27), 19–24. http://leadertoleader.org/leaderbooks/L2L/winter2003/hamel.html.

Hanson, R. J. and Krogh, F. T. (1992). A quadratic-tensor model algorithm for nonlinear least-squares problems with linear constraints. *ACM Trans. Math. Softw.*, **18**(2), 115–133.

Hastie, T. and Tibshirani, R. (1990). *Generalized Additive Models*. Chapman & Hall, New York.

Haugen, R. A. (1993). *Modern Investment Theory*. Prentice-Hall, Englewood Cliffs, NJ, 3rd edition.

He, X., Inman, J. J., and Mittal, V. (2008). Gender jeopardy in financial risk taking. *Journal of Marketing Research*, **XLV**(2), 414–424.

Henderson, S. G. (2003). private communication.

Henrion, M. and Fischhoff, B. (1986). Assessing uncertainty in physical constants. *American Journal of Physics*, **54**(9), 791–798.

Hersey, P. and Blanchard, K. H. (1996). *Management of Organizational Behavior. Utilizing Human Resources*. Prentice-Hall, Upper Saddle River, NJ, 7th edition.

Herzberg, F. (1959). Work and the nature of man. In F. Herzberg, B. Mausner, and B. Snyderman, editors, *The Motivation to Work*. John Wiley, New York.

Herzberg, F. (1966). *The Work and the Nature of Man*. World Publishing, Cleveland, OH.

Hirschi, T. (1969). *Causes of Delinquency*. University of California Press, Berkeley, 3rd edition.

Hirshleifer, J. (1988). *Price Theory and Applications*. Prentice-Hall, Englewood Cliffs, NJ, 4th edition.

Hofstadter, D. R. (1982). Number numbness, or why innumeracy may be just as dangerous as illiteracy. *Scientific American*, **246**(5), 20–34.

Hogg, R. V. and Klugman, S. A. (1984). *Loss Distributions*. John Wiley, New York.

Hogg, R. V. and Tanis, E. A. (1993). *Probability and Statistical Inferenece*. John Wiley, New York, 4th edition.

Horngren, C. and Foster, G. (1987). *Cost Accounting. A Managerial Emphasis*. Prentice-Hall, Englewood Cliffs, NJ, 6th edition.

Hosking, J. R. M. (1990). L-Moments: Analysis and estimation of distributions using linear combinations of order statistics. *Journal of the Royal Statistical Society*, **Series B**(52), 105–124.

House, R. J. (1971). A path goal theory of leader effectiveness. *Administrative Science Quarterly*, **16**(3), 321–339.

Huber, P. J. (1981). *Robust Statistics*. John Wiley, New York.

Hull, J. C. (1997). *Options Futures and other Derivative Securities*. Prentice-

Hall, Upper Saddle River, NJ, 3rd edition.

Hume, D. (1978). *A Treatise of Human Nature*. Clarendon, Oxford, 2nd edition. Original edition 1740.

Huntington, S. P. (2004). *Who Are We? The Challenges to America's National Identity*. Simon & Schuster, New York.

Hyer, R. N. and Covello, V. T. (2005). *Effective Media Communication During Public Health Emergencies*. World Health Organization, Geneva.

ICAF (2002). *Strategic Leadership and Decision Making*. Industrial College of the Armed Forces, Washington.

Inglehart, R. (1997). *Modernization and Postmodernization: Cultural, Economic, and Political Change in 43 Societies*. Princeton University Press, Princeton, NJ.

Inglehart, R. and Welzel, C. (2005). *Modernization, Cultural Change, and Democracy: The Human Development Sequence*. Cambridge University Press, New York.

International Auditing Practices Committee (2001). The relationship between banking supervision and bank's external auditors. Technical report, International Federation of Accountants.

Jaynes, E. T. (2007). *Probability Theory: The Logic of Science*. Cambridge University Press, Cambridge.

Jensen, M. C. and Meckling, W. H. (1976). Theory of the firm: Managerial behavior, agency costs and ownership structure. *Journal of Financial Economics*, **3**(4), 305–360.

Jensen, M. C. and Meckling, W. H. (1998). The Nature of Man. *Foundations of Organizational Strategy*.

Jobst, A. A. (2007). Constraints of consistent operational risk measurement and regulation: Data collection and loss reporting. *Journal of Financial Regulation and Compliance*.

Johnson, J. E. V. and Powell, P. (1994). Decision-making, risk and gender: Are managers different? *British Journal of Management*, **5**(2), 123–138.

Johnson, R. A., Kast, F. E., and Rosenzweig, J. E. (1964). Systems theory and management. *Management Science*, **10**(2), 367–384.

Judt, T. (2005). Europe vs. America. *The New York Review of Books*, **52**(2).

Kagan, R. (2004). *Of Paradise and Power*. Vintage Books, New York.

Kahan, D. M., Braman, D., Gastil, J., Slovic, P., and Mertz, C. K. (2007). Culture and identity-protective cognition: Explaining the white male effect in risk perception. *Journal of Empirical Legal Studies*, **4**(3), 465–505.

Kahneman, D., Slovic, P., and Tversky, A., editors (1982). *Judgment under Uncertainty: Heuristics and Biases*. Cambridge University Press,

New York.

Kane, E. J. (1997). Ethical foundations of financial regulation. NBER Working Papers 6020, National Bureau of Economic Research, Inc. Available at http://ideas.repec.org/p/nbr/nberwo/6020.html.

Kaufmann, D., Kraay, A., and Mastruzzi, M. (2006). Governance matters V: Governance Indicators for 1996-2005. Technical report, World Bank Institute, Wahington DC. Available at SSRN: http://ssrn.com/abstract=-929549.

Kern, S. (2004). *A Cultural History of Causality. Science, Murder Novels, and Systems of Thought*. Princeton University Press, Princeton, NJ.

Keynes, J. M. (1921). *A Treatise on Probability*. MacMillan, London.

Keynes, J. M. (1936). *The General Theory of Employment, Interest, and Money*. Harcourt, Brace & World, New York.

King, J. L. (2001). *Operational Risk. Measurement and Modelling*. John Wiley, New York.

Klugman, S. A., Panjer, H. H., and Willmot, G. E. (1998). *Loss Models. From Data to Decisions*. John Wiley, New York.

Knight, F. H. (1921). *Risk, Uncertainty, and Profit*. Houghton Mifflin, Boston. Available at http://www.econlib.org/library/Knight/knRUPtoc.html.

Kokoska, S. and Nevison, C. (1989). *Statistical Tables and Formulae*. Springer Verlag, New York.

Koontz, H. and Weihrich, H. (1988). *Management*. McGraw-Hill, New York, 9th edition.

Kotz, S. and Nadarajah, S. (2000). *Extreme Value Distributions: Theory and Applications*. Imperial College Press, London.

KPMG (2008). Risk management, methodology for implementing risk management. Brochure. http://www.kpmg.ch/docs/20081201_Broschuere_-RM_e_web.pdf.

Kuhn, T. S. (1996). *The Structure of Scientific Revolutions*. University of Chicago Press, Chicago.

Kynn, M. (2008). The "heuristics and biases" bias in expert elicitation. *Journal of the Royal Statistical Society: Series A (Statistics in Society)*, **171**(1), 239-264.

Laplace, P. S. (1825). *Essai philosophique sur les probabilits*. Bachelier, Paris.

Latter, T. (1997). The causes and management of banking crises. Technical report, Bank of England, Centre for Central Banking Studies, Handbooks in Central Banking no.12.

Law, A. M. and Kelton, W. D. (1991). *Simulation Modeling & Analysis*. McGraw-Hill, New York, 2nd edition.

Lawley, D. N. and Maxwell, A. E. (1971). *Factor Analysis as a Statistical Method*. Butterworths, London, 2nd edition.

Lawrence, P. and Lorsch, J. (1967). Differentiation and integration in complex organizations. *Administrative Science Quarterly*, (12), 1–30.

Leitch, M. (2008). *Intelligent Internal Control and Risk Management*. Gower, Aldershot.

Lempert, R. J., Popper, S. W., and Bankes, S. C. (2003). Shaping the next one hundred years: New methods for quantitative, long-term policy analysis. Technical report, RAND, Santa Monica, CA.

Lewin, K., Lippit, R., and White, R. (1939). Patterns of aggressive behavior in experimentally created social climates. *Journal of Social Psychology*, **10**, 271–301.

Lipset, S. M. (1996). *American Exceptionalism: A Double-Edged Sword*. W.W.Norton, New York.

Lombroso, C. (1876). *L'Uomo Delinquente*. Hoepli, Milano.

Lopez, J. A. (1999). Using CAMELS ratings to monitor bank conditions. *FRBSF Economic Letter*, (Jun).

Luhmann, N. (1991). *Soziologie des Risikos*. de Gruyter, Berlin.

Luhmann, N. (1994). *Risk: A Sociological Theory*. Aldine de Gruyter, New York.

Lundberg, F. (1909). Über die Theorie der Rückversicherung. *Transactions of the International Congress of Actuaries*, **1**(2), 877–955.

Mack, T. (2002). *Schadenversicherungsmathematik*. Number 28 in Schriftenreihe Angewandte Versicherungsmathematik. Verlag Versicherungswirtschaft, Karlsruhe.

Mackay, C. (1995). *Extraordinary Popular Delusions and the Madness of Crowds*. Wordsworth Reference, Ware, UK.

Mackie, J. L. (1974). *The Cement of the Universe. A Study of Causation*. Clarendon, Oxford.

Maslow, A. H. (1943). A theory of human motivation. *Psychological Review*, **50**, 370–396.

Matsumoto, M. and Nishimura, T. (1998). Mersenne twister: A 623-dimensionally equidistributed uniform pseudo-random number generator. *ACM Transactions on Modeling and Computer Simulation*, (8), 3–30.

Mayr, O. (1971). Adam Smith and the concept of the feedback system. *Technology and Culture*, **12**, 1–22.

McCormick, R. (2004). The management of legal risk by financial institutions. Technical report, IBA Draft Discussion Paper. http://www.federalreserve.gov/SECRS/2005/August/20050818/OP-1189/OP-1189_2_1.pdf.

McFarlin, D. B. and Sweeney, P. D. (2002). *Where Egos Dare. The Untold Truth About Narcissistic Leaders – and How to Survive Them.* Kogan Page, London.

Mcgregor, D. (1960). *The Human Side of Enterprise.* McGraw-Hill, New York.

Mehrabian, A. (1968). Communicating without words. *Psychology Today*, pages 53–55.

Merton, R. C. (1974). On the pricing of corporate debt: The risk structure of interest rates. *Journal of Finance*, (75), 643–669.

Merton, R. K. (1938). Social structure and anomie. *American Sociological Review*, **3**, 672–682. http://www.d.umn.edu/cla/faculty/jhamlin/2111/-Readings/MertonAnomie.pdf.

Metropolis, N. C. (1987). The beginning of the Monte Carlo method. *Los Alamos Science*, **15**, 125–130. (Special Issue, Stanislaw Ulam 1909–1984).

Metropolis, N. C. and Ulam, S. M. (1949). The Monte Carlo method. *Journal of the American Statistical Association*, **44**, 335–341.

Meyers, G. (2000). Coherent measures of risk; an exposition for the lay actuary. Technical report, CASACT, Arlington, VA.

Michael, J., Schucany, W., and Haas, R. (1976). Generating random roots from variates using transformations with multiple roots. *American Statistician*, **30**(2), 88–91.

Mikes, A. (2008a). Counting risk and making risk count: The organizational significance of risk management. *SSRN eLibrary*.

Mikes, A. (2008b). Risk management and calculative cultures. *SSRN eLibrary*.

Mikes, A. (2008c). Risk management at crunch time: Are chief risk officers compliance champions or business partners? *SSRN eLibrary*.

Miller, I. and Miller, M. (1999). *John E. Freund's Mathematical Statistics.* Prentice Hall, Upper Saddle River, NJ, 6th edition.

Mittelstrass, J., editor (2004). *Enzyklopädie Philosophie und Wissenschafts-theorie*, volume 1–4. J.B. Metzler, Stuttgart.

Modigliani, F. and Miller, M. H. (1958). The cost of capital, corporation finance and the theory of investment. *The American Economic Review*, **48**(3), 261–297.

Modigliani, F. and Miller, M. H. (1963). Corporate income taxes and the cost of capital: A correction. *The American Economic Review*, **53**(3), 433–443.

Morgan, M. G. and Henrion, M., editors (1990). *Uncertainty: A Guide to Dealing with Uncertainty in Quantitative Risk and Policy Analysis.* Cambridge University Press, Cambridge.

National Research Council (1989). *Improving Risk Communication.* National

Academy Press, Washington, DC.

Nicoletti, G., Scarpetta, S., and Boylaud, O. (2000). Summary indicators of product market regulation with an extension to employment protection legislation. Technical report, Organisation for Economic Co-operation and Development, Economic Department, Working Paper No. 226.

Nolan, J. P. (2010). *Stable Distributions - Models for Heavy Tailed Data*. Birkhäuser, Boston. In progress, Chapter 1 online at academic2.american.edu/~jpnolan.

Noll, P. and Bachmann, R. (1999). *Der kleine Machiavelli: Handbuch der Macht für den alltäglichen Gebrauch*. Pendo Verlag, Zürich, 3rd edition.

Nystrom, K. and Skoglundy, J. (2002). Quantitative operational risk management. Working paper, Swedbank, Group Financial Risk Control. Available at http://www.gloriamundi.org/picsresources/nsqopm.pdf.

O'Donnell, J. B. and Rechtman, Y. (2005). Navigating the standards for information technology controls. *CPA Journal online*.

OeNB and FMA, editor (2006). *Guidelines on Operational Risk Management*. Oesterreichische Nationalbank in cooperation with the Austrian Financial Market Authority, Vienna. http://www.oenb.at/en/img/-operational_risk_screen_tcm16-49652.pdf.

Office of Investor Education and Assistance (1998). A plain English handbook. How to create clear SEC disclosure documents. Technical report, U.S. Securities and Exchange Commission. http://www.sec.gov/news/extra/-handbook.htm.

Olkin, I. and Pratt, J. W. (1958). Unbiased estimation of certain correlation coefficients. *Annals of Mathematical Statistics*, (29), 201–211.

Ong, M. K. (1998). On the quantification of operational risk. A short polemic. In Arthur Andersen, editor, *Operational Risk and Financial Institutions*, pages 181–184. Risk Books, London.

Panetta, F., Faeh, T., Grande, G., Ho, C., King, M., Levy, A., Signoretti, F. M., Taboga, M., and Zaghini, A. (2009). An assessment of financial sector rescue programmes. BIS Papers 48, Bank for International Settlements, Monetary and Economic Department, Basle. http://www.bis.org/publ/-bppdf/bispap48.pdf.

Panjer, H. H. (1981). Recursive evaluation of a family of compound distributions. *Astin Bulletin*, **12**(12), 22–26.

Panjer, H. H. (2006). *Operational Risks: Modeling Analytics*. John Wiley, New York.

Papoulis, A. (1991). *Probability, Random Variables, and Stochastic Processes*. McGraw-Hill, Boston, 3rd edition.

Passmore, J. (1994). *A Hundred Years of Philosophy*. Penguin, London.

Paulos, J. A. (1988). *Innumeracy: Mathematical Illiteracy and Its Conse-quences*. Hill and Wang, New York.

Paulos, J. A. (1989). The odds are you're innumerate. *The New York Times*, (1st of January, 1989).

PCAOB (2007). Auditing standard no. 5: An audit of internal control over financial reporting that is integrated with an audit of financial statements. Technical report, Public Company Accounting Oversight Board, Washington, DC.

Pearl, J. (2000). *Causality. Models, Reasoning, and Inference*. Cambridge University Press, Cambridge.

Pézier, J. (2004). Do operational risks matter and if so which? Available http://www.ifk-cfs.de/papers/Pezier2.pdf.

Pfeffer, J. (1992). *Managing with Power. Politics and Influence in Organiza-tions*. Harvard Business School Press, Boston.

Pfeffer, J. and Sutton, R. I. (2000). *The Knowing-Doing Gap. How Smart Companies Turn Knowledge into Action*. Harvard Business School Press, Boston.

Pfeffer, J. and Sutton, R. I. (2006). The half-truth of leadership. *Stanford Business*, **74**(3), 14–19.

Popper, K. A. (1959a). *The Logic of Scientific Discovery*. Hutchinson, London.

Popper, K. A. (1959b). The propensity interpretation of probability. *British Journal of the Philosophy of Science*, **10**, 25–42.

Porter, M. L. (1990). *The Competitive Advantage of Nations*. Free Press, New York.

Posner, R. A. (2004). *Catastrophe. Risk and Response*. Oxford University Press, Oxford.

Power, M. (1994). *The Audit Explosion*. Demos, London.

Power, M. (1997). From ,risk society to audit society. *Soziale Systeme*, (3), 3–21.

Power, M. (2004). *The Risk Management of Everything. Rethinking the politics of uncertainty*. Demos, London.

Power, M. (2005). The invention of operational risk. *Review of International Political Economy*, **4**(12), 577–599.

Power, M. (2008). *Organized Uncertainty, Designing a World of Risk Man-agement*. Oxford University Press, Oxford.

Powojowski, M. R., Reynolds, D., and Tuenter, H. J. (2002). Dependent events and operational risk. *Algo Research Quarterly*, **5**(2), 65–73.

Press, W., Teukolsky, S., Vetterling, W., and Flannery, B. (1992). *Numerical Recepies in FORTRAN: The Art of Scientific Computing*. Cambridge

University Press, Cambridge, 2nd edition.

PricewaterhouseCoopers (2005). Global economic crime survey 2005. Technical report, PricewaterhouseCoopers.

PricewaterhouseCoopers (2007). The internal control system: A rapidly changing management instrument. Technical report, PricewaterhouseCoopers AG (Switzerland). http://www.pwc.ch/user_content/editor/files/-publ_ass/pwc_ics_changing_management_06_e.pdf.

RAMP (1998). *RAMP: Risk Analysis and Management for Projects*. Thomas Telford, London.

Rebonato, R. (2007). *Plight of the Fortune Tellers*. Princeton University Press, Princeton, NJ.

Rebonato, R. and Jäckel, P. (1999). The most general methodology to create a valid correlation matrix for risk management and option pricing purposes. Technical report, Quantitative Research Centre of the NatWest Group, London.

Reckless, W. C. (1961). *The Crime Problem*. Appleton-Century-Crofts, New York, 3rd edition.

Rejda, G. E. (1995). *Principles of Risk Management and Insurance*. Harper Collins, New York, 5th edition.

Reynolds, T. and Flores, A. (1986). *Foreign Law: Current Sources of Codes and Basic Legislation in Jurisdictions of the World*. Rothman, Littleton.

Rice, J. A. (1995). *Mathematical Statistics and Data Analysis*. Duxbury Press, Belmont, CA, 2nd edition.

Ropeik, D. and Gray, G. (2002). *Risk: A Practical Guide for Deciding What's Really Safe and What's Really Dangerous in the World Around You*. Houghton Mifflin, Boston.

Rosengren, E. (2004). Data for operational risk. Presentation at the National Academy of Science. Presentation. January 14th, 2004.

Rubinstein, R. Y. (1981). *Simulation and the Monte Carlo Method*. John Wiley, New York.

SAMHSA (2002). Communicating in a crisis: Risk communication guidelines for public officials. Technical report, U.S. Department of Health and Human Services, Substance Abuse & Mental Health Services Administration, Washington, DC.

Sapienza, P., Zingales, L., and Maestripieri, D. (2009). Gender differences in financial risk aversion and career choices are affected by testosterone. *PNAS*, **106**(36), 15268–15273.

Schein, E. (1980). *Organizational Psychology*. Prentice-Hall, Englewood Cliffs, NJ, 3rd edition.

Scheuer, M. E. and Stoller, D. C. (1962). On the generation of normal random vectors. *Technometrics*, **4**, 278–281.

Segrè, E. (1980). *From X-Rays to Quarks: Modern Physicists and Their Discoveries*. W.H. Freeman, San Francisco.

SFBC (2005). Circular: Audit of banks and securities firms (audit). Technical report, Swiss Federal Banking Commission, Bern. http://www.kpmg.ch-/docs/20050629-SFBC_Circ_05-01.pdf.

Shannon, C. E. (1948). A mathematical theory of communication. *Bell System Technical Journal*, **27**, 379–423 and 623–656. http://cm.bell-labs.com/-cm/ms/what/shannonday/shannon1948.pdf.

Shiller, R. J. (2001). Bubbles, human judgment, and expert opinion. Technical report, Yale Cowles Foundation Discussion Paper No. 1303. Available at SSRN: http://ssrn.com/abstract=275515.

Sigmund, K. (1995). *Games of Life. Explorations in Ecology, Evolution and Behaviour*. Penguin, London.

Smith, A. (1991). *An Inquiry into the Nature and Causes of the Wealth of Nations*. Everyman's Library Alfred A. Knopf, Inc., New York.

Sowey, E. ((2003). The getting of wisdom: Educating statisticians to enhance their clients numeracy. *The American Statistician*, **57**, 89–93.

Sparrow, M. K. (2000). *The Regulatory Craft: Controlling Risks, Solving Problems and Managing Compliance*. Brookings Institution Press, Washington, DC.

Spira, L. and Page, M. (2003). Risk management: The reinvention of internal control and the changing role of internal audit. *Accounting, Auditing & Accountability Journal*, **16**, 640–661(22).

Stanford University (1996). Internal control factors. Web page. http://www.stanford.edu/dept/Internal-Audit/docs/internal_controls.shtml.

Steiger, J. H. (1988). Tests for comparing elements of a correlation matrix. *Psychological Bulletin*, **87**, 245–251. Available at http://www.statpower.-net/Steiger_20Biblio/Steiger80.pdf.

Stempsey, W. E. (1998). Causation and moral responsibility for death. Online. Bioethics and Medical Ethics.

Stogdill, R. (1974). *Handbook of Leadership: A Survey of the Literature*. Free Press, New York.

Straub, E. (1988). *Non-Life Insurance Mathematics*. Springer Verlag, Berlin.

Sutter, R. (2005). U.S. tort costs and cross-border perspectives: 2005 update. Technical report, Towers Perrin – Tillinghast.

Sykes, G. and Matza, D. (1957). Techniques of neutralisation: A theory of delinquency. *American Sociological Review*, **22**, 664 – 670.

Szabó, I. (2001). *Höhere Technische Mechanik*. Springer Verlag, Berlin, 5th edition.

Taleb, N. N. (2005). *Fooled by Randomness. The Hidden Role of Chance in Life and in the Markets*. Random House, New York.

Taleb, N. N. (2007). *The Black Swan. The Impact of the Highly Improbable*. Random House, New York.

Taleb, N. N. and Martin, G. A. (2007). The risk of severe, infrequent events. *The Banker*, (September), 188–189.

Taylor, F. W. (1911). *Principles of Scientific Management*. Harper & Bros., New York.

Taylor, F. W. (1933). *The Human Problems of an Industrial Civilization*. MacMillan, New York.

The Joint Forum (2003). Operational risk transfer across financial sectors. Technical report, Bank for International Settlements. http://www.bis.-org/publ/joint06.pdf.

Timm, M. (1991). *How to Lie with Maps*. University of Chicago Press, Chicago.

Tirole, J. (2005). *The Theory of Corporate Finance*. Princeton University Press, Princeton and Oxford.

Toevs, A., Zizka, R., Callender, W., and Matsakh, E. (2003). Negotiating the risk mosaic: A study on enterprise-wide risk measurement and management in financial services. *The RMA Journal*, (March), 16–21.

Toint, P. L. and Tuyttens, D. (1992). LSNNO, a FORTRAN subroutine for solving large-scale nonlinear network optimization problems. *ACM Trans. Math. Softw.*, **18**(3), 308–328.

Traeger, L. (1904). *Der Kausalbegriff im Straf- und Zivilrecht: Zugleich ein Beitrag zur Auslegung des BGB*. Elwert, Marburg.

Tuft, E. G. (2003). PowerPoint is evil, power corrupts, PowerPoint corrupts absolutely. *Wired Magazine*, **11**(9).

Tukey, J. W. (1977). *Exploratory Data Analysis*. Addison-Wesley, Reading, MA.

Tversky, A. and Kahneman, D. (1974). Judgement under uncertainty: Heuristics and biases. *Science*, **185**, 1124–1131.

Tversky, A. and Kahneman, D. (1982). Judgements of and by representativeness. In Kahneman et al. (1982), pages 84–98.

Välikangas, L. and Hamel, G. (2001). Internal markets – emerging governance structures for innovation. Presentation.

Vaughan, E. and Vaughan, T. (1995). *Essentials of Insurance. A Risk Management Perspective*. John Wiley, New York.

Venables, W. and Ripley, B. (1999). *Modern Applied Statistics with S-Plus*. Springer Verlag, New York, 3rd edition.

Venn, J. (1876). *The Logic of Chance*. Macmillan, New York, 2nd edition.

Vineis, P. (2003). Causality in epidemiology. *Soziale Präventivmedizin*, **48**, 80–87.

von Bortkiewicz, L. (1898). *Das Gesetz der kleinen Zahlen*. Teubner Verlag, Leipzig.

von Mises, R. (1936). *Wahrscheinlichkeit, Statistik und Wahrheit - Einführung in die neue Wahrscheinlichkeitslehre und ihre Anwendung*. Springer Verlag, Wien, 2nd edition.

von Mises, R. (1939). *Probability, Statistics and Truth*. William Hodge, London. Translated by J.Neyman, D.Sholl and E.Rabinowitsch.

Wang, S. S. (1998). Aggregation of correlated risk portfolios: Models & algorithms. Technical report, Casualty Actuarial Society (USA): Committee on Theory of Risk. http://www.casact.org/cotor/wang.pdf.

Watzlawick, P. (1977). *How Real Is Real?: Confusion, Disinformation, Communication*. Random House, New York.

Weber, M. (1922). *Wirtschaft und Gesellschaft. Band 1*. J.C.B. Mohr, Tübingen.

Weber, M. (1930). *The Protestant Ethic and the Spirit of Capitalism*. Unwin Hyman, London. Translated by Talcott Parsons and Anthony Giddens.

Weber, M. (1988). Die protestantische Ethik und der Geist des Kapitalismus. In *Gesammelte Aufsätze zur Religionssoziologie. Band 1*, pages 1–2. Mohr-Siebeck, Tübingen. Original edition 1904/05.

Weber, M. (1991). Objektive Möglichkeiten und adäquate Verursachung in der historischen Kausalbetrachtung. In *Schriften zur Wissenschaftslehre*, pages 102–131. Philipp Reclam, Stuttgart. Published 1906.

Wilcox, J. (1976). The gambler's ruin approach to business risk. *Sloan Management Review*, (Fall), 33–46.

Woll, A. (1978). *Allgemeine Volkswirtschaftslehre*. Vahlen, München, 6th edition.

Yamai, Y. and Yoshiba, T. (2002). Comparative analysis of expected shortfall and value-at-risk: Their estimation error, decomposition, and optimization. *Monetary and Economic Studies*, (January), 87–122.

Yukl, G. A. (2006). *Leadership in Organizations*. Pearson Education, Upper Saddle River, NJ, 6th edition.

Zimmermann, D. W., Zumbo, B. D., and Williams, R. H. (2003). Bias in estimation and hypothesis testing of correlation. *Psicologica*, (24), 133–158.

Index

386   *Operational Risk Modelling and Management*

matching moments, 126
matching quantiles, 126
material weakness, 329
materiality, 325
matrix structure, 278
maximum entropy, 63
maximum likelihood ML, 126
Mayo, Elton, 238
maysir, 8
McGregor, Douglas, 233
mean, 124
mean excess function, 116, 219
measure of association, 153
measurement, 301
measurement scale, 171
Merrill Lynch, 8
Mersenne Twister, 144
Merton, Robert K., 262
method, 86
method of least squares, 128
Metropolis, Nicholas, 86
minimum chi-square, 126
minimum cost network flows, 214
misappropriation, 324
mixing data, 51, 95
mixing distribution, 118
mixture, 118, 125
model, 85
model distribution, 96
model structure, 92
modelling, 49
Modigliani-Miller, 22
moment matching equations, 126
moments about the origin, 124
monitoring
    in-process, 287
    process-independent, 287
monitoring cost, 246
monotonicity, 180
Monte Carlo, 85–90, 188
    example, 358
    method, 85
    simulation, 86
moral expectation, 26
Morgan Stanley, 8, 24

Morgan, J.P., 177
motivation, 232
multi-peril basket insurance, 42
mutual self-insurance pool, 42

narcissistic leader, 231
net capital rule, 24
net loss, 187
non-sampling risk, 325
NORTA, 159

objective function, 211
observability, 239
OCC, 335
Ockham, William, 159, 170, 293
operating effectiveness of controls, 326
operational risk, 28
    BCBS definition, 29
    capital requirement ORR, 34
OTS, 335
over-fitting, 313
overconfidence, 75, 77, 257
oversimplification, 269

P-P plot, 142
p-value, 139
Pólya urn, 79
Pólyia-Cox desiderata, 62
Panjer algorithm, 197
paradigm anchoring, 78
Pareto law, 114
Pareto principle, 148
Pareto, Vilfredo, 113, 114, 148
Pareto-optimality, 318
path diagram, 74
path-goal theory, 231
Pearson correlation coefficient, 153
Pearson, Karl, 139
peer-to-peer, 59
person-to-person lending, 59
Pillar 2, 335
Planck, Max, 78
Plato, 228
Poincaré, Henry, 81
Poisson, Siméon Denis, 105